AF566855

Absolute Radiometry

Electrically Calibrated Thermal Detectors of Optical Radiation

Absolute Radiometry

Electrically Calibrated Thermal Detectors of Optical Radiation

Edited by

F. Hengstberger

National Physical Research Laboratory
Council for Scientific and Industrial Research
Pretoria, South Africa

ACADEMIC PRESS, INC.
Harcourt Brace Jovanovich, Publishers
Boston San Diego New York
Berkeley London Sydney
Tokyo Toronto

ACADEMIC PRESS, INC.
1250 Sixth Avenue, San Diego, CA 92101

United Kingdom Edition published by
ACADEMIC PRESS INC. (LONDON) LTD.
24–28 Oval Road, London NW1 7DX

Library of Congress Cataloging-in-Publication Data

Absolute radiometry : electrically calibrated thermal detectors of optical radiation / edited by F. Hengstberger.
p. cm.
Bibliography: p.
Includes index.
ISBN 0-12-340810-5
1. Radiation—Measurement. 2. Electromagnetic waves—Measurement.
I. Hengstberger, F. (Franz)
QC475.A27 1989
539.2—dc19 88-2185
CIP

Printed in the United States of America

89 90 91 92 9 8 7 6 5 4 3 2 1

Contents

Contributors

Numbers in parentheses refer to the pages on which the authors' contributions begin.

L. P. BOIVIN (170), Division of Physics. National Research Council, Ottawa KIA OR6, Canada.

F. HENGSTBERGER (1, 145, 193, 234, 252), National Physical Research Laboratory, Council for Scientific and Industrial Research, PO Box 395, Pretoria 0001, South Africa

K. MÖSTL (118), Physikalisch-Technische Budesanstalt, Postfach 3345, D-3300 Braunschweig, Federal Republic of Germany

A. ONO (156), National Research Laboratory of Metrology, 1–4, Umezono 1-Chome, Sakura-Mura, Niihari-gun, Ibaraki, 305, Japan

Foreword

With the beginnings of absolute radiometry dating back to about 1893, this area of optical radiometry will soon be 100 years old. It is therefore fitting that the first book devoted to absolute radiometry (for the intended interpretation of the term, refer to the Preface) should be published at this time.

The idea for writing the book originated in a subcommittee on absolute radiometry of Technical Committee 2.2 (Physical Detectors of Optical Radiation) of the International Commission on Illumination (CIE). The subcommittee was chaired by myself, and it had the task of compiling a CIE Technical Report on absolute radiometry for nonexperts in this area of metrology. Although the report eventually produced—CIE publication no. 65 (1985), entitled *Electrically Calibrated Thermal Detectors of Optical Radiation* (*Absolute Radiometers*)—was merely a brief summary of the most important aspects of absolute radiometry, the committee had compiled extensive additional material on the subject. Since the contributors felt that it would be unfortunate just to forget about the additional information and because of their conviction that a book on the subject was needed in any case, it was agreed to seek a publisher for the material. This decision had the support of all the members of CIE TC 2.2, and official approval was granted by the CIE Council to use the material for this purpose.

The authors of the various chapters in this book are the original authors of the same subjects in the original CIE material. However, that material was merely used as the starting point for the work, and it has been extensively edited, updated, and expanded.

It is hoped that this work will fill the gap that has existed in absolute radiometry in the absence of a recognized reference work on the subject and that it will also be useful for researchers in related scientific disciplines.

F. Hengstberger

Preface

This book covers the subject of absolute radiometry, one of the techniques employed in optical radiation metrology for the absolute measurement of radiant power. As expressed by the subtitle of the book ("Electrically Calibrated Thermal Detectors of Optical Radiation"), the term *absolute radiometry* as used throughout the text is interpreted in the narrow sense of "the use of electrically calibrated detectors of optical radiation for the realization of an optical power scale." It is acknowledged that a wider interpretation of the term also exists, one that embraces all methods used for the realization of optical power scales. It is stressed right from the start that the wider interpretation does not apply in the context of this book, although other methods for realizing optical power scales are also discussed (Chapter 7), albeit in a much more condensed form.

The subject matter of absolute radiometry is put in perspective within the area of optical radiometry as well as relative to similar approaches used in other areas of metrology. As such, it is directed to metrologists, optical radiometrists, photometrists, meteorologists, space scientists, and professionals in all other disciplines concerned with accurate optical power measurements. The knowledge acquired in this area over the past century is summarized, including the instrumentation and materials used, methods of analyzing the temperature distribution in the detector elements, error sources, and experiments to measure the required corrections. Extensive bibliographies are appended to each chapter, and experimental results are used to support the theoretical derivations and conclusions.

Acknowledgments

Phil Boivin wishes to thank Gloria Dumoulin for her excellent typing of the chapter on environmental corrections and the frequent correspondence with the editor.

Franz Hengstberger is indebted to the South African Council for Scientific and Industrial Research (CSIR) for granting permission to participate in this book, both as author and as editor. He also wishes to express his gratitude to the late Franc Grum, coeditor of the Academic Press volumes on *Optical Radiation Measurements*, for his efforts in arranging the project with Academic Press. His further sincere thanks go to his secretary, Christine Isaacson, who typed all of his correspondence and chapters for the book and retyped the contributions of all the other authors. Without her support and hard work as well as the understanding and help of his wife, Margie, the project could not have been finished in time. The assistance of other CSIR colleagues, including Jean Turnbull for drawing the figures, Torsten Appenroth (electronics), Lorraine King (word processing), Althea Adey and Nicoline Basson (editing), Erik Dressler (RF metrology) Marek Marczak (AC–DC transfer), Lorene du Preez (ionizing radiations), Dick Turner (head, National Measuring Standards and Metrology Division), and George Ritter (director, General Physics), is also gratefully acknowledged. The helpful and friendly advice of R. Martinez-Herrero (Universidad Complutense, Madrid, Spain) and P. M. Mejias (Universidad Nacional de Educacion a Distancia, Madrid, Spain) in connection with partial coherence (Section 1.2.5) is much appreciated.

Klaus Möstl wishes to thank his wife, Claudia, for the accurate proof reading of his manuscript and for the drawing of his figures.

Akira Ono would like to express his thanks to his wife, Akiko, for her help in typing the manuscript.

All the authors are indebted to the International Commission on Illumination (CIE) for granting permission to use some of the material compiled by one of its technical committees as the starting point for their work.

1

The Absolute Measurement of Radiant Power

F. HENGSTBERGER
Council for Scientific and Industrial Research
Pretoria, South Africa

1.1 UNITS AND STANDARDS OF MEASUREMENT

1.1.1 Introduction

Physical *quantities* are used to characterize the properties of objects, states, or processes in a quantitative way. They may be scalars, vectors, or tensors and they are expressed either by symbols or as products of a number and a unit. Their most important property is their invariance against a change of unit. The part of a physical quantity that contains only its qualitative aspects (and no quantitative ones) is known as its *dimension*. The number of dimensions is much less than the number of quantities, since all quantities can be characterized either by means of a small set of basic dimensions or, alternatively, in terms of various derived dimensions. In any particular system of dimensions, it is possible to assign a base unit to each basic dimension. A base unit is a quantity selected from the set of quantities of identical dimension because of its particular value. Units are always scalar quantities. They are the basis for expressing the quantitative aspect of a quantity by means of the fraction or multiple of the unit (also known as the numerical value of the quantity in terms of the chosen unit) that is equal to the assessed quantity. This quantitative comparison of two quantities is the basic principle of a *measurement*.

The result of a measurement, which is generally subject to some error, is expressed as the product of a numerical value and a unit. The measurement itself is not restricted to the (direct or indirect) process of comparison, but may also include some processing of the acquired data (for example, the

ISBN 0-12-340810-5

application of corrections). The reported result of a measurement is not complete without a full statement of the influence parameters (for example, temperature, humidity, number of measurements taken) and of the estimated uncertainty.

Units of measurement are usually defined in abstract terms. Their concrete representations are known as *standards.* Before embarking on a discussion of this subject, it must be remembered that the word *standard* is ambiguous in English. When used in relation to measurement, it can have two meanings that are distinct in other languages. The first of these is a standard specification (German *Norm,* French *norme*), which is a document issued by institutions such as the International Organization for Standardization (ISO), the American National Standards Institute (ANSI), or the British Standards Institution (BSI). (At the end of chapter 1, an appendix gives the acronyms of associations, government agencies, equipment, and scientific terms.) These documentary standards describe recommended procedures, terminology, or specifications for certain applications. The second meaning of the word *standard* is a measurement standard (German *Normal,* French *étalon*), which is the physical embodiment of the unit of some quantity. It is a material object such as a thermometer, a clock, or a voltmeter that has been calibrated by a metrology laboratory for a specific type of measurement. The discussion in the current context is confined to the second meaning of the word *standard.* Measuring standards are used in calibration work to determine the relationship between the input and output quantities of a measuring instrument. Use is normally made of standards calibrated against a local reference standard. The latter is usually referred to as a higher-echelon standard and this measurement chain continues until it is eventually linked to the national standard held by a designated laboratory (e.g., the National Bureau of Standards in the United States, the National Physical Laboratory in the U.K., or the Physikalisch-Technische Bundesanstalt in the Federal Republic of Germany). This process establishes the *traceability* of a particular measuring instrument.

Historically, the earliest requirement for measurements and measurement standards arose through trade, religion, and science. Commodities had to be sold by weight, length, area, or volume; thus, fixed units for these quantities were already in use in ancient civilizations several thousand years ago. The human need to keep time and predict seasons—combined with the scientific and religious significance of studying the movement of the moon and the stars—resulted in astronomy becoming a highly developed science in many early civilizations. Ever since then, the measurement of time has remained one of the most accurate types of measurement. The need for standards and accurate measurements in industry was generated much later by modern mass-production techniques. These rely on the interchangeability of components made to the same specifications at different factories, locations, countries, and even continents. Many different types of measuring instruments are required to check the vast variety of products coming off the production line and the readings of these instruments have to be harmonized by a hierarchy of calibrations traceable to nationally and internationally agreed standards. Another relatively recent development is the need for traceable measuring standards to enforce legislation in health- and safety-related areas.

Metrology is the science of measurement, which includes all related scientific theories as well as the problem of practical measurements. It encompasses the definition and realization of units and measurement standards, the determination of physical constants and material properties, the theory and application of measuring instruments, the properties of measuring instruments, and problems of measurement and of interpretation of measurements.

The enconomic significance of a well-developed measurement system combined with its scientific importance has led to the establishment of sophisticated national metrology infrastructures in all industrialized countries and to the formation of the corresponding international umbrella organizations.

1.1.2 The International System of Units (SI)

The SI is the modern version of the original metric system of units, which was established by a diplomatic treaty (the Convention of the Meter) and signed by 17 nations on May 20, 1875. By 1979, the number of countries that had joined the Meter Convention had grown to 45, while the vast majority of other nations either used the metric system or were in the process of converting to it.

The treaty established an international weights and measures organization (depicted schematically in Fig. 1.1) and also resulted in the establishment of national measuring-standards laboratories in various countries.

In contrast with the original metric system, which represented a group of units derived from the meter, the SI of the 1980s is the result of a long process of creating a comprehensive, yet practical international system of units for science, technology, and education. It no longer attempts to derive all other units from the meter but from seven independent base units. It was approved by the Eleventh General Conference on Weights and Measures (CGPM) in 1960 with the objective of replacing the large number of units and metric systems of units that had by then been developed in various areas of technology. It is a coherent system, which means that it consists of only the base units and such derived units, which are formed by simple multiplication

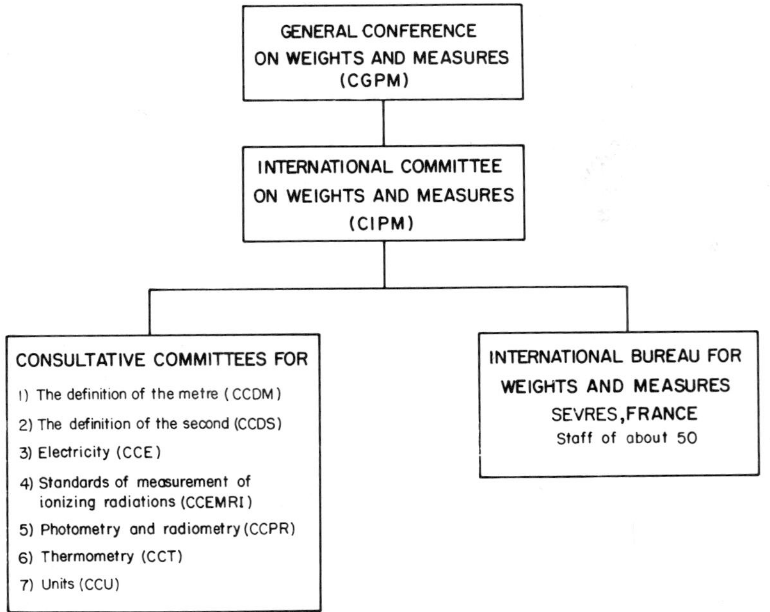

Fig. 1.1 Organs of the Meter Convention.

TABLE 1.1 The Seven SI Base Units

Quantity	Unit
Length	Meter
Mass	Kilogram
Time	Second
Current	Ampere
Temperature	Kelvin
Amount of substance	Mole
Luminous intensity	Candela

and/or division of the base units and the two supplementary units without involving factors other than the number one. This property and the choice of the base units ensure that there is only one SI unit for each physical quantity. Table 1.1 lists the seven SI base units.

Their definitions are as follows:

• The *meter* is the length of the path travelled by light in vacuum during a time interval of 1/299 792 458 of a second.

- The *kilogram* is the mass of the international prototype of the kilogram recognized by the CGPM and in the custody of the Bureau International des Poids et Mesures (BIPM) in Sevres, France.
- The *second* is the duration of 9 192 631 770 periods of the radiation corresponding to the transition between the two hyperfine levels of the ground state of the cesium-133 atom.
- The *ampere* is that constant current that, if maintained in two straight parallel conductors of infinite length, of negligible circular cross section, and placed one meter apart in vacuum would produce between these conductors a force equal to 2×10^{-7} newton per meter of length.
- The *kelvin*, unit of thermodynamic temperature, is the fraction 1/273.16 of the thermodynamic temperature of the triple point of water.
- The *mole* is the amount of substance of a system that contains as many elementary entities as there are atoms in 0.012 kg of carbon 12.
- The *candela* is the luminous intensity, in a given direction, of a source that emits monochromatic radiation of frequency 540×10^{12} hertz and that has a radiant intensity in that direction of (1/683) watt per steradian.

Although the units of a system of units consist only of base units and derived units, the CGPM introduced a third category called supplementary units into the SI. The reason for this step was the difficulty of reaching agreement on whether they were in fact base units or derived units. They are listed in Table 1.2.

TABLE 1.2 The Two SI Supplementary Units

Quantity	Unit
Plane angle	Radian
Solid angle	Steradian

- The *radian* is the plane angle between two radii of a circle that cut off on the circumference an arc equal in length to the radius.
- The *steradian* is the solid angle that, having its vertex in the center of a sphere, cuts off an area of the surface of the sphere equal to that of a square with sides of length equal to the radius of the sphere.

Further information on the SI can be found in Bureau International des Poids et Mesures (1973) and ISO (1973).

1.2 OPTICAL RADIOMETRY

1.2.1 Introduction

Radiometry is the science and technology of the measurement of electromagnetic energy (Grum and Becherer, 1979). The electromagnetic spectrum spans some 16 orders of magnitude in wavelength (and frequency), from a low 10^{-11} m in the gamma-ray region to a high 10^5 m in the very low frequency radio band (Fig. 1.2a). The instrumentation and techniques employed to measure electromagnetic energy differ widely within this vast spectral region. *Optical radiometry* is the subfield of radiometry concerned with measurements of optical radiation (Fig. 1.2b), which is electromagnetic radiation obeying the laws of optics (i.e., which can be reflected, dispersed, or imaged with optical elements such as lenses or mirrors) and covering a wavelength range from about 10^{-8} to 10^{-3} m (five orders of magnitude).

While the wave nature of electromagnetic radiation is predominant at the long-wavelength end of the electromagnetic spectrum, the short-wavelength parts are dominated by its quantum nature. With the optical region situated near the transition region between the two extreme manifestations of the dual nature of electromagnetic radiation, both wave and quantum aspects are important to varying degrees. Wave and quantum aspects are linked via Planck's constant h, which can be regarded as the product of two quantities, one characteristic of a wave, the other of a particle. If one selects energy (E) and frequency (ν) or momentum (p) and wavelength (λ) as the pairs of conjugate variables in this relationship, it can be expressed as

$$\mathrm{h} = E/\nu = p\lambda, \tag{1.1}$$

where E and p are particle attributes and υ and λ are wave properties. With a value for Planck's constant of 6.6×10^{-34} Js, it can be deduced that the wave and particle aspects are of about equal prominence in a frequency region around 2.5×10^{17} Hz (i.e., the X-ray region). The optical region is close to this transition region but nevertheless is clearly on the side of the wave properties and is thus dominated to a large extent by the wave nature of electromagnetic radiation.

With radiometry (originally only in the form of light measurements by means of the human eye) being one of the oldest branches of optics, the field is purely empirical in origin. Its beginnings date back as far as the Renaissance (Rozenberg, 1973) and some of its basic concepts were formulated in the eighteenth century by scientists such as Bouguer and Lambert. Its most notable contributions to physics include the studies of energy transfer by heat radiation, which culminated in the derivation of the blackbody radiation laws at the beginning of the twentieth century. In view of the early beginnings of

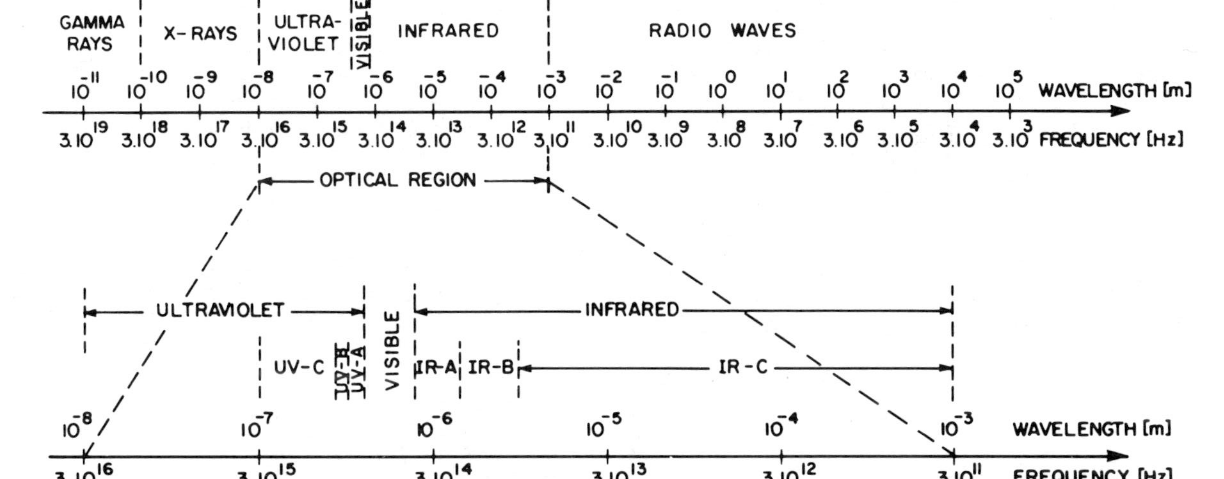

Fig. 1.2 (a) Electromagnetic wavelength and frequency spectrum. (b) Optical wavelength and frequency spectrum.

optical radiometry, it is understandable—yet nevertheless noteworthy—that it has featured in all physics textbooks for centuries as an independent part of optics, with no relation to the basic theories on which that field is based. Early attempts to include optical radiometry in the electromagnetic theory were unsuccessful and only relatively recently has significant progress been made in this respect. (See Sections 1.2.4 and 1.2.5.)

Although optical radiometry is one of the oldest branches of optics, measurements of the various radiometric quantities cannot be made with a very high degree of accuracy compared to the measurements of other physical quantities. While time or length for instance can often be measured with accuracies of parts per million or parts per billion or better, an accuracy of 1% in optical radiometry is often the best that can be achieved with the most advanced equipment and techniques. There are a number of reasons for this. First, radiant energy is distributed over wavelength, direction, time, position, and polarization and is subject to correlations in the field fluctuations, which can be characterized by means of the concept of coherence. In contrast, the accurately measurable quantities of length and time are not distributed in this complex way. Second, the whole radiometric measuring system is itself radiating, absorbing, scattering, and reflecting radiant energy, which results in a wide array of possible interactions and error sources.

1.2.2 Quantities and Units

This section contains the terminology, definitions of quantities, and units of quantities most commonly used in the field of optical radiometry. Wherever possible, use is made of the definitions and terminology proposed jointly (CIE, 1970) by the International Electrotechnical Commission (IEC) and the International Commission on Illumination (CIE). However, as of early 1988, a new edition of the CIE–IEC vocabulary is in preparation.

Although every effort has been made to include the most up-to-date terminology and definitions according to the new edition, there may be small discrepancies in wording compared to the version eventually appearing in print. The units of measurement used conform with the SI system of units. (See Section 1.1.2)

Radiometric quantities are purely physical in nature and are derived from the quantity "energy." The corresponding photometric quantities on the other hand involve the additional evaluation of the radiant energy in terms of a defined weighting function, usually the standard photometric observer (CIE, 1983). The two types of quantities are represented by the same principal symbol and may be distinguished by their subscripts. Radiometric quantities either have the subscript "e" or no subscript, while photometric quantities have the subscript "v."

TABLE 1.3 Radiometric Quantities

Quantity	Symbol	Defining Equation	Unit
Radiant energy	Q, Q_e		J (joule)
Radiant power	Φ, Φ_e	$\Phi = dQ/dt$	W (watt)
Radiant exitance	M, M_e	$M = d\Phi/dA$	$W\ m^{-2}$
Irradiance	E, E_e	$E = d\Phi/dA$	$W\ m^{-2}$
Radiant intensity	I, I_e	$I = d\Phi/d\omega$	$W\ sr^{-1}$
Radiance	L, L_e	$L = d^2\Phi/[d\omega\ dA \cos\theta]$	$W\ m^{-2} sr^{-1}$

Some quantities are functions of wavelength, in which case their designation must be preceded by the adjective *spectral.* Symbols for such quantities are followed by the bracketed symbol for wavelength (λ). As an example, the symbol for spectral radiance is $L(\lambda)$ or $L_e(\lambda)$. This is in contrast with the convention for the spectral concentration of a quantity X, which may also be preceded by the adjective *spectral.* In that case, however, the symbol is subscripted with the wavelength symbol, i.e., $dX/d\lambda = X_\lambda$.

The most frequently used fundamental radiometric and photometric quantities are listed in Tables 1.3 and 1.4, respectively, together with their symbols, defining equations, and units. Of the additional physical quantities used in the defining equations, $d\omega$ is the element of solid angle through which radiant power from the point source is radiated and θ is the angle between the line of sight and the normal to the surface considered. The unit of solid angle, the steradian, has the abbreviation sr. Although the photometric units in Table 1.4 are derived from the lumensecond (lms) to emphasize the correspondence to the derivation of the radiometric units from the joule (J = Ws), it must be stressed that the SI base unit for the photometric quantities actually is the candela. (See Section 1.1.2.) However, presenting the photometric units in terms of the candela makes it more difficult to spot the formal correspondence between the two sets of units.

TABLE 1.4 Photometric Quantities

Quantity	Symbol	Defining Equation	Unit
Luminous energy	Q_v	$Q_v = K_m \int v(\lambda) Q_\lambda\ d\lambda$	lm s
Luminous flux	Φ_v	$\Phi_v = dQ_v/dt$	lm (lumen)
Luminous exitance	M_v	$M_v = d\Phi_v/dA$	lx (lux)
Illuminance	E_v	$E_v = d\Phi_v/dA$	lx
Luminous intensity	I_v	$I_v = d\Phi_v/d\omega$	cd (candela)
Luminance	L_v	$L_v = d^2\Phi_v/(d\omega\ dA \cos\theta)$	$cd\ m^{-2}$

1.2.3 Terminology: A Glossary

Radiometer. Instrument used for measuring radiation in energy or power units.

Photometer. Instrument for measuring photometric quantities.

Radiometer [photometer] head. The part of a radiometer [photometer] containing the detector and means for spectral and spatial corrections of the detector response.

Detector (of optical radiation). Device in which optical radiation produces a measurable physical effect.

Photoelectric detector. A detector of optical radiation that utilizes the interaction between radiation and matter resulting in the absorption of photons and the liberation of electrons from their initial states, excluding electrical phenomena caused by temperature changes.

Thermal (radiation) detector. A detector of optical radiation in which a measurable physical effect is produced by the heating of the part that absorbs radiation.

Radiation thermocouple. Thermal radiation detector in which a single pair of thermovoltaic junctions is used as a sensor of temperature changes.

(Radiation) Thermopile. Thermal radiation detector in which more than one pair of thermovoltaic junctions are used as a sensor of temperature changes.

Bolometer. Thermal radiation detector in which the heating of the part that absorbs the radiation gives rise to a change in its electrical resistance.

Pyroelectric detector. Thermal radiation detector that utilizes the temperature dependence of the spontaneous electrical polarization of certain dielectric materials.

Note. This definition is also extended to include materials in which a long-lived but nonspontaneous electric polarization can be induced.

Detector input. The radiometric or photometric quantity that a detector is being used to measure or detect.

Detector output. The physical quantity, usually electrical, yielded by a detector in response to a detector input.

Note. This quantity may, for example, be current, voltage, or change in resistance.

Responsivity (sensitivity). Quotient of the detector output (quantity Y) by the detector input (quantity X).

$$\text{Symbol:} \quad s, \qquad s = Y/X$$

Note 1. If there exists a detector output (quantity Y_w) without irradiation and the measured total detector output Y_t, the detector output (quantity Y) caused by the detector input (quantity X) is $Y = Y_t - Y_w$.

Note 2. The responsivity depends on the relative spectral distribution. Moreover, it may depend on the polarization and the direction of the incident radiation, on the homogeneity of irradiation, and on the temperature of the detector as well as on the electric circuit. If the detector output quantity and the detector input quantity are not proportional, the responsivity also depends on the value of the detector input quantity.

Spectral responsivity. Quotient of the detector output (quantity $dY(\lambda)$) by the detector input quantity $dX_e(\lambda) = X_e(\lambda)d\lambda$ at the wavelength λ.

$$\text{Symbol:} \quad s(\lambda), \qquad s(\lambda) = dY(\lambda)/dX_e(\lambda)$$

Response time: rise [fall] time. The time required for the detector output Y to rise [fall] from a stated low [high] percentage to a stated higher [lower] percentage of the maximum value when a steady state of radiation is instantaneously applied [removed]. The levels used must be stated. It is usual to consider the 10%[90%] and 90%[10%] levels.

Note. If the output rises [falls] exponentially with time, the time required for it to change from its initial value by the fraction $(1 - 1/e)$ of the final change is called the *time-constant*.

Noise-equivalent detector input. The value of the detector input that produces a detector output equal to the rms noise input within a stated bandwidth and at a stated frequency.

$$\text{Symbol:} \quad P_N$$

Note 1. It is usual to consider a one-Hz bandwidth and this value is implied unless stated otherwise.

Note 2. If the detector input is a radiation flux, the noise-equivalent detector input is referred to as the *noise-equivalent power* (NEP).

Note 3. If the detector input is a uniform irradiance, the noise-equivalent detector input is referred to as the *noise-equivalent irradiance* (NEI).

Detectivity. Reciprocal of the noise-equivalent detector input P_N.

$$\text{Symbol:} \quad D, \qquad D = 1/P_N.$$

Normalized detectivity. A normalized value of the detectivity, related to some relevant parameters of the detection system, such as the sensitive area of the detector and the measurement bandwidth. (Usually the square root of the sensitive area and bandwidth is used.)

$$\text{Symbol:} \quad D^*, \qquad D^* = D(A\Delta f)^{1/2} = (1/NEP)(A\Delta f)^{1/2},$$

where A is the detector area and Δf the bandwidth. This implies that the responsivity and noise output of the detector are frequency-independent in the frequency range under consideration.

Note 1. The relative spectral distribution of the detector input and the frequency should be stated.

Note 2. As the noise-equivalent irradiance is given by

$$NEI = NEP(A^{-1}),$$

the normalized detectivity is

$$D^* = NEI^{-1}(A^{-1/2})(\Delta f)^{1/2}.$$

Linearity. The linearity of a detector is the property that the output quantity of the detector is proportional to the input quantity. Then the responsivity is constant over a specified range of inputs.

Note 1. A detector is usually linear only over a certain range of input levels. Outside this range, it may become nonlinear. The range must be stated.

Note 2. The linearity range of a detector may be affected by the use of unsuitable electronic circuitry.

Note 3. The linearity range of some detectors also depends on wavelength and irradiance.

1.2.4 Classical Radiometry

1.2.4.1 Introduction

Optical radiometry evolved historically parallel to, but independent of geometrical optics. Both branches of optics deal with the concept of the beam of optical radiation as a photon flux. While geometrical optics is concerned with the influence of the environment on the trajectory of this beam, optical radiometry, in the classical sense, is concerned with the beam dynamics, thus involving all the complexities of the conservation laws. Since the concepts of optical radiometry arose from purely empirical origins, later attempts to present them as an approximation from, e.g., wave optics have not always been completely successful. However, three simplifying assumptions are fairly generally accepted as applying to classical optical radiometry:

(a) The wavelength is negligibly small compared with the dimensions of interest in the optical system.

(b) The time-constants of detectors used in a measurement are much longer than one period of the oscillating field.

(c) The contributions to the energy flow from the various parts of a source are calculated by simple additive superposition.

1.2.4.2 Rays and Radiance

Rays of radiation in optical radiometry have the same meaning as in geometrical optics, where the concept follows from the so-called eikonal equation

$$(\nabla S)^2 = n^2, \tag{1.2}$$

which is an approximate form of the wave equation for $\lambda \to 0$. The wave equation itself is derived directly from Maxwell's equations (Born and Wolf, 1964). In this model, rays can be interpreted as the normals to the geometrical wavefront and they also indicate the direction of energy flow in the electromagnetic field (Grum and Becherer, 1979).

In order to describe the energy transfer in a beam of radiant energy in a homogeneous isotropic medium, it is useful to consider a so-called elementary beam of rays as depicted in Fig. 1.3 (Nicodemus, 1963).

Such a beam is defined as consisting of a single central ray connecting the centers of the two area elements dA_1 and dA_2 plus all the rays passing through both of them. The solid angles subtended by each area element when viewed from the center of the other are $d\omega_1 = \cos\theta_1\, dA_1/D^2$ and $d\omega_2 = \cos\theta_2 dA_2/D^2$. θ_1 and θ_2 are the angles between the central ray and the

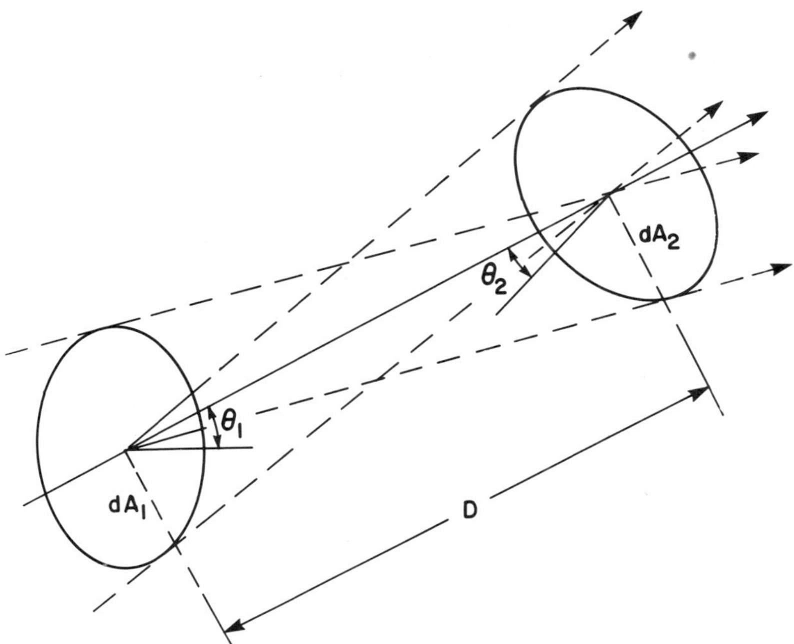

Fig. 1.3 Elementary beam of radiation.

respective surface normals, and D is the distance between the centers of the two area elements. It can then be shown (Nicodemus, 1963) that the quantities $d\omega_2 \cos\theta_1\, dA_1$ and $d\omega_1 \cos\theta_2\, dA_2$ are equal. These invariant products of a projected area element and the solid angle subtended at this element by a second area element are known as the *geometric extent* (symbolized by G) of the elementary beam of rays.

$$dG = d\omega_1 \cos\theta_2\, dA_2 = d\omega_2 \cos\theta_1\, dA_1 \tag{1.3}$$

From the definition of radiance given in Section 1.2.2, it is evident that radiance can now also be expressed as

$$dL = d\Phi/dG. \tag{1.4}$$

Since the same rays pass through both area elements, classical optical radiometry interprets this as implying that the same power is passing through them and that the radiance along a ray in a passive, lossless, uniform, and isotopic medium is therefore invariant according to Eq. 1.4, i.e.,

$$L_1 = L_2 = L, \tag{1.5}$$

L_1 and L_2 being the radiances in the direction of the central ray at dA_1 and dA_2, respectively.

If radiation has to pass through regions of different refractive indices (as, for example, lenses or other optical elements), then it can be shown (Nicodemus, 1963) that it is not radiance (L) that is conserved, but a quantity called *basic radiance*. It is defined as L/n^2, where n is the refractive index of the medium and, therefore,

$$L_1/n_1^2 = L_2/n_2^2. \tag{1.6}$$

However, if the ray emerges again (without attenuation loss) into the same medium, it does not matter that it has transversed different media on its way. In that case, the conservation of basic radiance also implies the conservation of ordinary radiance from the original medium to the final medium with the same refractive index.

Since the conservation of radiance or basic radiance holds for a lossless medium, it is convenient to characterize the loss of radiant energy due to, e.g., absorption, reflection, or scattering in terms of a reduction in radiance. From the invariance of the geometric extent and Eq. 1.4, it is evident that any attenuation of an element of flux ($d\Phi$) along the path of a ray implies a corresponding attenuation of the radiance (dL). Without going into detail about the particular effect causing the loss, one can define a quantity called *propagance*, which characterizes the attenuation along the beam path, as (Nicodemus, 1976)

$$\tau^* = d\Phi_p/d\Phi_i = L_p/L_i, \tag{1.7}$$

where $d\Phi_p$ and L_p are the propagated quantities at the end of the beam path and $d\Phi_i$ and L_i are the initial quantities at the start of the beam path.

If wavelength is taken into account in these relationships, then the various equations apply separately for each wavelength since both the radiance and the refractive index may vary with it.

1.2.4.3 Transfer of Radiant Power

In many instances in optical radiometry, it is not the distribution of radiance that has to be inferred, but the total radiant power in a beam impinging on a defined area at a given position. Assuming again that both ends of the beam path are in the same medium and considering the wavelength dependence of the quantities involved, the total initial (Φ_i) and propagated radiant power (Φ_p) follow as

$$\Phi_i = \iiint_{A\omega\Delta\lambda} [L_{p,\lambda}(x, y, \theta, \phi, \lambda)/\tau^*(x, y, \theta, \phi, \lambda)]d\lambda \cos\theta \, d\omega \, dA \quad (1.8a)$$

$$\Phi_p = \iiint_{A\omega\Delta\lambda} [L_{i,\lambda}(x, y, \theta, \phi, \lambda)\tau^*(x, y, \theta, \phi, \lambda)]d\lambda \cos\theta \, d\omega \, dA. \quad (1.8b)$$

Using these equations, it is possible to calculate the radiant power passing through a defined area from the corresponding radiance distribution over another area and the propagance of the beam path. However, although the formalism is apparently simple, it is by no means a trivial matter to acquire all the data required for the calculation. Given the coordinates (x, y) and the directions (θ, ϕ) of all the rays in a beam at a given reference surface, it is necessary to calculate the corresponding coordinates and directions at the intersection of each ray with a second surface. Sometimes this can be achieved by using a table or formula, but in many cases it will constitute a ray-tracing problem, which is best solved by using the techniques developed for this purpose in geometrical optics.

The ray concept and the radiance-conservation laws will now be used to describe an important case in optical radiometry, namely the transfer of radiant power from a source surface to a detector surface. It will be assumed that the intervening medium is homogeneous, lossless, and isotropic and that there are no imaging or focusing elements between the two. The situation is depicted in Fig. 1.4.

Omitting the parameter list of the various quantities, the radiant power transferred from an arbitrary area element dA_1 to another area element dA_2 is given by

$$d^3\Phi_{12} = L_{\lambda 1} \, d\lambda \, d\omega(dA_1 \cos\theta_1), \quad (1.9)$$

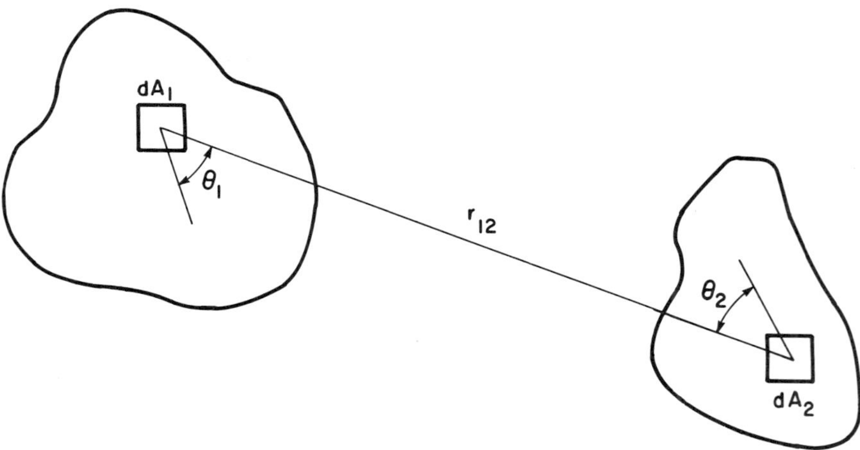

Fig. 1.4 Geometry of radiant-power transfer between two arbitrary surfaces.

where $L_{\lambda 1}$ is the spectral radiance of dA_1 in the direction of dA_2, $d\omega$ is the solid angle subtended by the second area element when viewed from the first area element, and $(dA_1 \cos \theta_1)$ is the area element dA_1 projected on a plane perpendicular to L_1. If one also expresses $d\omega$ in terms of the area element dA_2, projected on a plane perpendicular to L_1, and the distance r_{12} between the two area elements, Eq. 1.9 can be rewritten as

$$d^3\Phi_{12} = L_{\lambda 1}\, d\lambda\, dA_1\, dA_2 \left(\frac{\cos \theta_1 \cos \theta_2}{r_{12}^2} \right) \tag{1.10}$$

and the total radiant power received at surface 2 from surface 1 is

$$\Phi_{12} = \int_{\Delta\lambda} \int_{A_1} \int_{A_2} L_{\lambda 1} \left(\frac{\cos \theta_1 \cos \theta_2}{r_{12}^2} \right) d\lambda\, dA_1\, dA_2. \tag{1.11}$$

In this triple integral, all the integrated quantities will generally be functions of the variables over which the integration is performed. If no further simplifying assumptions can be made about these quantities, analytical solutions are often not possible and one has to resort to numerical integration methods. This is so in spite of the fact that the contributions from the various source elements were found by simple addition (thus neglecting the coherence properties of the source), that polarization was neglected, and that it was assumed that there were no interreflections between the source and the detector. The transfer of radiant power from partially coherent sources will be treated in Section 1.2.5, while procedures for dealing with interreflections can be found in Siegel and Howell (1972).

When Eq. (1.11) is applied to two small, coaxial area elements A_1 and A_2, the radiance of A_1 within the small solid angle subtended by A_2 can be approximated by a constant value L_1 and one obtains the so-called inverse-square law of radiometry, namely

$$\Phi_{12}/(A_2 \cos \theta_2) = E_{2n} = L_1(A_1 \cos \theta_1)/r_{12}^2 = I_1/r_{12}^2 \tag{1.12}$$

where $I_1 = L_1(A_1 \cos \theta_1)$ is the radiant intensity of A_1 in the direction of A_2 and E_{2n} is the normal irradiance ($E_2/\cos \theta_2$) at A_2. (See Section 1.2.2 for the relevant definitions.) The approximation is accurate to better than 1% if the source or detector dimensions normal to the beam direction amount to less than one-tenth of the source-detector distance.

If it can be assumed that the radiance distribution of the source is *Lambertian* (i.e., if the spectral radiance of the source is the same at all points and for all directions), then $\int_{\Delta\lambda} L_{\lambda_1}\, d\lambda = L_1$ can be separated from the two area integrations and Eq. (1.11) can be rewritten as

$$\Phi_{12} = L_1 \int_{A_1} \int_{A_2} (\cos \theta_1 \cdot \cos \theta_2/r_{12}^2) dA_1\, dA_2. \tag{1.13}$$

The double integral on the right-hand side is the geometric extent (G) defined in Eq. (1.3), a purely geometric quantity, which can be evaluated for any particular measuring geometry. A great variety of other geometric quantities are used to facilitate the calculations concerned with the interchange of radiant power between surfaces in the fields of heat-transfer and illuminating engineering (e.g., form factor, shape factor, view factor, exchange coefficient, configuration factor). The relationships between these factors and between them and geometric extent are important (Nicodemus, 1978) in view of the extensive tables and formulas available in the literature (Siegel and Howell, 1972; Sparrow and Cess, 1966). Using the results published for a great variety of geometries, the calculation of the geometric extent is often considerably simplified. However, since only relatively few geometries are required in the majority of cases in absolute radiometry, the subject will not be pursued any further in this context. Instead, a single example of an important measurement geometry will be presented.

Example. Beam between two coaxial disks

Most of the optical systems used in absolute radiometry use circular elements and, in particular, circular baffles (for the reduction of stray light) and circular detector apertures. Therefore, the transfer of radiant power between two coaxial disks separated by a distance D along the axis between their

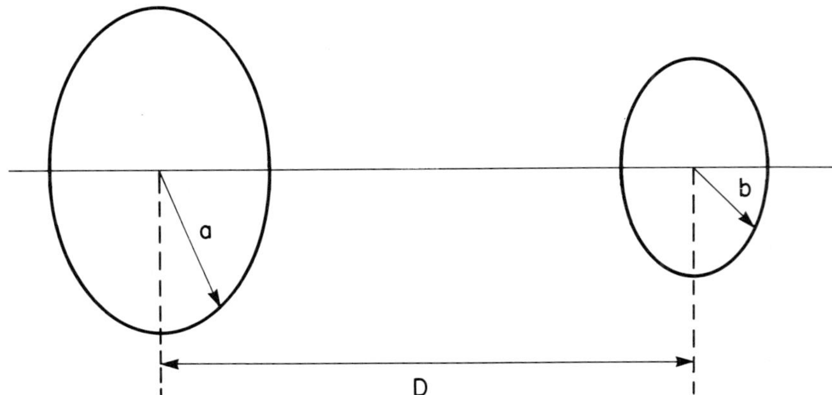

Fig. 1.5 Geometry for radiant-power transfer between two parallel coaxial disks.

centers (Fig. 1.5) is a fairly common beam configuration. For that case, the geometric extent is given by (Nicodemus, 1978)

$$G = (\pi^2/2)(a^2 + b^2 + D^2) - [(a^2 + b^2 + D^2)^2 - 4a^2b^2]^{1/2} \quad (1.14)$$

and the radiant power (Φ_{12}) transferred from the disk with radius a to the disk with radius b under the assumption of a Lambertian radiance distribution of the source is

$$\Phi_{12} = L_1 G = (M_1/\pi)G, \quad (1.15)$$

M_1 being the radiant exitance of the source. (See Section 1.2.2.)

1.2.5 Optical Radiometry with Sources in Any State of Coherence

In cases where the basic premises of classical radiometry (see Section 1.2.4) are no longer valid (as, for example, in the measurement of highly coherent laser radiation, or, more generally, in the measurement of partially coherent sources), a more complex wave-optics approach is required to understand the processes involved and the implications for radiometric measurements. When the classical electromagnetic fields from two or more separate oscillators are superimposed, their electric and magnetic field vectors add, producing a field vector, the effect of which on the detector is interpreted as a measure of the incident radiant power. The oscillators may reinforce or attenuate each other and (depending on their relative phase, amplitude, and frequency stability) the regions over which the interference phenomena occur as well as the interference patterns will vary with time. The degree to which the oscillators in a source are able to interfere over varying spatial and temporal extents is characterized by the concept of coherence. It should also be remembered that the radiation field from a given source may be incoherent at one distance and highly coherent at another. An example of this is the radiation from stars, which is highly incoherent close to the source yet is highly coherent at astronomical distances. Both interference and diffraction phenomena require a wave-optics description and are quite outside the scope of geometrical optics and classical radiometry.

Considerable progress has since been achieved with the interpretation of classical radiometric quantities in terms of wave optics. This has come about through the discovery of an intimate connection between the foundations of classical radiometry and the theory of coherence. The first break-through was made by Walther (1968), who succeeded in deriving an expression for radiance in terms of the cross-spectral density function of coherence theory. This is a function that is measurable in principle and for which theoretical models are known for typical classes of sources encountered in practice. It soon transpired, however, that the complete answer required a lot of further theoretical work to remove some originally unexplained inconsistencies. These included the facts that the radiance could become negative for some values of the arguments in the expression and that other (nonequivalent) expressions could be found that also correctly predicted the total power radiated by a planar source. It was later shown in a general proof (Friberg, 1979) that no radiance function for a finite, planar source in an arbitrary state of coherence could be defined on the basis of a linear transform of the cross-spectral density function, which would be consistent with all the implicit and explicit postulates of radiometry. These postulates were that the radiance function should be nonnegative and have zero value at all points in the source plane outside the source area and that the radiant intensity calculated from it according to classical radiometry should agree with the radiant intensity calculated independently on the basis of physical optics. A similar difficulty existed for the radiant exitance and doubts were expressed (Wolf, 1978) as to whether these quantities were therefore measurable in the strict physical sense. On the positive side, radiant intensity in the far zone of the source could be expressed in terms of the cross-spectral density distribution across the source as (Wolf, 1978)

$$I(\mathbf{s}) = (2\pi k)^2 \cos^2 \theta \tilde{W}^{(0)}(k\mathbf{s}_\perp, -k\mathbf{s}_\perp), \tag{1.16}$$

where $\tilde{W}^{(0)}(\mathbf{f}_1, \mathbf{f}_2)$ is the spatial (four-dimensional) Fourier transform of the cross-spectral density function $W^{(0)}(\mathbf{r}_1, \mathbf{r}_2)$ of the electromagnetic field in the source plane and $\mathbf{s}_\perp$ denotes the two-dimensional vector obtained by the projection of the three-dimensional unit vector $\mathbf{s}$ onto the same source plane. $\tilde{W}^{(0)}(\mathbf{f}_1, \mathbf{f}_2)$ is given by

$$\tilde{W}^{(0)}(\mathbf{f}_1, \mathbf{f}_2) = (2\pi)^{-4} \iint W^{(0)}(\mathbf{r}_1, \mathbf{r}_2) \exp[-i(\mathbf{f}_1\mathbf{r}_1 + \mathbf{f}_2\mathbf{r}_2)] d^2r_1\, d^2r_2 \tag{1.17}$$

whereby the integration extends twice independently over the source. This generalized radiant intensity was found to be always nonnegative and to represent the angular distribution of radiant power in the far zone. Thus, it had the same physical interpretation as the radiant intensity of classical radiometry.

It has since also been shown (Foley and Wolf, 1985) that Walther's generalized radiance for spatially highly incoherent sources and in the limit of large wave numbers can be reduced to a function with all the basic properties of traditional radiance.

Since then, a radiance function that satisfies all the postulates of classical radiometry has been derived (Martinez-Herrero and Mejias, 1984, 1986). As it does not depend linearly on the cross-spectral density function, it is not in violation of Friberg's (1979) theorem and could thus represent the generalized radiance concept needed for making further progress in this area. The propagation of generalized radiance through arbitrary optical systems and for any kind of source remains one of the outstanding matters awaiting clarification (Walther, 1978). Another issue of potential significance is the possibility of the existence of sources of different states of coherence, yet all producing the same radiant intensity (Collet and Wolf, 1978; Martinez-Herrero and Mejias, 1981, 1982). Important new applications could arise if such sources could be implemented in practice.

It therefore seems likely that modern coherence theory will vindicate all the intuitive concepts and quantities in use in classical radiometry by providing them with the long-sought link with physical optics. One can also expect that the generalized expression for the basic quantities of classical radiometry and the associated propagation laws and invariants will enable the formulation of modern radiometric concepts applicable to sources in any state of coherence. These will not be subject to the constraints of geometrical optics, but the price for their wider validity will be a considerably increased complexity. The exact implications of the generalized radiance concept for practical radiometry are still being discussed (Welford and Winston, 1987).

The concepts described so far have only been concerned with single components of the field vector (scalar theory of partial coherence). In order to extend them to coherence between all components, one has to replace the cross-spectral density function with a 2×2 matrix of four cross-spectral densities. The Stokes parameters are then one possible set of four linear combinations of the four matrix elements of the cross-spectral density (Shumaker, 1983).

1.2.6 Applications

As everything on earth is constantly subjected to irradiation by a wide array of radiation, among which electromagnetic and specifically optical radiation play an important role, the measurement of optical radiation (optical radiometry) has a wide range of applications. Due to the importance of the human sense of vision, important application areas have traditionally centered on radiometric applications in the visible part of the optical spectrum. These have included mainly light measurements (photometry) and color measurements (colorimetry) in fields such as illuminating engineering, photography, television, film, visual information displays, and signalling.

In the ultraviolet region, important applications involve measurements in health-related areas such as tanning, phototherapy, photochemotherapy, disinfection, and sterilization and for industrial purposes such as organic synthesis, photopolymerization, curing of inks (e.g., in the printing industry) and coatings, photoresist exposure in the electronics industry, and nondestructive testing.

Infrared applications include thermal design and analysis and sensors for military applications.

Radiometric measurements are also important in areas such as astronomy and astrophysics, meteorology and atmospheric physics, remote sensing, solar energy, lasers, material science, agriculture, and biology.

1.3 ABSOLUTE RADIOMETRY

1.3.1 Introduction

In North America, absolute radiometers are sometimes referred to as *electrically calibrated radiometers* (ECRs) or *electrical substitution radiometers* (ESRs), while the expression *compensation pyrheliometer* is common for them in the field of meteorology. However, not all compensation pyrheliometers are absolute instruments. Some of them measure in terms of relative scales based on particular constructions, and results obtained with them may differ by several percent from the absolute power scale. (See Section 1.3.4.) A very descriptive but somewhat long-winded alternative term has been introduced by the CIE (1985): *electrically calibrated thermal detectors of optical radiation.* This term has been used as the subtitle of this book, while the shorter term *absolute radiometers* will be used exclusively throughout the text.

Although there is no complete consensus on the definition of these terms, the difference between an absolute radiometer and a calorimeter is another matter in need of clarification. Until about the beginning of the twentieth century, the distinction was relatively straightforward. In absolute radiometers, the radiant power or energy was determined in terms of the substituted electrical power or energy, while in calorimeters it was calculated from the mass, specific heat, and time rate of change of the temperature or the temperature rise of the calorimeter. Since then, matters have become somewhat less clear in that most calorimeters now also employ the electrical-substitution principle. As of the late 1980s, it would seem fair in optical radiometry to define a calorimeter as an instrument primarily used for measuring radiant energy, while an absolute radiometer is primarily used for radiant-power measurements. Although the two techniques have some features in common, they often differ quite substantially in the construction of the measuring head, the execution of the measurements, and the processing of the acquired data. It is emphasized again at this point that the term *absolute radiometry* in the context of this book is interpreted as "the use of electrically calibrated thermal detectors of optical radiation for the realization of an optical power scale. (See also Preface.)

1.3.2 The Operating Principle of an Absolute Radiometer

Absolute radiometry is one of the few techniques that can be used to measure the radiant power or radiant power per unit area (irradiance) in a beam of optical radiation in absolute units. As has been shown in Section 1.2.2, the SI unit for radiant power is the watt (symbol: W) and for irradiance it is the watt per square meter (symbol: W/m^2). All of the techniques employed for the direct measurement of optical radiation in terms of these units use either a detector or a source as the measurement standard. In the same way as the expansion of a metal with temperature (e.g., in mercury-in-glass thermometers) is no absolute measure of temperature (in degrees Celsius or Fahrenheit) without a prior calibration against a suitable temperature standard, the outputs of optical radiation detectors by themselves are no absolute measure of the incident radiant power or irradiance. Depending on the type of detector used, the output could be a change in one or more physical parameters (for example, temperature, pressure, length, electrical resistance, voltage, or current). Absolute radiometry (as well as a few other methods to be described in Chapter 7) can be used to establish an absolute measuring scale for radiant power and irradiance to calibrate the great number and variety of ordinary (nonabsolute) optical radiation detectors and radiation sources used in all areas of optical radiometry.

In a nutshell, an absolute radiometer is a thermal-radiation detector with a built-in electrical heating element. The thermal-radiation detector is used to convert radiant energy into heat by the process of absorption (see Chapter 2) and to provide an output signal, which is proportional to the resulting temperature rise of the detector. The detector can also be heated electrically until the signal (and thus, by implication, the temperature rise) equals that due to the unknown radiant power. That substituted electrical power is measured and the unknown amount of radiant power is equated to it. The ideal absolute radiometer and the ideal substitution process are depicted schematically in Figs. 1.6a and 1.6b.

In the ideal absolute radiometer of Fig. 1.6, the radiant power is converted to heat in exactly the same place in the detector as the electrical power. In practice, however, this objective has proven very difficult to achieve. It is true that some materials—e.g., metal blacks (see Chapter 2)—are both absorbers of radiant energy and electrically conducting and can thus be used for the dual purpose of absorber and electrical heater in an absolute radiometer. On the other hand, this still does not mean that the conversion of both forms of energy takes place with the same conversion efficiency, at the same depth in the material or that the resulting radial and axial temperature distribution in the radiometer element (see Chapter 4) can be made the same for both forms of heating. Most absolute radiometers built to date use a resistive heater

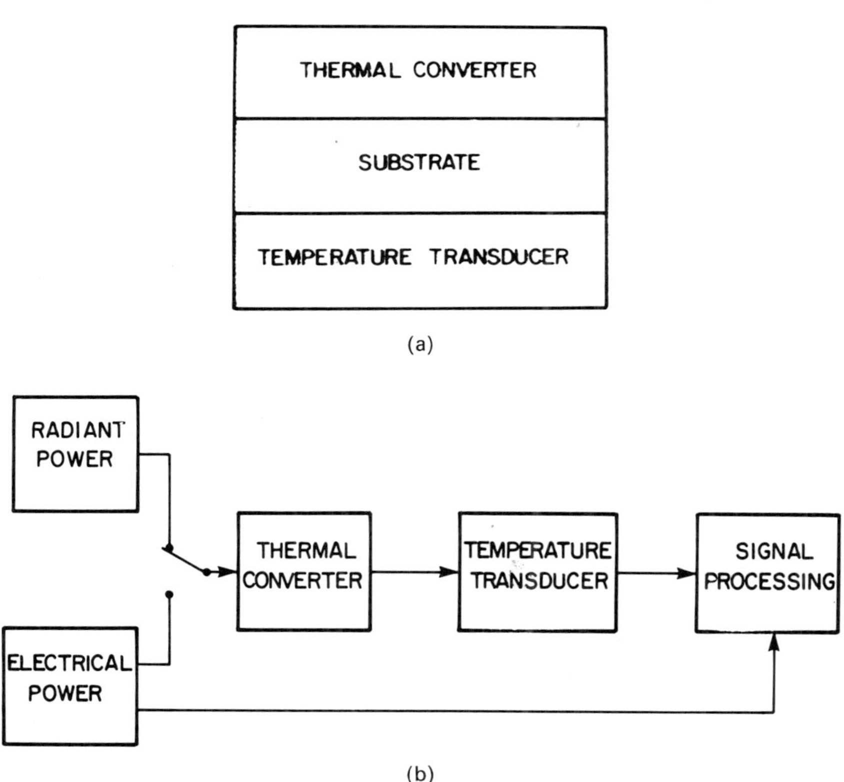

Fig. 1.6 (a) Schematic diagram of an ideal absolute radiometer. (b) Schematic diagram of the substitution process in an ideal absolute radiometer.

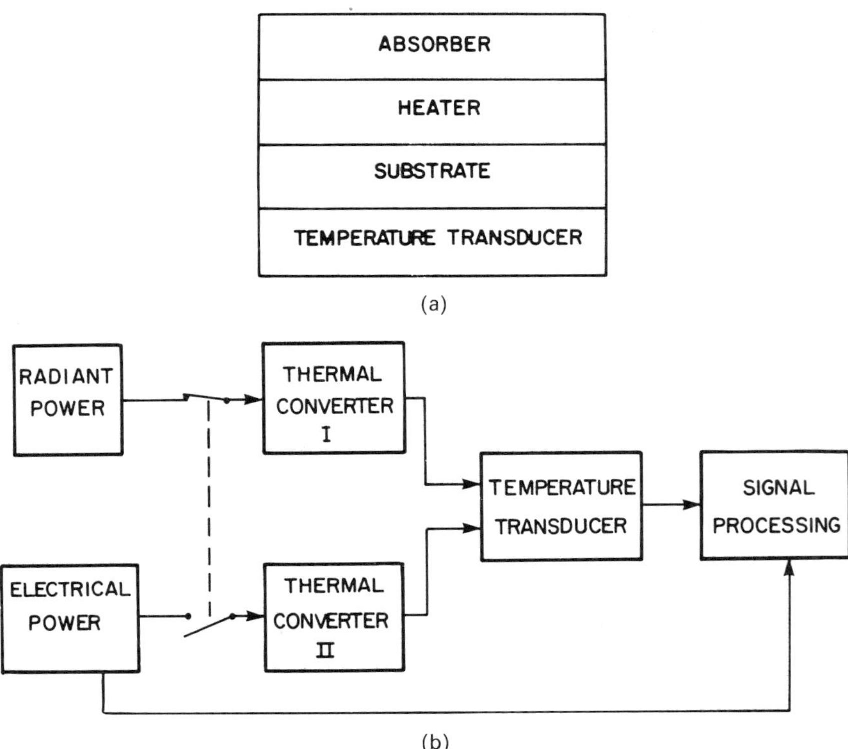

Fig. 1.7 (a) Schematic diagram of a real absolute radiometer. (b) Schematic diagram of the substitution process in a real absolute radiometer.

located under an electrically nonconducting absorber layer. This case is depicted schematically in Figs. 1.7a and 1.7b.

In the more realistic model of Fig. 1.7, it is clear that the measured electrical power can only be equated approximately to the unknown radiant power. For the highest accuracy, corrections must be applied for the different conversion efficiencies of the two forms of energy, for the difference in the temperature distribution in the radiometer element for the two forms of heating, as well as for parasitic power sources linked to the two kinds of heat supply.

1.3.3 Terminology in Use in Absolute Radiometry: A Glossary

Some of the terms frequently used in absolute radiometry have not been defined formally in any other context. Since variations in terminology have led to considerable difficulties in interpreting and comparing results by different authors, definitions for the most important terms are listed next.

Detector element. The part or parts of the radiometer head consisting of the thermal-radiation detector, the electrical heating element, and the absorber.

Note. Other structural or functional components—such as a substrate, insulating layers, or a thermal diffuser—may also form part of the detector element of an absolute radiometer.

Radiometer aperture. Aperture placed close to the detector element to define the size of the measured beam area.

Note. If the distance between source and radiometer has to be established, it has to be measured relative to the plane of the radiometer aperture.

Diffraction correction. Ratio of the radiant power passing through an arbitrary area to the radiant power passing through the same area, with a baffle or aperture placed between this area and the radiation source.

Note 1. It is assumed that the baffle or aperture has nonreflecting, narrow edges, that the aperture location and dimensions are such that no part of the source is obscured from any part of the receiving area, and that there is no straylight from any part of the environment obscured by the baffles. In that case, the difference in the amount of radiant power passing through the area with and without the insertion of a baffle or aperture is solely due to diffraction.

Note 2. In the case of the diffraction correction for baffles and apertures located inside the radiometer head, the radiant-power source used for determining the basic correction factor should be the reference source defined at the end of this glossary.

Absorption correction. Ratio of the radiant power incident on a detector from a radiation source in a nonabsorbing medium to the incident radiant power from the same radiation source in an absorbing medium.

Refraction correction. Square of the ratio of the optical distances between a source and a detector in a medium with refractive index n_1 with and without the insertion of a plane parallel plate with refractive index n_2 and thickness t.

Lead-heating correction. Ratio of the electrical power dissipated in the heater leads of an absolute radiometer that reaches the detector element, to the corresponding electrical power dissipated directly in the heater element.

Reflection correction. Fraction of the radiant power incident on the detector element that is not converted to heat by absorption in the absence of scattered, diffracted, or case-heating components originating inside the radiometer head.

Note 1. It is assumed that the detector element is opaque.

Note 2. The radiant power source used for determining the basic correction factor should be the reference source defined at the end of this glossary.

Case-heating correction. Ratio of the difference in net power supplied to the detector element from its thermal environment in the radiometer head by way of emission, convection, and conduction when the shutter is open and closed to the radiant power passing through the radiometer aperture in the absence of reflected, scattered, diffracted, or case-heating components originating inside the radiometer head.

Note. The radiant-power source used for determining the basic correction factor should be the reference source defined at the end of this glossary.

Scattering correction. Ratio of the radiant power reaching the detector element of an absolute radiometer by diffuse reflection at apertures, edges, or other surfaces inside the radiometer head to the radiant power passing through the radiometer aperture in the absence of reflected, scattered, diffracted, or case-heating components originating inside the radiometer head.

Note. The radiant-power source used for determining the basic correction factor should be the reference source defined at the end of this glossary.

Nonequivalence correction. Ratio of the total power converted to heat in the detector element of an absolute radiometer in the radiant heating mode to the total electrical power supplied to it in the electrical heating mode, with the two powers producing identical detector outputs.

Note 1. The radiant power source used for determining this correction should be the reference source defined at the end of this glossary.

Note 2. The correction is due to the different temperature distributions in the detector element under radiant and electrical heating.

Correction for nonuniform responsivity. Ratio of the radiant power from the reference source defined at the end of this glossary passing through the radiometer aperture to the radiant power passing through a differently irradiated radiometer aperture, with the two producing identical detector outputs.

Note. The correction is due to nonuniformities in the responsivity of the absolute radiometer across the surface of the detector element.

Dual detector correction. Ratio of the total electrical power converted to heat in the detector element used for the measurement of the radiant power to the total electrical power converted to heat in the compensating detector element, with the two producing identical detector outputs when heated simultaneously with electrical power.

Note. The correction is due to different detector temperature coefficients, different thermal resistances to the environment, and so on. It is only applicable if two detectors in a compensating configuration are used simultaneously, one being exposed to the radiant power to be measured and the other being heated electrically. (See Section 1.3.6.)

Feedback correction. Ratio of the substituted electrical power measured with a feedback-controlled absolute radiometer with zero error signal at the input of the error amplifier to the electrical power measured with a finite error signal at the input of the error amplifier.

Reference source. Source for which the basic set of instrumental correction factors for the radiometer head of an absolute radiometer is determined.

Note. Unless specifically stated otherwise, the reference source is a source with a spectral irradiance distribution equivalent to CIE Illuminant A (CIE, 1986), producing a spatially uniform irradiance distribution over the radiometer aperture, positioned at a distance of one meter from the radiometer aperture and in line with the normal through the center of the meter aperture, fully within the clear field of view of the radiometer head, and with the dimensions of the radiating area not exceeding 50 mm in any direction.

Set of basic instrumental correction factors. Set of instrumental correction factors determined for the reference source just defined and listed in the specification sheet of the absolute radiometer.

Note. In order to facilitate the comparison of the instrumental correction factors of different absolute radiometers, it is recommended that the default reference source just specified be used wherever possible. If the set of basic instrumental correction factors has to be given for another source, this source should be clearly specified.

1.3.4 Historical Review of Absolute Radiometry

1.3.4.1 General

In this section, the history of absolute radiometry will be reviewed with special emphasis on those developments having a major influence on the technology and methodology in use in the field as of 1988. Naturally, it is impossible to discuss every single contribution or publication in this field in the available space. For this reason, a list of additional references is given after the main references to cover further relevant material. However, no claim for completeness is made regarding these supplementary references. In spite of the care taken in surveying the literature in the field, it is possible that isolated important contributions published outside the mainstream of Western literature may have been missed. Among the main identifiable application areas of detector-based radiometry, namely:

(a) traditional photometry and radiometry
(b) solar radiation and space applications
(c) laser applications

(Geist, 1976), there will be a clear emphasis on the first. The reason for this is that absolute radiometry by its very name is concerned primarily with absolute measurements of radiant power. The radiation scales used for solar measurements have been based on nonabsolute instruments for most of this century, while laser-related measurements have a much shorter history, starting only after the invention of the laser in 1959. It must also be remembered that many laser measurements rely on calorimetric methods, which are not of central interest in this context. As also pointed out by Geist (1976), the development in the three forementioned application areas took place along separate lines with very little interaction between the user communities.

1.3.4.2 The Historical Roots of Absolute Radiometry

The emergence of absolute radiometers at the turn of the twentieth century depended on the prior invention of thermal radiation detectors, which in turn depended on a reasonably mature temperature-measurement technology.

Thermal radiation detectors were accessible to people long before the invention of the thermometer in the form of the heat sensors in the human skin. Use was made of these detectors by a number of investigators both in antiquity and during the Middle Ages to study the heating effect of the sun and of artificial radiation sources with the help of such rudimentary optical instruments as the burning glass and the mirror (Putley, 1982). Sactorius in 1612 appears to have been the first to use an air thermometer to study the heating effects of the sun and the moon. The first laboratory experiments with artificial radiation sources and involving the use of a liquid-in-glass thermometer instead of the biological heat sensors in the human skin were apparently carried out at the Accademia del Cimento in Florence, Italy, in 1660 (Putley, 1982). Liquid-in-glass thermometers continued to be used in many studies of thermal radiation ("calorific" or "frigorific" rays) throughout the seventeenth and eighteenth centuries. The need to compensate for temperature drifts of the thermal environment was first pointed out by de Saussure and Pictet during the last quarter of the eighteenth century, but it was Leslie who developed the first truly differential air thermometer in 1797 (Putley, 1982). The first blackened conical cavity absorber attached to a liquid-expansion thermometer was built by Count Rumford in 1804, who also developed a differential air thermometer similar to Leslie's (Putley, 1982). However, a much more sensitive temperature sensor was required before thermal radiation detectors could be used as precise measuring instruments for optical radiation in general. The development of just such a sensor was made possible by the discovery of the thermoelectric effect by Seebeck in 1826 and the complementary heating or cooling process by Peltier in 1834. The thermoelectric effect was put to use for optical-radiation measurements within a very short time after its discovery when Nobili constructed a radiation thermocouple in 1830. This was soon followed by a further refinement, the introduction of the radiation thermopile by Melloni in 1833. Although it consisted of blocks of bismuth and antimony and thus had a fairly large heat capacity, it proved extremely useful for the type of investigations conducted over the next 50 years because of its superior responsivity. The invention of the bolometer, a sensitive temperature sensor based on the change of electrical resistance with temperature, by Langley in 1880 was a further important milestone. Its responsivity considerably exceeded that of contemporary thermopiles, which had not been improved much since their invention by Melloni. However, the following decade also saw a radical improvement in the performance of thermopiles, so that there was eventually little to choose between these two types of thermal radiation detectors.

The first absolute measuring scales for radiant power made use of the calorimetric principle. In an ideal calorimeter of the isoperibol type (Gunn, 1973), a test piece of mass m and specific heat c_m is thermally completely isolated from its environment. If the temperature of the test piece is raised by an amount dT, the supplied heat energy dQ can be calculated from

$$dQ = mc_m\, dT = C\, dT, \tag{1.18}$$

where C is the thermal capacity of the test piece. For a continuous supply of radiant power Φ to the test piece, it follows that

$$\Phi = dQ/dt = C\, dT/dt. \tag{1.19}$$

Equation 1.19 implies that the incident radiant power is equal to the time rate of change of the temperature of the test piece multiplied by its thermal capacity, which in general is a function of temperature.

In practice, of course, it is difficult to realize a calorimeter that is completely isolated from its thermal environment. For small temperature differences $(T - T_0)$ between the test piece and the environment, the heat losses from the test piece can be expressed in the form $\mathrm{b}(T - T_0)$ and the basic calorimetric equation then has the form

$$\Phi = dQ/dt = C\, dT/dt + \mathrm{b}(T - T_0). \tag{1.20}$$

In this case, the calculation of the absorbed radiant power also requires a knowledge of the calorimeter's thermal conductance to the environment (constant b).

The first pyrheliometer based on the calorimetric principle was constructed by Pouillet in 1837. It used the rate of rise in the temperature of a mass of water contained in a thin metal box. Further versions of such instruments were developed; especially those by Crova (around 1877) and Violle (around 1879) were fairly widely used. A very interesting calorimetric method invented by Ångstrom (1886, 1890) cancelled the loss term to the environment by means of a second, compensating element and an effective measuring technique. The calorimeter consisted of two circular copper disks of 30-mm diameter and 5 to 7-mm thickness, which were blackened on the side exposed to the radiation. A thermocouple junction was attached to each plate and, thus, the temperature difference between the two plates could be read with a mirror galvanometer. The two plates were alternately exposed to the incident radiation in such a way that one plate was cooling down while the other one was being heated by the incident radiation. Since the cooling plate was in effect compensating for the heat loss from the heated plate, the incident radiant power could be expressed by the simple relation

$$\Phi = 2C\,\Delta T/(a\,\Delta t), \tag{1.21}$$

where C is again the thermal capacity of the calorimeter plate, ΔT the measured temperature difference, a the absorptance of the absorbing surface, and Δt the exact time required to reverse the temperature difference between the two plates.

Although calorimetric methods for establishing radiant-power scales had thus reached a certain level of sophistication by the end of the nineteenth century, they had some definite drawbacks. They required a knowledge of the material properties of the calorimeter, involved measurements on a dynamic (instead of a steady-state) system, and were neither accurate nor sensitive enough for the many challenging measuring tasks at hand at the time. These included nothing less than the establishment and/or confirmation of the basic radiation laws. As happens with conspicuous frequency in science, the problem was solved simultaneously by two researchers working independently of each other, namely Kurlbaum in Germany and Ångstrom (the person who invented the just-mentioned compensated calorimeter) in Sweden.

1.3.4.3 The Invention of the Electrical-Substitution Principle

Since references to Kurlbaum's role as one of the two founders of absolute radiometry have almost disappeared in the modern literature on this subject, it would seem important to consider Kurlbaum's and Ångstrom's roles in this invention on the basis of the documented historical record. In this regard, there is clear evidence that both Ångstrom and Kurlbaum accepted the fact of their independent and practically simultaneous discovery of the electrical-substitution principle for radiant-power measurements. Ångstrom states in a footnote of his 1899 paper in *Astrophysical Journal* that

> F. Kurlbaum has, in a short notice in the *Berichten der Thätigkeit d. Phys-Techn. Reichsanstalt* in 1891 and 1892 (published in Nov. 1892) and in the *Zeitschrift für Instrumentenkunde*, March 1893, p. 122, suggested a similar principle for the measurement of radiation in absolute measure, which he developed further in *Wied. Ann.*, **51**, 591, 1894. He later employed his method for the determination of the radiation of a blackbody (*Wied. Ann.*, **65**, 746, 1898). We each, therefore, had almost simultaneously, but evidently independently, an idea in many respects identical. But as Kurlbaum himself remarks (*Wied. Ann.*, **51**, l.c.) we suggested entirely different methods of carrying out the idea.

Kurlbaum implied similar sentiments in his 1894 article, although he could undoubtedly claim the first published account of the method. His note in the 1892 reference, cited in Ångstrom's 1899 footnote, translates as follows:

> The temperature rise of the bolometer, which is caused by the incident radiation, can also be produced by means of an electric current. Consequently, radiation may be compared with an electric current. The quantities determining the current are measurable absolutely and one will therefore be able to express the radiation in absolute measure.

However, because of the short time between Kurlbaum's 1892 note and Ångstrom's publication in 1893 and the fact that Ångstrom's publication on the other hand described his technique in more detail, the fairest conclusion would seem to be to credit both men equally with the invention of the absolute radiometer. This view also seems to be shared by the authors of many of the earliest publications on this subject (Callendar, 1911; Guild, 1937) who generally cite the work by both men with equal weight. In spite of this, references to Kurlbaum's work began to fade more and more from the reference lists in the second half of this century. This is probably partly due to the fact that most of Kurlbaum's publications were in German, (which made them accessible to a smaller number of scientists in the field) and partly due to the emphasis on solar-radiation measurements in Ångstrom's work. Solar measurements were the dominant application for absolute radiometers in the first half of the twentieth century and scientific applications (such as Kurlbaum's measurements of the Stefan–Boltzmann constant and their later use as radiometric references in metrological laboratories) were not such active research areas over that period and were therefore cited less frequently.

In his absolute radiometer, Ångstrom (1893) used a thin blackened platinum strip both as the heater and as the main structural component

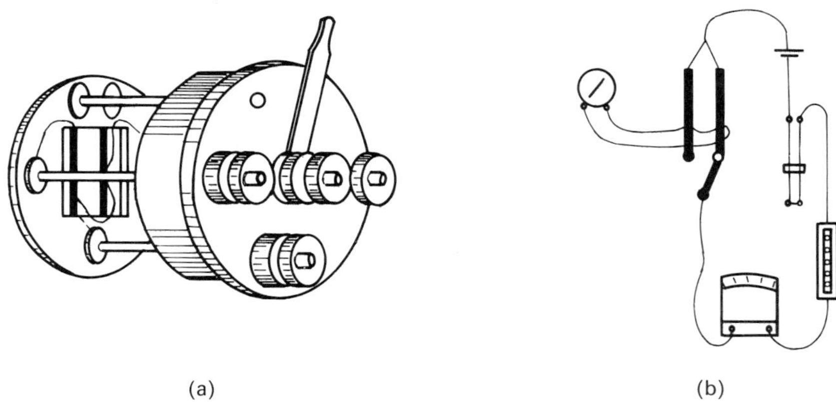

Fig. 1.8 (a) Electric compensation pyrheliometer. (b) Electric circuit diagram. (Ångstrom, 1899)

(substrate) of his detector elements. A thin mica sheet was originally glued to the back of the strip to enable him to attach a copper/German silver thermocouple to the middle of the rear surface. Although Ångstrom did not specify the exact method used to blacken the detector element in his 1893 publication, it is likely that he used the same method he described in his earlier and later publications (Ångstrom, 1886, 1899). There, the platinum strips were coated galvanically with a layer of zinc and then treated with a 1% solution of platinum chloride to obtain a black appearance. The absorption factor was further increased by a thin coating of lampblack, produced by a stearin candle under a copper mesh. Ångstrom estimated the absorption factor thus obtained as 98%. Two strips, closely matched for electrical resistance, were fastened to an ebonite frame as depicted in Fig. 1.8a, with the two thermocouples connected in opposition in a compensating configuration. The circuit diagram is shown in Fig. 1.8b.

Ångstrom mentioned three possible operating modes for his absolute radiometers. In the first, one of the two detector elements was heated by radiation and the second one was heated electrically until the thermocouple output was the same as when neither of the two elements was heated. The second variant involved exposing one of the two detector elements to the incident radiation, while shielding the other one. After noting the thermocouple output via the deflection of the galvanometer, the exposed detector element was also screened. It was then heated electrically until the galvanometer deflection reached the same value as before. That current (I) was recorded and used to calculate the incident irradiance (E) according to

$$aAE = RI^2, \tag{1.22}$$

where a was the strip absorptance, A the strip area, and R the previously determined resistance of the strip. With the area of the detector element simply being the product of the strip length (l) and the strip width (b), the irradiance (E) due to the measured source followed as

$$E = \frac{RI^2}{(lba)}. \tag{1.23}$$

In the third operating mode, Ångstrom irradiated one strip while shading the second one and observed the resulting galvanometer deflection D. Under the assumption that his instrument response was linear, he could then express the relationship between his observed deflection, the strip area A, the strip absorptance a, and the incident irradiance E in the form

$$albE = \mathrm{k}D, \tag{1.24}$$

where k is a constant. Next, he sent an arbitrary heating current I through the irradiated strip of resistance R, which resulted in an increased galvanometer deflection D_1. From the assumed linearity he could then get another equation of the form

$$albE + I^2R = \mathrm{k}D_1. \tag{1.25}$$

From these two equations he deduced the incident irradiance as

$$E = \frac{I^2RD}{(D_1 - D)alb}. \tag{1.26}$$

Testing his three variants on a single source, Ångstrom found that the results agreed to within approximately 2%. For increased accuracy, he suggested a way in which the change of strip resistance with different heating currents could also be taken into account.

In later models, Ångstrom (1899) insulated the thermocouple from the platinum strip with shellac plus silkpaper and used constantan/copper or nickel/copper thermocouples. He also specified the thickness of his platinum strips at that time as 0.001 to 0.002 mm.

Ångstrom's compensation pyrheliometers were recommended for universal adoption as measuring standards in the worldwide network of solar-radiation-monitoring stations by the International Meteorological Organization in 1905. Although the instruments were capable of being used absolutely, they were mostly used in a relative manner by meteorological observatories, with their calibration factors having been derived by comparison with standard pyrheliometers, which were maintained first at Upsala and later at Stockholm. The incident irradiance was then calculated by multiplying the square of the compensation current with the calibration factor. This, presumably, assured a greater consistency for the measured values gathered by the network, which was at that stage more important than the absolute accuracy of the Ångstrom Pyrheliometric Scale. It was later discovered that the exposed strip in Ångstrom pyrheliometers is shaded at the ends by the diaphragm closest to the strip, while the electrically heated strip was heated uniformly over its full length. This "edge-effect" is estimated to have caused measurements to be low by approximately 2% (Drummond, 1970).

In Kurlbaum's (1892, 1894) original instrument, a platinum bolometer strip of 0.001-mm thickness was used as the heater, substrate, and bolometric sensor at the same time. The strip was incorporated in a Wheatstone bridge circuit, with the other three bridge resistors being thick manganin wires. In view of this arrangement, Kurlbaum's absolute radiometers were of the uncompensated type.

The surface of the detector element was blackened with platinum black, which was deposited electrolytically from platinum electrodes in a solution of 1 part of platinum chloride plus 0.008 parts of lead acetate in 30 parts of water (Kurlbaum, 1899). In order to increase the area of the detector element without increasing the width of the bolometer strips (which would have reduced their resistance to an inconvenient value), Kurlbaum (1898) later used two grids of bolometer strips as his detector element. The first grid was placed in front of the other in such a way that the gaps in the front grid were filled in by the strips of the second grid. One of his instruments consisted of 10 strips in front and 9 strips behind, each 35 × 2 × 0.001 mm, with 1-mm gaps between the strips. Another was made up of 5 front strips and 4 rear strips, each 35 × 4 0.001 mm, with gaps of 3 mm between adjacent strips. The total areas of these detector elements were 1003 and 1129 mm², respectively. The relatively large area had the advantage that it could be measured with greater accuracy. A schematic diagram of Kurlbaum's (1898) absolute radiometer is depicted in Fig. 1.9.

In order to measure an unknown irradiance, Kurlbaum first determined the resistance of his detector element (R_1) for a certain bridge current I_1 from the values of the other bridge resistors needed to balance his Wheatstone bridge. Then the detector element was exposed to the radiation to be measured and the bridge was balanced again. This gave a new value R_2 for the resistance of the detector element. Finally, the shutter was closed again and the bridge current was increased until the bridge was balanced at the same value of resistance (R_2). With an absorptance a for the detector element, an area A of the limiting aperture, and the new bridge current I_2, the unknown irradiance followed as

$$E = \frac{R_2 I_2^2 - R_1 I_1^2}{(aA)}. \tag{1.27}$$

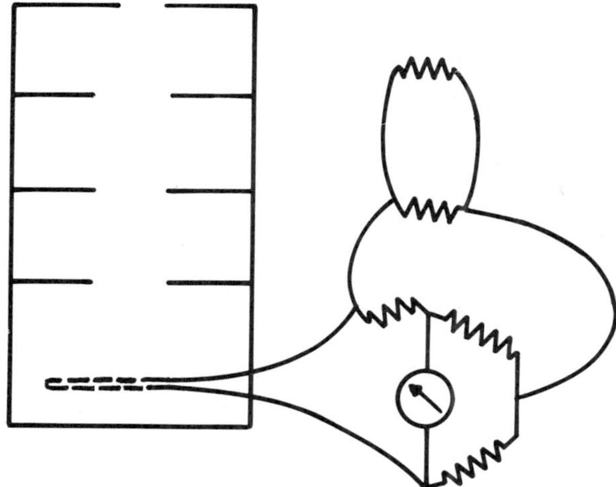

Fig. 1.9 Absolute radiometer. (Kurlbaum, 1898)

From practical experience, Kurlbaum later concluded that it was easier to produce equal galvanometer signals by radiant and electrical heating rather than to balance the bridge each time. In that case, however, the right-hand part of the Eq. (1.25) had to be multiplied by the current ratio I_2/I_1.

Kurlbaum's highly regarded measurements of the Stefan–Boltzmann constant, which were performed on blackbodies of 0 and 100°C (Kurlbaum, 1898), yielded a value of $5.32 \times 10^{-8}\ \mathrm{Wm^{-2}\,K^{-4}}$, which is about 6% lower than the best estimate of this constant in 1985 (Quinn and Martin, 1985). Determinations done before Kurlbaum by calorimetric methods had given values for the constant ranging between 4.23 and $5.06 \times 10^{-8}\ \mathrm{Wm^{-2}\,K^{-4}}$ (Kurlbaum, 1898 using a value for the electrothermal equivalent of 4.185W/[gcal/s]).

Kurlbaum (1899) also did a detailed study of the absorptance of platinum black and lampblack as a function of the layer thickness. He did this by comparing the emittance of surfaces covered with these absorbers to the emittance of a blackbody cavity at the same temperature. He also investigated the difference between the surface and interior temperatures of radiating bodies, which is an important factor in such emittance measurements (Kurlbaum, 1900). For relative measurements, where no electrical compensation was required, Kurlbaum used the bolometer grids in a different configuration. This involved the use of a second, compensating grid detector in the bridge and connecting the front and rear grids of the irradiated and compensating detectors in opposite bridge arms, with a resulting doubling of the responsivity as well as thermal compensation (Lummer and Kurlbaum, 1892; Kurlbaum, 1899).

1.3.4.4 Absolute Radiometry During the First Half of the Twentieth Century

Callendar (1911) considered the relative merits of Ångstrom's and Kurlbaum's constructions and concluded that Kurlbaum's instrument offered definite advantages except for the fact that it was uncompensated. Drawing on his experience with platinum thermometers, his "absolute recording bolometer" (which was first exhibited at a meeting of the Physical Society of London in 1905) used two (unequal) platinum grids in opposite arms of a Wheatstone bridge to achieve thermal compensation. (See Fig. 1.10.)

In order to overcome the disadvantage that the measurements of the radiation and of the equivalent heating current were not simultaneous, Callendar experimented with several other bolometric variants, all of which were analyzed and described in considerable detail (Callendar, 1911).

Arguably, Callendar's most innovative developments were his so-called radio-balances in which the temperature rise due to the absorbed radiant power was compensated or "balanced" against the Peltier cooling effect due to the passage of a current through a thermocouple. In his disk radio-balance (Fig. 1.11), he used two detector elements in a dual compensated configuration.

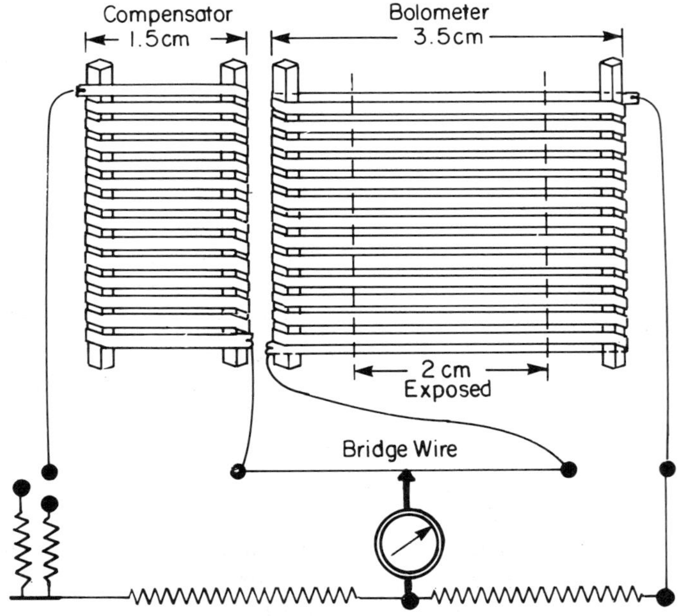

Fig. 1.10 Absolute recording bolometer. (Callendar, 1911)

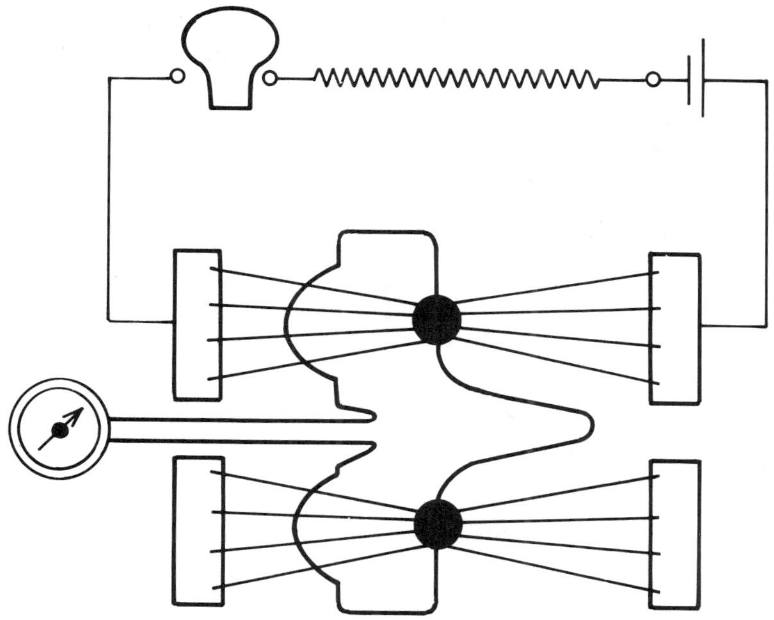

Fig. 1.11 Disk radio-balance. (Callendar, 1911)

Callendar noted that there was an upper limit to the incident radiant power, which could be directly compensated by Peltier cooling. This is due to the Joule heating caused by the passage of the current through the thermocouple wires. The power balance for equilibrium between incident radiant power and Peltier cooling is given by

$$aAE = PI - I^2R = PI(1 - I/I_0), \tag{1.28}$$

where a is the absorptance, A the area of the radiometer element, E the incident irradiance, P the Peltier coefficient of the thermocouple, and R the effective resistance of the Peltier thermocouple. $I_0 = P/R$ is the neutral current, which produces neither heating nor cooling. The Peltier coefficient P is related to the thermoelectric power dV/dT via the relation

$$P = T(dV/dT), \tag{1.29}$$

where V is the voltage produced by the thermocouple and T the absolute temperature.

In order to reduce the uncertainty arising from the absorptance of the black absorber and to be in a position to measure small quantities of power due to the conversion to heat of nonoptical radiations (e.g., the emission from radioactive substances), Callendar also constructed radio balances in which he used cup-shaped detector elements. These used thermopiles for increased responsivity and two Peltier junctions connected in opposition, thus cooling the cup exposed to the radiation and heating the other one.

Although electrical-substitution radiometers were well established and relatively mature instruments by the beginning of the second decade of the twentieth century, calorimetric radiation scales continued to be used in parallel. One of these, in the form of the Smithsonian Pyrheliometric Scale, relied on the silver-disc phyheliometers developed by Abbot (1911) to build up a wide international network of meteorological measuring stations for solar radiation. Their ultimate absolute reference was a water-flow pyrheliometer (a calorimetric instrument) maintained by the Astrophysical Observatory of the Smithsonian Institution. It was established in 1932 after the introduction of an improved water-flow pyrheliometer based on the electrical-substitution principle, that the original scale was in error by about 2.5% (Drummond, 1970).

Coblentz and Emerson (1916) conducted an intensive study of many variants of their absolute radiometer, which basically consisted of a blackened strip of metal in front of a bismuth-silver thermopile with a continuous receiving surface of tin. As there was only one detector element, the absolute radiometer was uncompensated. The thermopile was connected to a galvanometer for the indication of the temperature rise of the blackened metal strip, which served both as the absorber and as self-supporting electrical heating element. (See Fig. 1.12.)

Apart from the airgap separating the blackened absorber–heater strip from the thermopile detector (Coblentz, 1914), another noteworthy feature of these instruments was the use, for the first time, of potential wires in the heater circuit. This eliminated the uncertainty associated with the change of the heater resistance with the heater temperature. The devices were mainly employed in determinations of the Stefan–Boltzmann constant with the aid of a 1000°C blackbody furnace, and the total observed spread of the measured values was close to 4% (2% if 2 of the 13 variants were excluded). Although the exact cause for the spread of the results was not identified, the major part of the blame was apportioned to the absolute detectors. However, more recent experience with measurements of the Stefan–Boltzmann constant (Blevin and Brown, 1971; Quinn and Martin, 1985) has shown the crucial importance of diffraction phenomena, especially with such a relatively long-wavelength spectrum as that of a 1000°C blackbody. These could conceivably have accounted for a sizable portion of the observed scatter and the absolute radiometers developed at the NBS were perhaps not as unreliable as they seemed on the basis of these measurements. In spite of the scatter, however, the mean value of $5.72 \times 10^{-8}\,\mathrm{Wm^{-2}K^{-4}}$ obtained for the Stefan–Boltzmann constant was within less than 1% of the best 1985 estimate.

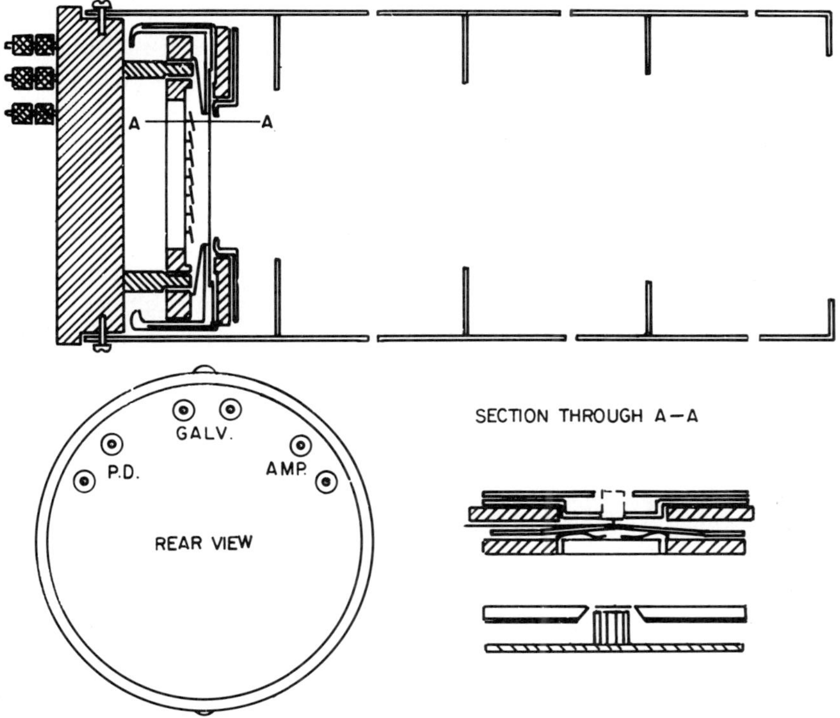

Fig. 1.12 Absolute thermopile. (Coblentz and Emerson, 1916)

Based on its unfavorable assessment of absolute radiometers, the NBS proceeded to base its radiometric scale, first, on the radiant exitance of a blackbody according to the Stefan–Boltzmann law and, later, on the radiance of a blackbody as given by Planck's law (Coblentz and Stair, 1933; 1970; Kostkowski et al., 1970).

In order to check the accuracy of the radio-balances of Callendar (1911), which had until then been used to maintain the radiation scale at Britain's National Physical Laboratory, Guild (1937) embarked on the construction of a new type of absolute radiometer. He wanted to make it as different as possible from Callendar's device in order to facilitate the identification of possible systematic errors in the radiometric scales realized with the two instruments. Therefore, he used relatively heavy copper disks, 45 mm in diameter and 6 mm thick, as the substrates for his detector elements. Two of the copper-disk substrates were cemented to the opposite sides of an ebonite ring, with the inner disk surfaces separated by about 10 mm. A thermopile consisting of seven copper–eureka thermocouples in parallel was connected between the two inner disk surfaces. The copper body separating the eureka wires represented the second leg of the thermocouples formed in this way. An insulated electrical heating element was attached to the outer surface of one of the disks. It was made either of 0.007-mm-thick platinum foil, divided into a zigzag pattern of conducting strips by a series of fine cuts, or of coiled copper wire attached to the disk by means of a special cement. The pattern of the two heating elements is depicted in Fig. 1.13a.

One current lead and one potential lead was attached to each end of the heating element. The outer surfaces of both disks were then coated with lampblack by dry-smoking over the flame of a turpentine lamp. Each detector unit, which consisted of one copper disk with an electrical heating element and one without, was mounted in the ebonite ring and housed in a massive tubular brass case. The arrangement is depicted in Fig. 1.13b.

In order to make his instrument as different as possible from Callendar's, Guild not only used a completely different construction but also a different measuring technique, his so-called temperature-drift method. He regarded it as a particular challenge since previous instruments of that kind such as Pouillet's pyrheliometer, Abbot's silver-disk pyrheliometer, and the water-stir pyrheliometer at the Smithsonian Institution had not succeeded as accurate absolute instruments. Guild concluded that their main drawback was their reliance on direct calorimetric principles instead of the more accurate electrical-substitution principle. He also analyzed the thermal behavior of his absolute radiometer theoretically and succeeded in working out an effective strategy to obtain accurate results with the drift technique.

On exposing the absolute radiometer to radiant or electrical heating, the rate of change in its temperature is greatest at the beginning and decreases continually until a steady condition of equilibrium is reached between the supplied power and the heat losses. The drift rate in a given time interval is therefore not constant but depends on the state of the system at the start of the heating or cooling interval. In order to overcome this difficulty, Guild preheated the instrument for a certain period based on his theoretical investigation, which ensured that he could operate in a cyclic condition for heating and cooling independent of the exact initial conditions. When the instrument had been preheated sufficiently with radiant power, a galvanometer reading (d_0) was taken, and the heating was continued for another 20 seconds and then stopped by the insertion of a shutter between the radiation source and the radiometer. The decreasing galvanometer signal was then observed until it again reached the previously recorded value and the time at that instant was noted. Exactly 100 s later, another reading was taken (d_1) and the cooling was continued for about another 20 s before the shutter was opened again. The transition of the signal through d_1 was timed; exactly 100 s later, the signal d_2 was recorded. After another 20 s, the shutter was closed again; 100 s after the signal had reached the level d_2, another signal (d_3) was

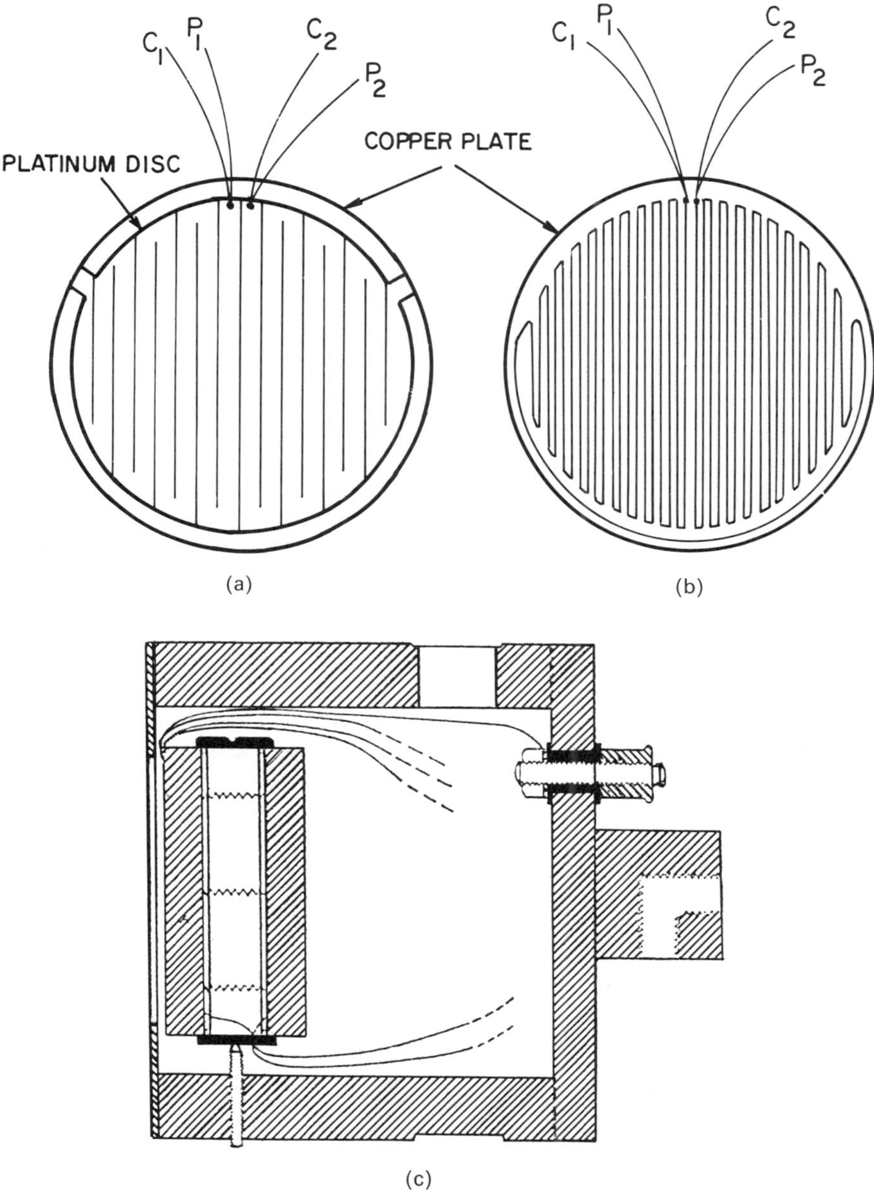

Fig. 1.13 (a) Foil heater. (b) Wire heater. (c) Absolute radiometer. (From Guild, 1937; by permission of the Royal Society, London)

recorded. On theoretical considerations, Guild used twice the observed signal rise plus the falls preceding and following it, i.e.,

$$R_1 = (d_0 - d_1) + 2(d_2 - d_1) + (d_2 - d_3) \tag{1.30}$$

as the significant quantity for each such group of observations. Twenty seconds after the observation d_3, the electrical heating of the absolute radiometer was switched on. Based on previous trials and experience, it was usually possible to set the electrical power to within better than 5% of the absorbed radiant power. An identical heating and cooling sequence was then followed as in the case of the radiant power, yielding another significant quantity E_1. The incident radiant power Φ was then calculated according to

$$\Phi = (R_1/E_1)IV/a, \tag{1.31}$$

where I and V were the measured heater current and voltage, respectively, and a was the absorptance of the lampblack absorber.

Apart from the correction for the reflection loss from the absorber, Guild also investigated two further potential error sources in his absolute radiometer. The first of these was the contribution of the heat generated in the current leads, and the second dealt with the nonequivalence between the radiant and electrical powers, which cause equal detector signals.

When Guild finally compared his two instruments with Callendar's radiobalance, he had the satisfaction of finding agreement to 0.1% between instruments intentionally constructed and operated in ways as different as possible. He also compared the radiometric scales realized by Britain's National Physical Laboratory (NPL) and the United States National Bureau of Standards (NBS) with four carbon-filament lamp standards of total irradiance purchased by the NPL from the NBS in 1929. Although their irradiances were rather low for the absolute radiometers, he used the radiobalance together with a very sensitive galvanometer and appropriate modifications to the measuring procedure and obtained results that agreed with the NBS values to better than the estimated uncertainty of the comparison (0.2%). Since the irradiance levels of the NBS lamps were in the order of 0.06 mW/cm^2 at the specified distance of 2 m, it can be inferred that the noise-equivalent irradiance in these measurements must have been in the order of a few hundred nW/cm^2, which is exceptional with instruments designed to operate at about a thousand times higher irradiance levels. In a further comparison, Guild used an Abbot silver-disk pyrheliometer calibrated at the Smithsonian Institution and found that this instrument read 2.2% higher than the NPL scale. This was within 0.1% of the Smithsonian Institution's own assessment of the amount by which its scale was in error. Finally, Guild also made a comparison with an Ångstrom pyrheliometer calibrated at Upsala. This comparison suffered from the fact that the absorber layer of the pyrheliometer was in a poor condition when Guild received it. After removing it and depositing a fresh layer, he found the instrument to read 0.45% lower than the NPL scale.

1.3.4.5 Absolute Radiometry Between 1950 and 1974

Rutgers (1951) used a mica foil of 0.01-mm thickness and 30 × 10 mm dimensions as the substrate for his absolute radiometer. An evaporated manganin layer, about 0.001 mm thick and with a resistance of 5 to 10 Ω, served as the heating element. The bolometric detector was an aluminum layer of about the same thickness and resistance, which was evaporated onto the opposite side of the mica substrate. The detector element was mounted between two metal clamps and was coated on its front surface with a mixture of lampblack, alcohol, and some shellac by means of a glass spray gun. The thickness of the absorber layer amounted to 0.006 to 0.010 mm. Two identical detector elements were mounted back to back on a rigid heat-sink and inserted in a brass case with two brass tubes opposite the strips to admit the radiation to be measured.

An important event for absolute radiometry occured in 1956, in the run-up to the third International Geophysical Year (IGY) of 1957. In that year, the World Meteorological Organization (WMO) accepted a proposal for the establishment of a single international pyrheliometric scale, the International Pyrheliometric Scale (IPS) of 1956 (IGY, 1958). It had its origin in the frustrations inherent in the interpretation of solar measurements, which were being reported in a variety of different revised and unrevised versions of pyrheliometric scales, whose differences to each other and to the absolute radiometric scale were difficult to establish. The IPS of 1956 was based on a compromise, in which a difference of 3.5% was assumed between the Ångstrom scale of 1905 and the revised Smithsonian scale of 1913. Although the new IPS of 1956 was originally implemented quite enthusiastically, it did not succeed in the longer term. This was partly due to the fact that it was an artificial scale based on artifacts without a defined relationship with the basic units of measurement and partly because the definition was ambiguous (Fröhlich, 1973). In order to overcome the ever more confusing state of affairs with solar-radiation measurements, quite a number of meteorology laboratories began to pursue vigorous programs in absolute radiometry, which to them offered the only realistic prospect of obtaining universally meaningful results. This development has contributed substantially to the progress achieved in absolute radiometry since then.

Eppley and Karoli (1957) made use of a silver substrate, 10 mm in diameter and 0.01 mm thick, to which they glued a bifilar spiral of 0.025-mm-thick platinum wire. The platinum spiral served both as a heating element and as a bolometric detector and had an electrical resistance of 100 to 200 Ω. The side of the silver disk opposite the detector–heater coil was coated with either lampblack or goldblack and the mounted detector element was placed in a housing. The resistance of the platinum coil was measured with a four-lead Mueller temperature bridge and there was thus no compensating element to eliminate temperature drifts. For this reason, elaborate temperature lagging and thermostatting of the measuring system was essential.

The instrument was used in measurements of the Stefan–Boltzmann constant, yielding a value of $5.77 \times 10^{-8}\,\mathrm{Wm^{-2}K^{-4}}$ (i.e., nearly 2% above the best 1985 estimate). Diffraction at the relatively small apertures is a likely reason for some of this discrepancy.

An absolute radiometer similar to the instrument constructed by Ångstrom (1893) was built at the Mendeleev Institute of Metrology (USSR) by Kartachevskaia (1961, 1962). The instrument, which had already been constructed during 1949, consisted of two manganin strips of 20-mm length, and 1.5-mm width covered with platinum black. A copper-constantan thermocouple was connected between the two strips to indicate their temperature difference. This instrument was the first to be used for a measurement of the maximum luminous efficacy of radiation (K_m), an important factor linking photometry and radiometry.

An important basis for the development of modern absolute radiometry was laid by Gillham (1962), who verified the NPL radiometric scale anew with three absolute radiometers of his own design. Apart from the advances inherent in the design of his radiometers, Gillham's other major contributions include his extremely detailed analysis of the error sources of absolute radiometers and the development of sophisticated techniques for measuring the various corrections.

Before embarking on the construction of his new absolute radiometers, Gillham carefully reevaluated Guild's instruments. He extended the reflectance measurements on the absorber layer to 1.3 μm and was apparently also not convinced by the evidence presented by Guild that the correction for the nonequivalence between radiant and electrical heating was indeed negligible. For this reason, he developed an experimental technique to measure the correction (see Chapter 6), which yielded a value of 0.4% for the smoked carbon absorber coating as used by Guild.

Further, he also changed the measuring sequence, alternating the two forms of heating without any cooling periods in between. He was thus in effect recording the differences $E - R$ and not E and R separately as in Guild's method. To find E/R, he had to make an additional recording of radiant heating cycles alternated with cooling. When recording the differences $E - R$, he pointed out the need for a rapid and accurate changeover from radiant to electrical and back to radiant heating as any break or overlap of duration t would result in an error of t/τ, where τ is the time-constant of the radiometer. For Guild's instruments, τ amounted to four minutes. While noting that the new method gave satisfactory agreement with Guild's original technique, Gillham did not analyze the new method in terms of the theoretical criteria that had led to Guild adopting his particular technique in the first place.

Gillham indicated that a gradually emerging difference between the two Guild radiometers had been observed at the NPL, with the wire-grid radiometer eventually measuring about 0.5% higher than the platinum-grid one. This difference seems to have been the main reason for Gillham's construction of the three new models, one of which was intentionally made very similar to Guild's radiometers. The other two were designed primarily with the objective of making measurements simpler and quicker. All three of the new models shared with the Guild radiometers the common feature of a circular detector element with a metal substrate and were provided with thermocouples for indicating its average temperature. The provision of a heating element that ensured a temperature distribution similar to the one due to the absorbed radiation was another common feature. However, they differed by their complete symmetry, incorporating two fully functional detector elements, and by their emphasis on achieving a uniform temperature throughout the instrument. They were provided with two current leads on each end of the heating element and had their thermojunctions in series rather than in parallel for increased responsivity. The metal disk was glued, in each case, to a circular disk of thin mica of somewhat larger diameter, which had the electrical heater deposited on its front surface. The mica disk was clamped to a metal ring of large enough diameter to ensure that the heat losses by conduction through the mica were much smaller than the total heat loss from the detector element. The heating element was deposited onto the mica substrate by first sputtering a thin film of tin oxide onto the surface to ensure good adhesion. A layer of gold about 20 μm thick was then evaporated over the tin oxide through a mask, which ensured that the electrical heating pattern matched the area heated by the radiation as closely as possible. The new Guild radiometer is depicted in Fig. 1.14.

The second type of absolute radiometer was designed for speedier and easier operation than the drift radiometers. In order to achieve a time-constant of about 10 s, the aluminum disk was shrunk to a diameter of 10 mm and a thickness of 0.5 mm. The detector element of the new radiometer is shown in Fig. 1.15a and its heating element in Fig. 1.15b.

As is evident from Fig. 1.15a, the thermopile detector was not connected directly between the two detector elements but rather to the heat-sink located between the two detector elements. The separate thermopiles of the two detector elements were connected in opposition to obtain compensation. The absorber coating used was specularly reflecting and was deposited on top of

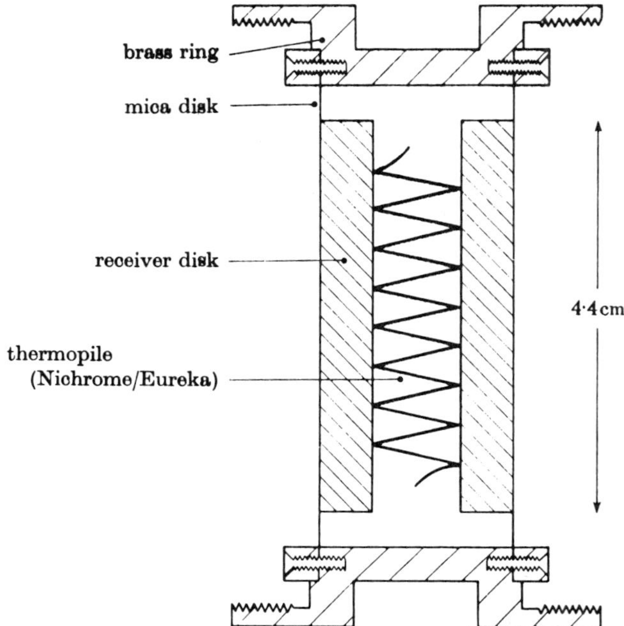

Fig. 1.14 New Guild radiometer. (From Gillham, 1962; by permission of the Royal Society, London)

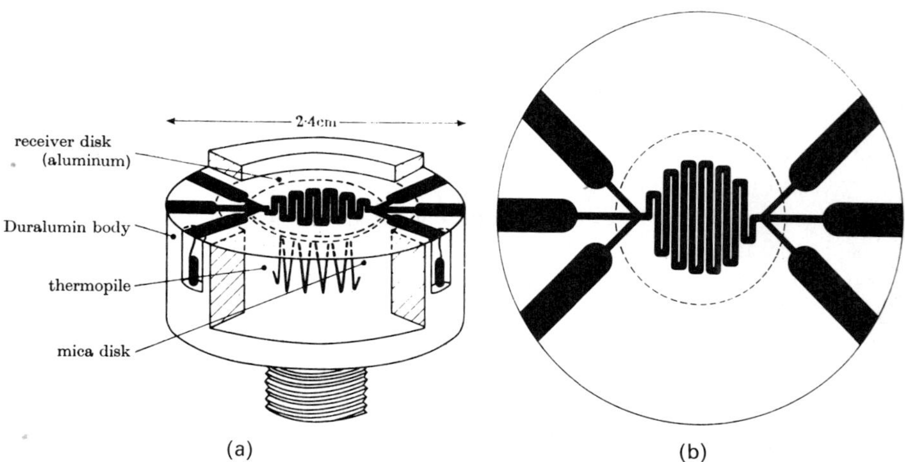

Fig. 1.15 (a) Disk radiometer. (b) Heater element. (From Gillham, 1962; by permission of the Royal Society, London)

the heating element by sputtering about 50 alternate layers of tin oxide and platinum to a total thickness of about 1 μm. This type of absorber was electrically nonconducting and had a reflectance that increased smoothly from about 6.5% at 0.5 μm to about 8.5% at 1.4 μm and 13% at 2 μm.

The third type of absolute radiometer constructed by Gillham was the blackbody radiometer, which was quite similar in most respects to his disk radiometer. The only major difference was that the detector element was in the form of a cavity as depicted in Figs. 1.16a and 1.16b.

The cavity was assembled from three separate tubular sections, each of which had its own radial thermopile attached to it. These three sections were

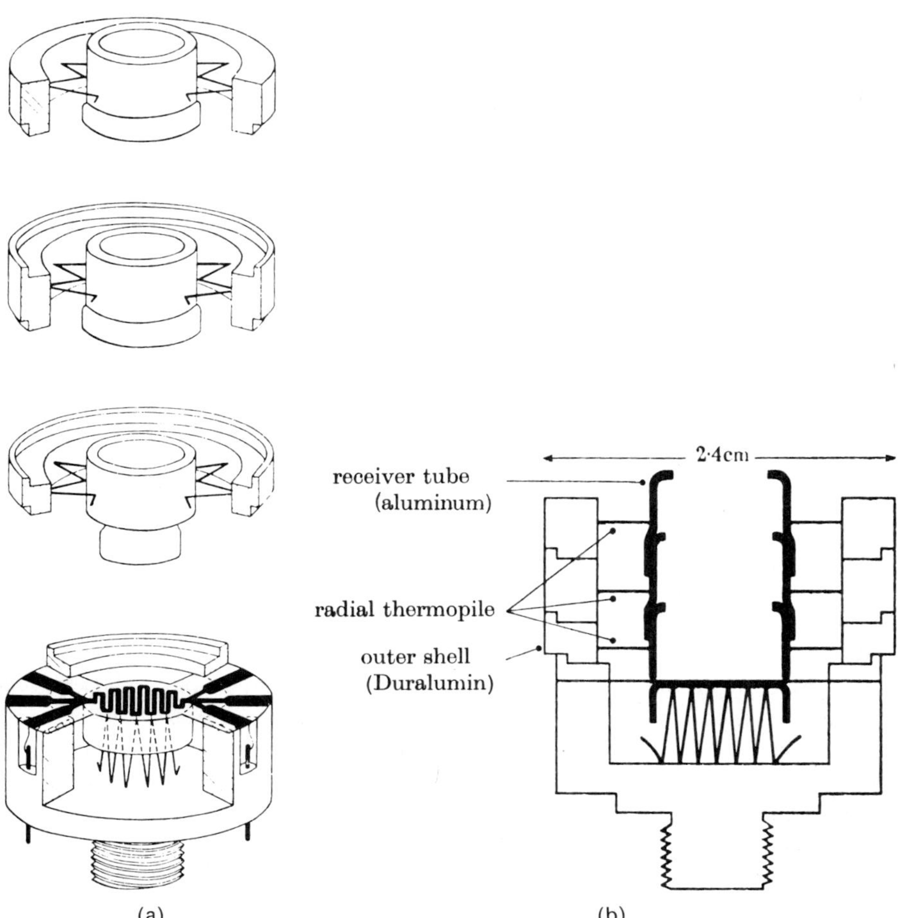

Fig. 1.16 (a) Blackbody radiometer before assembly. (b) Blackbody radiometer after assembly. (From Gillham, 1962; by permission of the Royal Society, London)

painted internally with matt black paint, assembled into a unit, and mounted on top of a disk thermopile, which formed the bottom of the cavity. In order to achieve a uniform responsivity over the whole surface area of the cavity, the number of thermojunctions per cylindrical section was varied and the final adjustment consisted of shunting the different thermopiles with appropriate shunt resistors before connecting them in series.

After completing the construction of the new instruments, Gillham characterized them fully in terms of their absorber reflectances, the thermal resistance of their absorbers, the lead-heating correction (according to his new technique involving the double current leads on each side of the heating element), the case heating correction, and the edge reflection from the limiting apertures.

The three new models, each consisting of two fully functional absolute radiometers, were finally compared with Guild's absolute radiometers. The results of the comparison showed that the three new instruments (equivalent to six absolute radiometers) agreed to about 0.2%, while Guild's platinum-grid instrument differed from them by about 0.5% and the wire-grid instrument by about 1%, discrepancies that were considered significant at the uncertainty level in the comparison. However, it is difficult to infer a difference between Guild's 1937 and Gillham's 1962 radiometric scales from this result. The Guild radiometers used in the comparison had been modified with new absorber layers, had apparently suffered a displacement of their heater leads, and were no longer used in their original drift sequence. From other indirect evidence cited by Gillham, it seems unlikely that Guild's original 1937 scale differed from Gillham's by more than 0.5%.

While Gillham's work provided an important basis for the development of modern absolute radiometry, a further significant impetus for the emergence of a considerable number of new programs in absolute radiometry at national measuring-standards laboratories was given by Preston's (1963) proposal for "a radiometric method of perpetuating the unit of light." It had been proven very difficult and time-consuming to achieve a reasonable level of accuracy in light measurements based on the 1948 SI definition of the candela (the "unit of light"). That definition was based on the luminance of a blackbody radiator at the freezing point of platinum, a concept that proved to be rather less reproducible in its practical realization than had been originally assumed. Preston showed that a fixed material standard for the candela was not required if the maximum luminous efficacy of radiation was fixed instead by definition at a wavelength of 555 nm. In that case, it would be possible to calculate the illuminance of any radiation source from the irradiance measured through a filter with a spectral transmittance similar to the spectral luminous efficiency function $V(\lambda)$ of the average human eye, as long as the spectral transmittance of the filter was known. In that way, photometry would be firmly based on radiometry and photometric units could be derived with radiometric techniques. Preston also demonstrated the technique by determining K_m experimentally as equal to 685 lm/W. An absolute radiation scale based on an absolute radiometer was a key element in the proposed new procedure, which proved to be one of the main stimulators for absolute radiometry in the following decades. Gillham (1964) was the first to follow up Preston's work, arriving at a value of 686 lm/W for K_m.

In 1965, Ooba reported that absolute radiometry had been pursued in Japan since 1959 with the aim of basing photometric standards on a radiometric basis. The absolute radiometers used in that work (Ooba, 1965) were closely similar to those reported earlier by Eppley and Karoli (1957). They incorporated a 0.02-mm-thick silver substrate 10 mm in diameter plus a bifilar spiral of platinum wire, insulated from the silver disk by a thin layer of shellac and of 0.025-mm thickness.

The front surface of the silver disk was first coated with an insulating layer, whereafter a layer of goldblack was evaporated. The detector element was finally mounted in a brass ring with two threads of silk (Fig. 1.17a) and the platinum spiral was connected like a three-wire platinum resistance thermometer to a modified Wheatsone bridge (Fig. 1.17b).

The four constructed radiometers (uncompensated) showed a maximum scatter of 0.5%. In order to check the equivalence between the two forms of heating, one of the radiometers was coated with goldblack on both the front

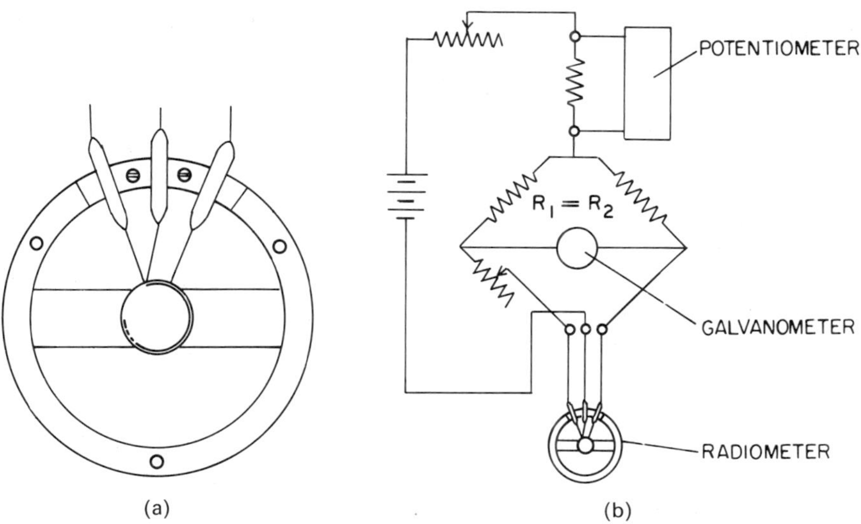

Fig. 1.17 (a) Detector element. (b) Circuit diagram. (From Ooba, 1965; by permission of the Bureau International des Poids et Mesures, Sevres, France)

and rear surfaces. Values obtained by irradiating either the front or rear surface did not differ significantly and the nonequivalence correction was therefore assumed to be negligible.

Both the need for a more accurate radiometric scale in meteorology and at metrology laboratories and the possible redefinition of photometric units in radiometric terms were given as the reason for the development of a new scale for optical radiation in Australia (Blevin and Brown, 1967). Because of uncertainties in the value of the Stefan–Boltzmann constant and in the thermodynamic temperature of various freezing points of metal, which were useful for the operation of blackbody radiators, it was decided to base the new radiometric scale on absolute radiometers.

Two different types of absolute radiometers were constructed: one based on a thermopile as the thermal-radiation detector and the other on a bolometer. In both cases, the most recent work in the field was taken as the point of departure.

The absolute thermopile was largely a copy of Gillham's (1962) disk radiometer, with the exception of the absorber layer, which was goldblack of 0.001-mm thickness (Blevin and Brown, 1966). The substrate for the detector element consisted of an aluminum disk, 10 mm in diameter and 0.5 mm thick. A 0.05-mm–thick mica disk was glued to its front surface with shellac to insulate the evaporated heating element. A thin layer of lacquer insulated the heating element from the goldblack absorber. The temperature change of the detector element was sensed with the aid of 28 copper-constantan thermocouples connected in series. The reference junctions were connected to the radiometer case behind the detector element and two radiometers were used as in Gillham's case for thermal compensation. Two pairs of absolute thermopiles were constructed in total.

Blevin and Brown's absolute bolometers were improved versions of Rutgers's (1951) instrument. They consisted of 20-mm–wide strips of 0.025-mm–thick Melinex sheet, 40 mm in length and with a nickel layer evaporated on both sides to serve as heater and bolometric detector, respectively. The detector element was clamped between copper jaws at each end. The goldblack absorber was insulated from the heated layer by a 0.01-mm–thick layer of glossy black paint. A metal heat-sink was mounted 1 mm behind the detector elements to improve the uniformity of the responsivity across them and they were then mounted in pairs in a thick-walled aluminum enclosure as depicted in Fig. 1.18. Three bolometric radiometer elements were made in all.

The absolute radiometers were compared with the aid of gas-filled incandescent lamps combined with a 14-mm–thick plate of crown glass to eliminate radiation at wavelengths longer than 2.8 μm. The lamps were operated at a color temperature of 2856 K and were positioned at a distance of 1 m from the absolute radiometers, resulting in irradiance levels of

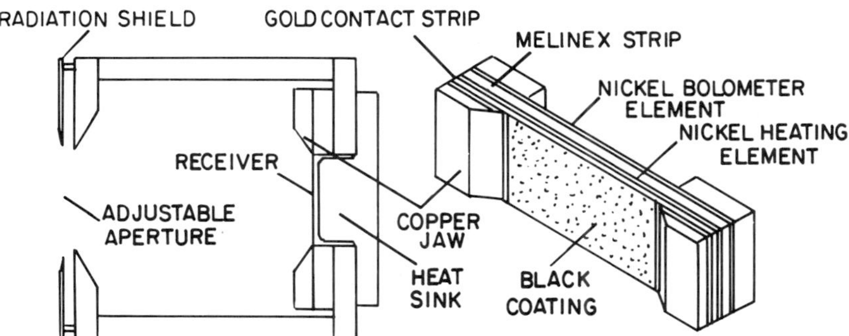

Fig. 1.18 Absolute bolometer. (From Blevin and Brown, 1967; by permission of *Australian Journal of Physics*)

approximately 1 $mWcm^{-2}$. Corrections were made for absorption by water-vapor bands as a function of the pathlength and of the vapor pressure. The spread of the seven instruments was approximately 0.2 %, with the mean of the absolute thermopiles and of the absolute bolometers agreeing to better than 0.05 %.

A prototype of an absolute radiometer was constructed at the Institut Royal Meteorologique de Belgique (IRMB) by Crommelynck (1967). The main motivation for the project was to contribute to the establishment of a universal and absolute pyrheliometric scale, but the potential use of absolute radiometers for photometric as well as other radiometric purposes was cited as an additional attraction. The detector element consisted of a square manganin strip of 0.05-mm thickness and 40-mm length, which was used as the heating element and was blackened at the front surface with Black Pearson's Optical Black. A commercial thermopile was attached to the rear surface. In order to increase the effective absorptance of the device, the detector element was placed at the center of a hemispherical mirror. The ratio of the mirror opening to the mirror radius was 0.1 and the use of the mirror increased the absorptance for a source of 2900 K color temperature by 5 %. Two identical detector elements were mounted side by side in separate hemispherical reflectors and connected in a compensating configuration (Fig. 1.19).

Circuits for controlling the electrical heater power by means of electronic feedback were also developed (Crommelynck and Vandenborre, 1967; Crommelynck and Van de Velde, 1972).

At the West German Physikalisch-Technische Bundesanstalt, Bischoff (1968) continued the institute's long tradition in absolute radiometry with the construction of a new absolute radiometer (Figs. 1.20a and 1.20b). Its detector element consisted of a 0.02-mm-thick circular mica disk of 55-mm

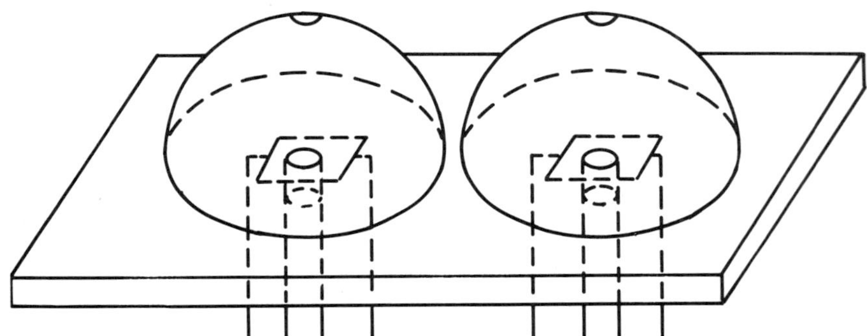

Fig. 1.19 Experimental absolute radiometer. (Crommelynck, 1967)

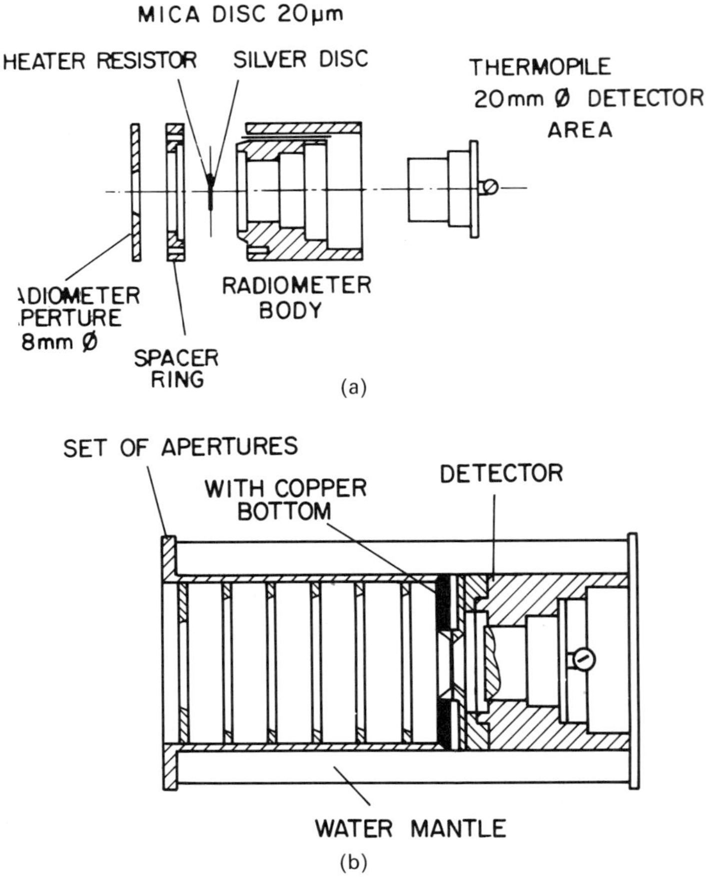

MICA DISC 20µm

HEATER RESISTOR

SILVER DISC

THERMOPILE 20mm Ø DETECTOR AREA

ADIOMETER PERTURE 8mm Ø

SPACER RING

RADIOMETER BODY

(a)

SET OF APERTURES WITH COPPER BOTTOM

DETECTOR

WATER MANTLE

(b)

Fig. 1.20 (a) Detector element. (b) Assembled absolute radiometer. (From Bischoff, 1968; by permission of *Optik*)

diameter, with an evaporated heating element (15-mm diameter) on the front and a 0.15-mm-thick silver disk (20-mm diameter) attached to the rear surface. The heating element was covered with an unspecified absorber layer of 20-mm diameter and with a spectral reflectance of 2.66 % at 546 nm. The thermal radiation detector was a commercial thermopile, which was mounted 2 mm behind the rear surface of the silver disk. That surface was also coated with matt black paint to facilitate the maximum heat transfer by radiation to the thermal radiation detector behind it. A steel aperture of 2.5 cm^2 was mounted 5 mm in front of the detector element, which was used together with a second, identical detector element in a compensating configuration. Each detector element together with a set of apertures was housed in a cylindrical tube with a water mantle to ensure a stable thermal environment.

In order to check the consistency of radiometric scales maintained by metrology laboratories and also to evaluate the potential of radiometric methods for establishing the unit of light, the Consultative Committee on Photometry (CCP) of the International Committee on Weights and Measures (CIPM) decided in 1962 to ask the United Kingdom's National Physical Laboratory to organize an international comparison of radiometric scales (Betts and Gillham, 1968; Bonhoure, 1969). The method of comparison chosen consisted of circulating batches of three gas-filled incandescent lamps of the type Osram W 40/G to the participating laboratories to be measured according to a precisely defined procedure. The lamps were run at a color temperature of 2700 K and were used together with a 1/8″-(3-mm) thick filter of white plate glass placed at an angle of 22.5° with the optical axis halfway between the lamp and the detector. At a measuring distance of 1 m, the resulting irradiance was about 0.8 $mWcm^{-2}$. A correction was applied by the participants for absorption by atmospheric water vapor and the lamps were measured at the NPL both before being sent to participants and after being received back from them.

Eight metrology laboratories, including the NPL, participated in the comparison: Deutsches Amt für Messwesen und Warenprüfung der DDR (DAMW), German Democratic Republic; Electrotechnical Laboratory (ETL), Japan; Mendeleev Institute of Metrology (VNIIM), USSR; National Bureau of Standards, United States; National Research Council (NRC), Canada; National Standards Laboratory (NSL), Australia; Physikalisch-Technische Bundesanstalt (PTB), Federal Republic of Germany; and the National Physical Laboratory, United Kingdom. Except for the DAMW, the scales of all participating laboratories were based on recent and independent work. Those of the ETL, VNIIM, NPL, NSL, and PTB were realized with absolute radiometers, while those of the NBS and NRC relied on blackbody radiators. Their total spread was 3 %, but six of the eight participants agreed to just over 1 %. The participants whose results differed most from the mean either had not based their scales on recent or independent work (DAMW) or had suffered from difficulties with lamps undergoing large changes during transport (NBS). However, in spite of the larger-than-expected scatter of the overall results, the scales realized by means of absolute radiometry had a spread of less than 1.2%. These results were encouraging enough for participating laboratories to proceed with their programs in absolute radiometry and certainly did not discourage further metrology laboratories from joining the increasing international activity in this area.

As a consequence of the importance of accurate radiometric measurements for the U.S. space program (e.g., for the thermal equilibrium testing of spacecraft, for the calibration of instruments used for remote sensing of earth resources, and for other scientific purposes), a series of sophisticated and novel absolute radiometers were developed at the Jet Propulsion Laboratory (JPL) in Pasadena, California (Kendall and Berdahl, 1970). The most essential feature of these was the use of cavity-type detector elements.

The first of the designs, the so-called standard cavity radiometer, was developed primarily for measuring the irradiance from solar simulators in space chambers. It consisted of a massive thermal guard of copper, gold-plated on both sides, which enclosed the silver measuring cavity spun from a 0.13-mm-thick silver sheet and held in position by a thin glass tube attached to the apex of the cavity (Fig. 1.21a).

A silver cylinder with a wall thickness of 0.5 mm and in thermal contact with the guard was located between the two. The heater winding and the platinum-wire temperature sensor attached to the outside surface of the thermal guard were incorporated in a servo-controlled feedback system, which kept the guard temperature constant at a preselected value. The platinum-wire temperature sensor, wound around the intermediate silver cylinder, was accurately calibrated for the precise measurement of the actual guard temperature. A further accurately calibrated platinum-wire thermometer was wound around the cylindrical outer surface of the cavity detector element. It was incorporated in a Wheatstone bridge circuit and also served as a heater, via a variable bridge current, to keep the cavity temperature exactly the same as the guard temperature. The inside of the cavity was coated with Parson's black lacquer with a thermal resistance of 2.5 KW^{-1} cm^2 in air and an emittance of 0.925 at 26°C and of 0.946 at 136°C. The corresponding effective cavity emittance thus varied from 0.987 to 0.9911 over the same temperature range. While the instrument had a thermal time-constant of about 30 minutes, it was possible to obtain full accuracy after one minute because of the efficient operation of the manual feedback loop with electronic gain.

The "standard cavity radiometer" was used to measure the Stefan–Boltzmann constant by inserting it into an opening in the wall of an evacuated

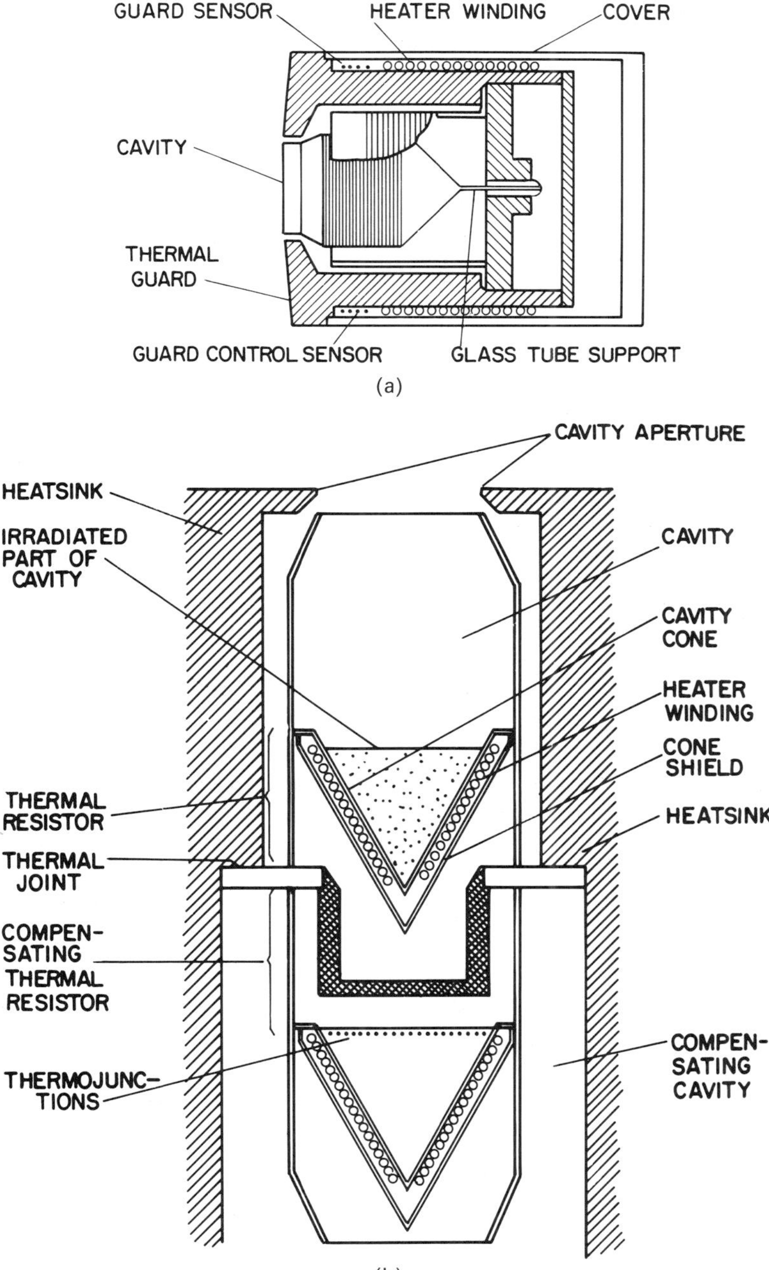

Fig. 1.21 (a) Standard cavity radiometer. (b) Primary cavity radiometer. (From Kendall and Berdahl, 1970; by permission of *Applied Optics*)

cavity, which was cooled to 77 K with liquid nitrogen. The radiometer, which was operated at a number of different set points from 26 to 135°C, was a nett radiator in the radiation interchange with the cooled cavity. The electrical power needed to maintain the cavity detector element at a given temperature was then related to the Stefan–Boltzmann constant via the Stefan–Boltzmann radiation law. Additional factors needed in the calculation included the area of the radiometer aperture, the emittances of the cavity detector element and of the cooled cavity, the thermodynamic temperatures of the cavity detector element and of the cooled cavity, and the instrumental corrections of the absolute radiometer. A value of $5.687 \times 10^{-8}\ \mathrm{Wm^{-2}K^{-4}}$ was found for the constant, which differed by only 0.3% from the best estimate in 1985, well within the estimated uncertainty of the experiment. Apart from the careful execution of the experiments and the quality of the absolute radiometers, it is clear in retrospect that the excellent result was also due to the chosen measurement geometry. It involved no intermediate apertures and thus avoided the large diffraction effects that had bedevilled practically all of the previous measurements of the Stefan–Boltzmann constant.

The second type of absolute radiometer developed at the JPL and referred to as "primary cavity radiometer" was designed for use in irradiance measurements under ordinary ambient conditions. When used together with a view-limiting attachment with an acceptance angle of about 5 degrees, it lent itself particularly well to solar irradiance measurements. With the addition of a temperature-controlled heat-sink, it could, however, also be used in a space simulator environment. A schematic diagram of the primary cavity radiometer detector element is shown in Fig. 1.21b.

The assembled radiometer incorporated an (optional) view-limiting tube and a so-called muffler tube, which also had the effect of shielding the cavity detector element from convection. The latter was again internally coated with Parson's black lacquer and the heat dissipated in the detector element was conducted to the heat-sink via a thermal resistor. A second cavity viewing the heat-sink provided thermal compensation. The material used for the cavities and the thermal resistors was a silver sheet of 0.13-mm thickness. A thermopile was connected across the thermal resistors and the heater winding attached to the conical portion of the cavity was completely surrounded by a heat shield in order to restrict most of the heat transfer from the cavity to the path via the thermal resistor. Both the heat-sink and the absolute radiometer inside it were supported in a dewar flask to limit the heat exchange with the environment. The primary cavity radiometer was operated in the conventional way, with the calibration constant for converting the thermopile output to irradiance established by electrical substitution. Corrections for all known instrumental imperfections were computed and the results were adjusted accordingly. A total of eight primary cavity radiometers were built and their total observed spread when compared was 0.15%.

A comparison of the two types of absolute radiometers in a field trial at Table Mountain in California showed that they agreed with each other within 0.5%. This was considered as quite satisfactory in view of the difficulties involved in using the standard cavity radiometers in an outside environment. The JPL cavity radiometers were also compared with Ångstrom pyrheliometers on several occasions and the latter were consistently shown to measure lower by an average of 2.3%.

In the meantime, work of great significance continued to be produced at the NSL in Australia. It included a more detailed investigation of the problem of water-vapor absorption (Blevin and Brown, 1971) and a theoretical analysis of diffraction errors caused by apertures in the measuring system (Blevin, 1970). This work provided an essential basis for Blevin and Brown's highly acclaimed determination of the Stefan–Boltzmann constant against a gold-point blackbody (Blevin and Brown, 1971), which was the first such measurement to come within 0.1% of the theoretical value. For these measurements, the Gillham-type disk radiometer had been modified with the addition of a gold-coated hemispherical mirror to increase its absorptance even further and reduce the effective thermal resistance of the absorber coating (Fig. 1.22).

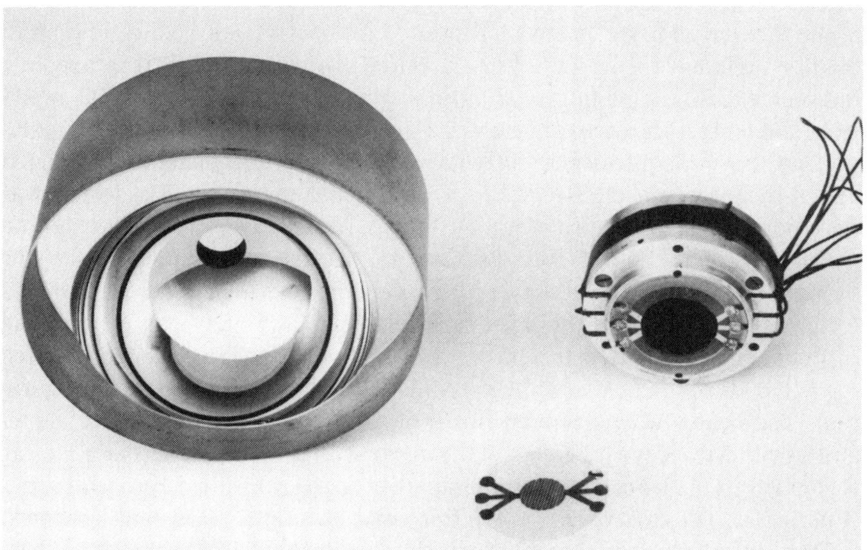

Fig. 1.22 Photograph of NSL absolute radiometer with hemispherical reflector.

Research on techniques for realizing the candela by radiometric means was also pursued at the NSL and, in a preliminary report to the Consultative Committee on Photometry (CCP) of the International Committee on Weights and Measures (Brown and Blevin, 1971), a value for K_m of 682 lm/W was quoted as being compatible with the NSL radiometric scale and the 1952 world mean candela.

At the National Physical Research Laboratory (NPRL) in South Africa, two Guild-type drift radiometers were built for some pilot research on the same subject (Kok, 1972). The radiometers, which were in all essential respects identical to the two types constructed by Guild, were used in conjunction with a 1000-W luminous-intensity standard lamp calibrated at the NPL. Using the most recent NPL value for K_m of 686 lm/W (Gillham, 1964), the radiometric results obtained at the NPRL agreed with the NPL value assigned to the lamp within 0.7% for one radiometer and within 0.4% for the other. These preliminary results were so encouraging that they led to the commencement of a full-scale research project on the construction of an advanced absolute radiometer.

After a long period in which the U.S. NBS had concentrated solely on blackbody sources as the basis for its radiometric scale, the first absolute radiometers since Coblentz's (1916) instruments were developed at the NBS over the period 1968–1971 by Geist (1971). The main aim of the program was the creation of instruments that could be employed to realize, maintain, and transfer a scale of spectral irradiance. It also included a critical investigation of the theoretical foundations of absolute radiometry and the formulation of the first coherent theory of the instrumental corrections applied to absolute radiometers. Fundamental contributions were made to the theoretical analysis of the temperature distribution in a detector element and to the formulation of the nonequivalence correction. Absolute radiometers constructed under the program consisted of an anodized aluminum disk, which was suspended in a concentric ring by means of a radial copper-constantan thermopile. The holder ring also served as the cold junction for the thermopile. A square heater layer of graphite mixed with lacquer was applied to the center portion of the front surface of the aluminum disk by spraying through a mask. Three platinum wires of 0.076-mm diameter were connected to both ends of the heater with silver-filled lacquer in the Gillham configuration. The heater was covered with Eppley-Parsons Optical Black Lacquer and a cylindrical cavity, coated on the inside with the same absorber material, was glued to the front of the aluminum disk with silver-filled epoxy adhesive (Fig. 1.23a). The cavity was drawn from silver sheeting of 0.25-mm thickness.

One of the radiometers was specifically modified for field use to participate in the Third International Pyrheliometer Comparison at Davos, Switzerland, and Locarno, Italy, in September 1970. For that purpose, two detector

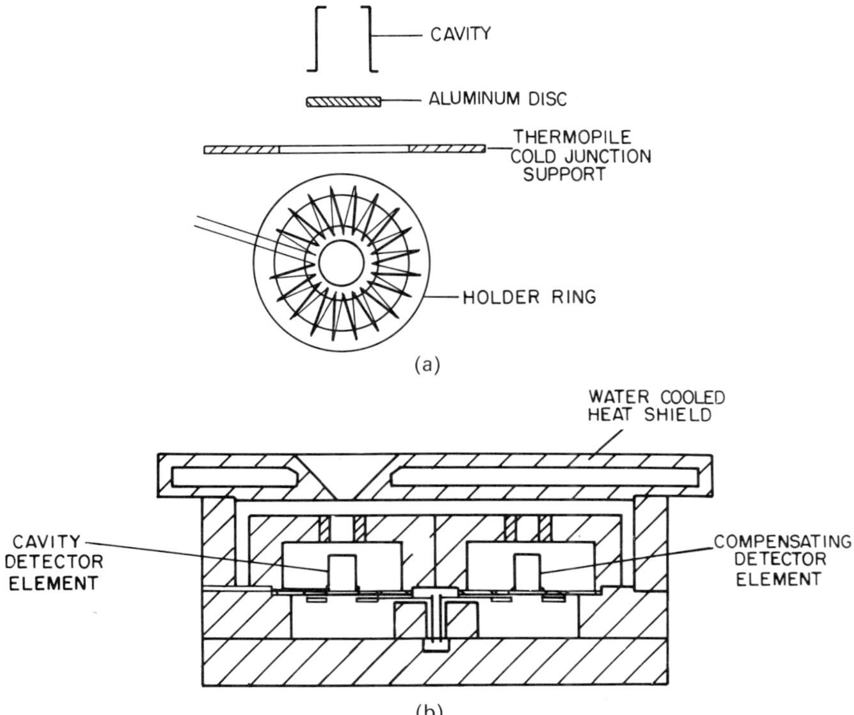

Fig. 1.23 (a) Detector element. (b) Assembled absolute radiometer. (From Geist, 1971; by permission of the U.S. National Bureau of Standards)

elements were incorporated in a special housing (Fig. 1.23b) with a water-cooled heat shield, a precision aperture insert (aperture diameter: 2.6 mm), a view-limiting aperture, and a support plate, which allowed it to be mounted together with one of the JPL radiometers (Kendall and Berdahl, 1970) on a single equatorial mount.

The comparison (Fröhlich et al., 1973) involved 28 Ångstrom pyrheliometers, 3 silver-disk radiometers, and 2 absolute radiometers. A total of 200 measurements were taken in which the irradiances ranged from 56 to 100 $\mathrm{mW cm^{-2}}$. It was the intention that the two Ångstrom pyrheliometers, numbers 210 (Davos) and 158 (Stockholm), would serve as references in the comparison, since they represented the International Pyrheliometric Scale of 1956. A discrepancy of 1.1% was discovered between them and traced to changes in no. 158. Thus, the final results, which were computed relative to no. 210 alone, showed a change ranging from +1.6 to −2.2% for the 28 participating Ångstrom pyrheliometers and from −0.4 to −0.7% for the two silver-disk pyrheliometers previously calibrated in terms of the IPS of 1956.

Moreover, the IPS of 1956 was shown to differ by 1.8 % from the absolute radiometric scale represented by the JPL and NBS instruments, thus substantiating earlier observations to that effect (Blevin et al., 1969; Kendall and Berdahl, 1970). The difference was again confirmed during a series of further extensive comparisons from 1973 to 1975.

At about the same time, a different section at the NBS developed instruments for the accurate maintenance of the NBS scale of laser power and laser energy (West et al., 1972). A new comparison between Geist's (1971) absolute thermopile and West et al.'s (1972) C-series calorimeter gave agreement to 0.09 % (Geist et al., 1973). The C-series calorimeters were of the isoperibol type (Gunn, 1973) and could measure pulse energies between 0.01 and 20 J and CW power levels between 0.04 mW and 1 W. The maximum pulse energy density was limited to 0.1 $\mathrm{Jcm^{-2}}$.

An extremely sophisticated cryogenic absolute radiometer had also been constructed in the meantime in the NBS temperature section (Ginnings and Reilly, 1973) to measure the thermodynamic temperatures of certain fixed points relative to the triple point of water via the Stefan–Boltzmann law. A schematic view of the detector element is depicted in Fig. 1.24. It consisted of a gold tube 150 mm in length and 19 mm wide, with a wall thickness of 0.05 mm. One end was closed with a gold cone and the other end was terminated with a gold aperture of 10-mm diameter. It was coated internally with electrolytically deposited platinum black and was connected through a stainless steel tube (137.5 mm long by 20.7 mm wide by 0.1 mm thick) to a copper shell maintained at 2 K. The temperature of the detector element was sensed with an uncased germanium resistance thermometer as was the temperature of a compensating twin cavity. Electrical heaters with radiation shields were attached to the part of the cavity exposed to the incident radiation. The temperature of the 2-K shell was controlled with a servo system and the instrument had a noise-equivalent power of about 0.2 nW.

While the technical sophistication of the cryogenic absolute radiometer was impressive, the objective of using it for the accurate measurement of thermodynamic temperatures was not achieved. The main reason for this was that the problem of diffractions in the complicated aperture system between radiometer and blackbody source could not be resolved with the necessary accuracy.

At the JPL, a new series of absolute radiometers referred to as "active cavity radiometers" (ACRs) were developed by Willson (1973). They were designed for automatic and remote operation in any environment and were specifically aimed at astrophysical, meteorological, and solar-engineering radiation measurements. The detector element was a conical cavity connected by a low thermal impedance to a heat-sink, which was insulated from the outside environment (Fig. 1.25a). The cavity opening of 2 $\mathrm{cm^2}$ was

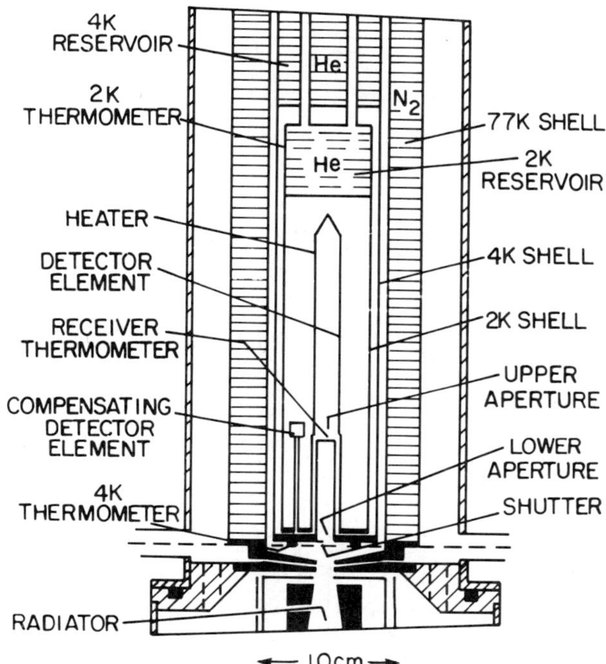

Fig. 1.24 The cryogenic absolute radiometer. (From Ginnings and Reilly, *Temperature: Its Measurement and Control in Science and Industry*, 1973; reprinted by permission. Copyright Instrument Society of America 1973)

reduced by a precision aperture of 1 cm^2 and the cavity walls were coated with an absorber material, which produced an effective cavity absorptance of 0.996 for solar radiant flux. An automatic thermostat circuit (Fig. 1.25b) maintained the cavity at a preselected temperature by controlling the voltage applied to the cavity heater via an electronic feedback loop. The incident radiant power was derived from the difference in the power dissipation in the electrical heater when the detector element was exposed to the radiation source to be measured and when it was not exposed.

An in-depth error analysis was carried out on the instrument and various models were constructed to cater for irradiance levels from 10 $mWcm^{-2}$ up to one solar constant (137 $mWcm^{-2}$) and also for 10, 20, and 30 solar constants. The one-solar-constant instrument was compared with the IPS of 1956; the results confirmed that the latter yielded values about 2.2% lower than values measured on the absolute scale.

A valuable addition to the range of thermal detectors available for absolute radiometers was the pyroelectric detector. It was based on the pyroelectric

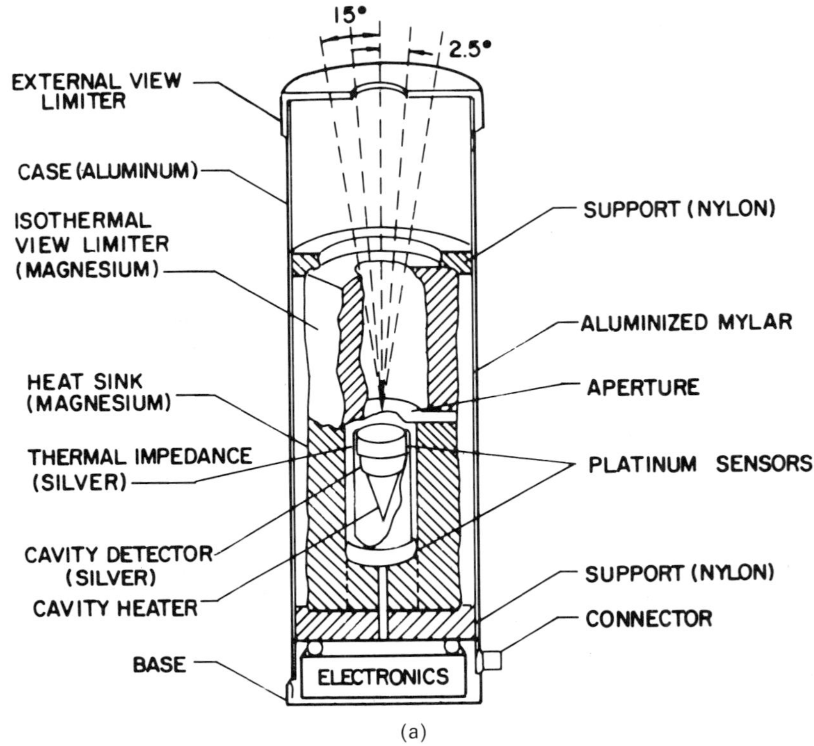

(a)

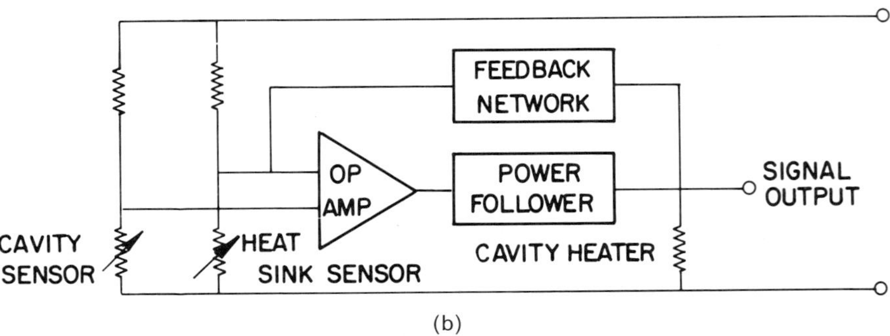

(b)

Fig. 1.25 (a) Active cavity radiometer. (b) Electric circuit. (From Willson, 1973; by permission of *Applied Optics*)

effect, which had been known since the eighteenth century but had not been considered for use in infrared detectors until well into the twentieth century. The first practical realization had to wait even longer and its use in the electrical substitution mode was pioneered at the NBS during the early 1970s (Phelan and Cook, 1973). It was based on the material polyvinylfluoride (PVF), which was employed in the form of a 0.01–0.02-mm thick circular sheet, clamped between two holder rings. Both sides were covered with evaporated nickel films of about 10-nm thickness, which overlapped over an area of 1 cm^2 at the center of the sheet (Fig. 1.26).

Contact to the nickel-covered area on the front surface was established with the aid of electroplated gold strips to which a pair of leads was attached on each side with conductive epoxy. The device was then poled with about 400 V between the two sides of the sheet at a temperature of 380 K, thus only activating the overlapping region. Using the nickel layer on the front surface as an electrical heater and supplying it with an AC signal, it was possible to establish the detector response to the absorbed radiant power and thus to establish the necessary calibration factor. Since the nickel film was also used as the absorber, the absorption correction was about 0.3 at both 633 and 10,600 nm. Two further problems were the capacitive coupling between the heater and the detector circuits and the difference in the shape of the chopped radiant heating and the AC electrical heating.

The idea was further developed at the NBS and combined with a waveform independent synchronous amplification technique (Geist, 1972). By the end of 1973, a prototype of a pyroelectric absolute radiometer had been constructed, in which the radiant and electrical powers were supplied to the detector in an alternating form (Geist and Blevin, 1973).

The heater circuit was electrically shielded from the pyroelectric detector to eliminate capacitive and inductive coupling, and a layer of goldblack evaporated onto the front surface of the detector served as both heater and

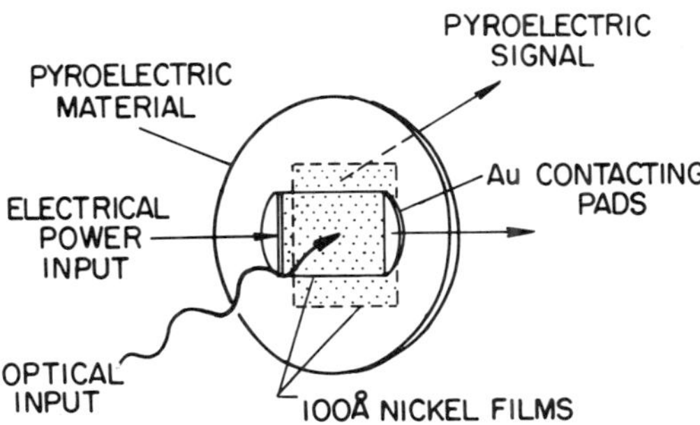

Fig. 1.26 Detector structure. (From Phelan and Cook, 1973; by permission of *Applied Optics*)

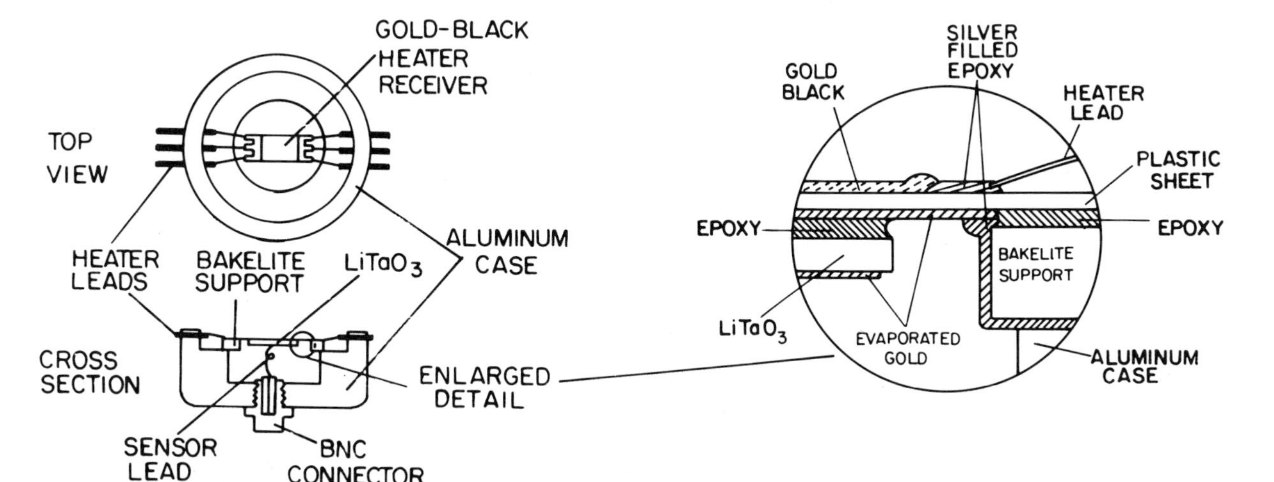

Fig. 1.27 (a) Detector for pyroelectric null radiometer. (b) Detector detail. (From Geist and Blevin, 1973; by permission of *Applied Optics*)

absorber. Heater leads in a Gillham configuration were used and the sensor was a 0.5-mm-thick disk of lithium tantalate. The plastic substrate that insulated the heater from the electrical shield had a thickness of 0.013 mm, and the epoxy layer between the plastic sheet and the pyroelectric detector was about 0.05 mm thick (Fig. 1.27).

The instrument was operated at a frequency of 12.5 Hz and achieved a noise-equivalent power level of 100 nW with an integration time of 10 s. It was compared with Geist's absolute thermopile detector and the agreement within about 2% was considered quite satisfactory in view of the new principle involved, the relatively immature technology, and some outstanding problem areas concerned with the device characterization. One of these was the influence of the black coating on the amplitude and phase of the pyroelectric signal, which was analyzed in considerable detail (Blevin and Geist, 1974).

By 1975, the NBS section responsible for the laser power and energy scale had also constructed four pyroelectric absolute radiometers based on the new principle and had added an electronic servo system, which adjusted the electrical power continuously to match the absorbed radiant power (Fig. 1.28). The instruments showed good agreement with each other and also with the C-series calorimeters (Hamilton et al., 1975).

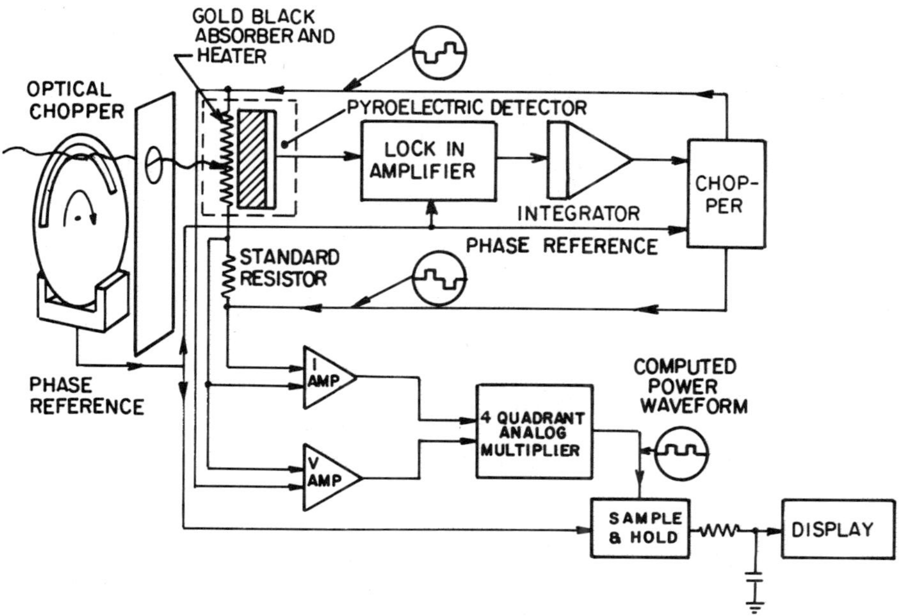

Fig. 1.28 Electronic circuit for autobalancing pyroelectric radiometer. (Hamilton et al., 1975)

1.3.4.6 Absolute Radiometry Over the Period Since 1975

The quest for a radiometric realization of the photometric units entered a decisive phase when Australia's NSL decided to implement the transfer of the Australian measurement scale for photometry onto a radiometric base (Brown, 1975). This step increased the pressure for an international redefinition of the candela and stimulated a further intensification of research efforts in this area. A new determination of the constant K_m in the course of this work, this time using a group of lamps calibrated by the BIPM in terms of the mean international candela of 1952, gave a value of $683\,\mathrm{lm W^{-1}}$. A formal proposal for a redefinition of the candela and the lumen was subsequently made jointly by Blevin (NSL) and Steiner (NBS) and was presented to the eighth session of the Consultative Committee on Photometry and Radiometry of the International Committee for Weights and Measures in 1975 (Blevin and Steiner, 1975). The proposal was accepted in principle pending further verification of and agreement on the value of the constant K_m and a further CCPR session to deal with a possible redefinition of photometric units was scheduled for 1977.

The dissatisfaction with the pyrheliometric radiation scale as a basis for solar-radiation measurements continued to grow in the meteorological community and further laboratories got involved in the realization of the absolute irradiance scale. One of these was the Physikalisch Meteorologisches Observatorium (PMO) in Davos, Switzerland (Brusa and Fröhlich, 1975). Two different absolute radiometers (PMO2 and PMO3) were constructed, both based on cavity-type detector elements (Fig. 1.29).

The heat flow from the heated cavity to the environment was channelled through a thermal resistor and the temperature difference between the active and the passive (compensating) cavities was measured. The temperature sensor was a resistance thermometer in one case and a thermopile in the other. The electrical heating element was attached to the irradiated portion of the cone-shaped cavity. Both absolute radiometers were operated with an electronic servo system, which maintained the temperature difference between the active and passive cavities at a constant value. The incident radiant power was derived from the difference in the electrical-power levels dissipated in the active cavity when the shutter was closed and when it was open. The 10–90% time-constants of the feedback-controlled instruments were 1.7 and 3.7 s, respectively, and the diameter of the radiometer aperture was 5 mm. Considerable attention was paid to the characterization of the instruments in terms of their instrumental corrections.

The PMO instruments were later further improved and the experimental characterization of the instruments was extended to all correction factors, including those that had previously been estimated by calculation (Brusa, 1983; Brusa and Fröhlich, 1986).

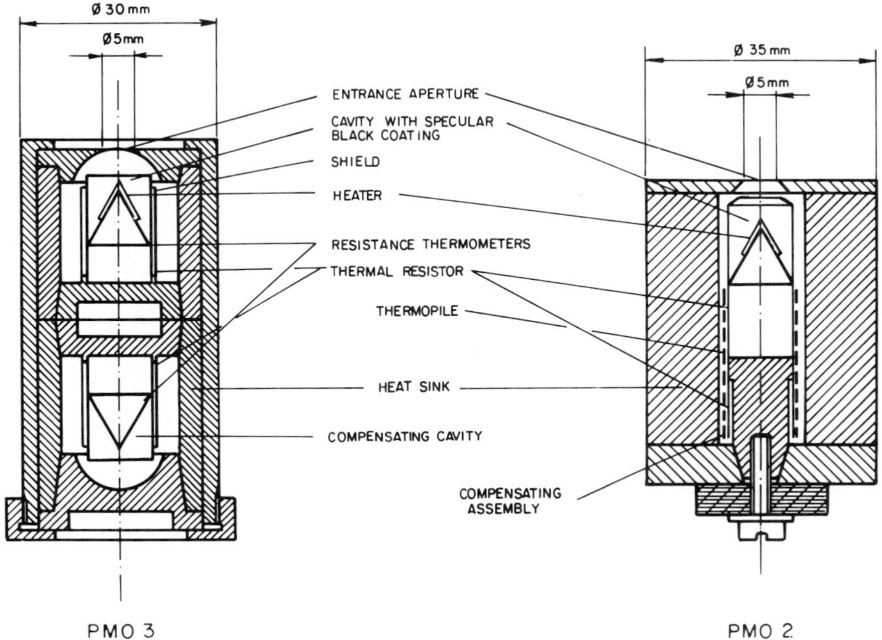

Fig. 1.29 Detector assembly for PMO radiometers. (From Brusa and Fröhlich, 1975; by permission of the World Radiation Centre)

The first commercial version of an absolute radiometer based on the pyroelectric device developed at the NBS was produced in the United States by Laser Precision Corporation in 1975. It was selected as one of the most significant new products of the year by *Industrial Research* magazine in the United States and received one of the magazine's I-R 100 awards for it. A photograph of the latest version of this instrument is depicted in Fig. 1.30.

Detailed studies were made of the required instrumental corrections (Doyle et al., 1976), and the most recent (as of early 1988) microprocessor-based model (Laser Precision Corporation, 1985) features an uncertainty specification (two sigma level) of plus or minus 1%, an autonulling servo-loop, autoranging, autozeroing of background, and various test and calibration modes. It works at a chopping rate of 15 Hz, has a radiometer aperture area of 0.5 cm^2, and uses a lithium tantalate crystal as the pyroelectric material.

During 1976, a Laser Precision electrically calibrated pyroelectric radiometer (ECPR) was compared with a cavity-type absolute radiometer at the PTB and a C-series calorimeter at the NBS using radiation from a helium-neon laser at 632.8 nm. The ECPR measured 0.4% higher than the PTB instrument and 0.5% lower than the NBS calorimeter. A further comparison

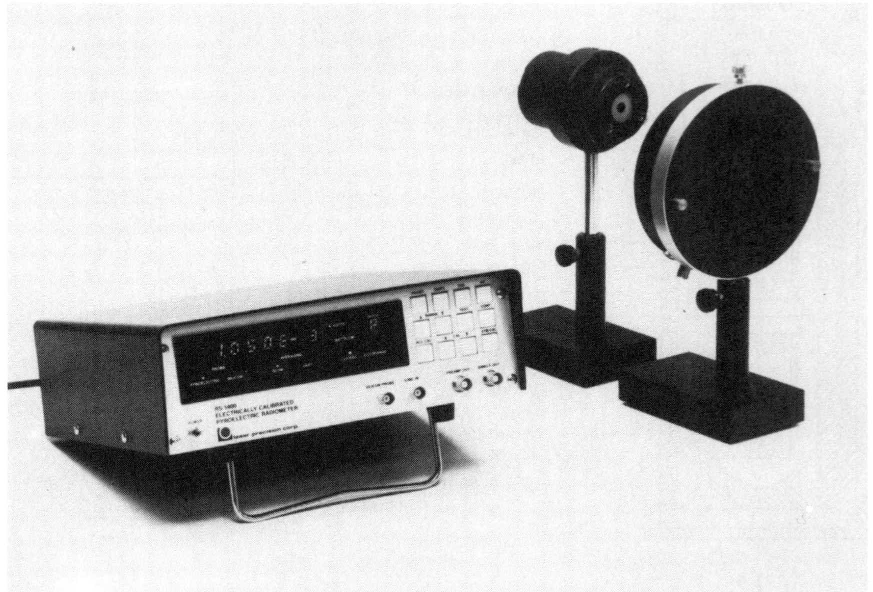

Fig. 1.30 Photograph of electrically calibrated pyroelectric radiometer by Laser Precision Corporation. (By permission of Laser Precision Corporation)

was made against the absolute radiometer at the PMO using broadband solar irradiance. Here the ECPR measured 0.3% lower. These values provided independent confirmation for the specified uncertainty estimate for the ECPRs, which have had a marked impact on improving the overall accuracy of many radiometric measurements in diverse application areas and will very likely continue to do so for many years to come.

First results with the new absolute radiometers, which had been under development at South Africa's NPRL since 1971, were reported by Hengstberger (1975). Since the most successful absolute radiometers during the previous few decades had been based on thermopile detectors, while instruments based on bolometers were apparently considerably less sensitive, the new instruments were intentionally designed with bolometric detector elements. This was motivated by the knowledge that nonabsolute bolometers were performing just as well as thermopiles and that a competitive broadening of the instrumental base of absolute radiometry would thus be desirable.

The developed detector elements are depicted in Fig. 1.31. They consisted of two mica substrates of approximately 0.025-mm thickness glued together with silver-filled epoxy (Fig. 1.31a). Before being glued together, gold layers of 100-nm thickness were evaporated on the two mica substrates, and the conductor patterns depicted in Figs. 1.31b and 1.31c were etched into them

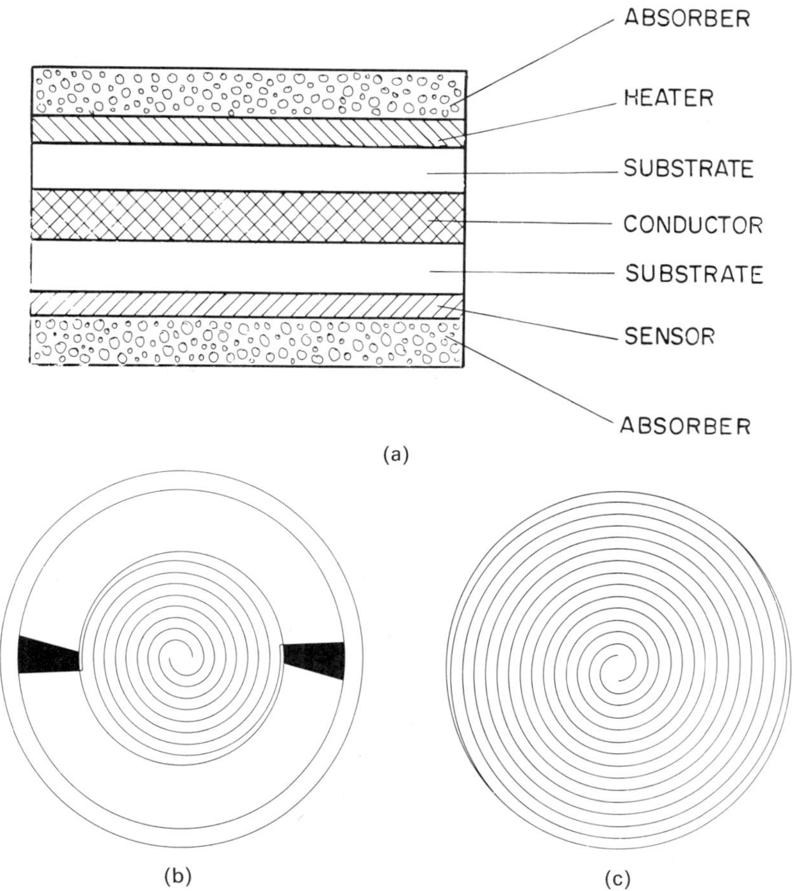

Fig. 1.31 (a) Schematic diagram of detector element. (b) Heater pattern. (c) Sensor pattern. (Hengstberger, 1975)

by a photographic exposure technique. Both patterns were in the form of bifilar spirals, one covering the whole substrate and the other only the central part. The area of the smaller spiral was identical to the area of the radiometer aperture (1 cm^2). It had an electrical resistance of about 180 Ω and was used as the electrical heating element. The other spiral, with an electrical resistance of about 400 Ω, was used as a bolometric detector. The assembled detector elements, with the gold spirals on the outside, were eventually covered with 3M Black Velvet Paint on both sides and mounted with quartz fibers in aluminum holder rings and connected with thin gold wires to insulated terminals in these rings. Two such rings were then mounted back-to-back on an aluminum heat-sink with a reflecting hemisphere facing both the front and the back of the detector elements. The front hemisphere had an opening for the radiometer aperture. The bolometric sensors were incorporated in a Wheatstone bridge circuit (Fig. 1.32) in a compensating configuration. The bridge was originally used with a DC supply and a galvanometer for measuring the bridge output. However, this mode of operation was soon changed to an AC bridge technique (1000 Hz), with a six-decade ratio-transformer across the middle resistor in the bridge reference arm for zeroing the bridge and a lock-in amplifier for measuring the bridge output. The AC technique had the advantage of immunity to drifts of thermal potentials in the circuit and also lent itself to narrow-band signal amplification at the bridge operating frequency.

The performance of the NPRL instrument was further improved (Hengstberger, 1977a) by increasing the thickness of the gold bolometer elements to 1000 nm and by removing the original sensitivity to vibrations through an improved mechanical design of the components of the radiometer head (Fig. 1.33).

The absolute radiometers were fully characterized by exact measurements of their instrumental correction factors and incorporated in sophisticated measuring systems designed for the realization of photometric and radiometric measuring scales (Hengstberger et al., 1977). A detailed study was undertaken of various experimental variants for measuring the lead-heating correction using a Gillham-type lead configuration (Hengstberger, 1977b), and an improved theory of the instrumental corrections for absolute radiometers was derived (Hengstberger, 1977c). Following the Australian precedent, the South African photometric scale was officially placed on a radiometric basis during 1976 using a preliminary value of the constant K_m of 682.5 $lm W^{-1}$.

At the ninth session of the Consultative Committee on Photometry and Radiometry (CCPR) of the International Committee on Weights and Measures in 1977 (CCPR, 1977), many national metrology laboratories reported new measurements of the constant K_m. Based on these values and the expressed preferences of the laboratories, a value of 683 $lm W^{-1}$ was chosen as the best estimate for the true value of the constant. This value was then used in a new, radiometric definition of the photometric units, which was recommended for ratification subject to the usual procedures of the organs of the Metre Convention applicable in such cases. The new definition of the candela, which was eventually ratified at a meeting of the General Conference on Weights and Measures (in French, the Conference General des Poids et Mesures, or CGPM) in 1979, reads (CGPM, 1979):

> The candela is the luminous intensity, in a given direction, of a source that emits monochromatic radiation of frequency 540×10^{12} hertz and that has a radiant intensity in that direction of (1/683) watt per steradian.

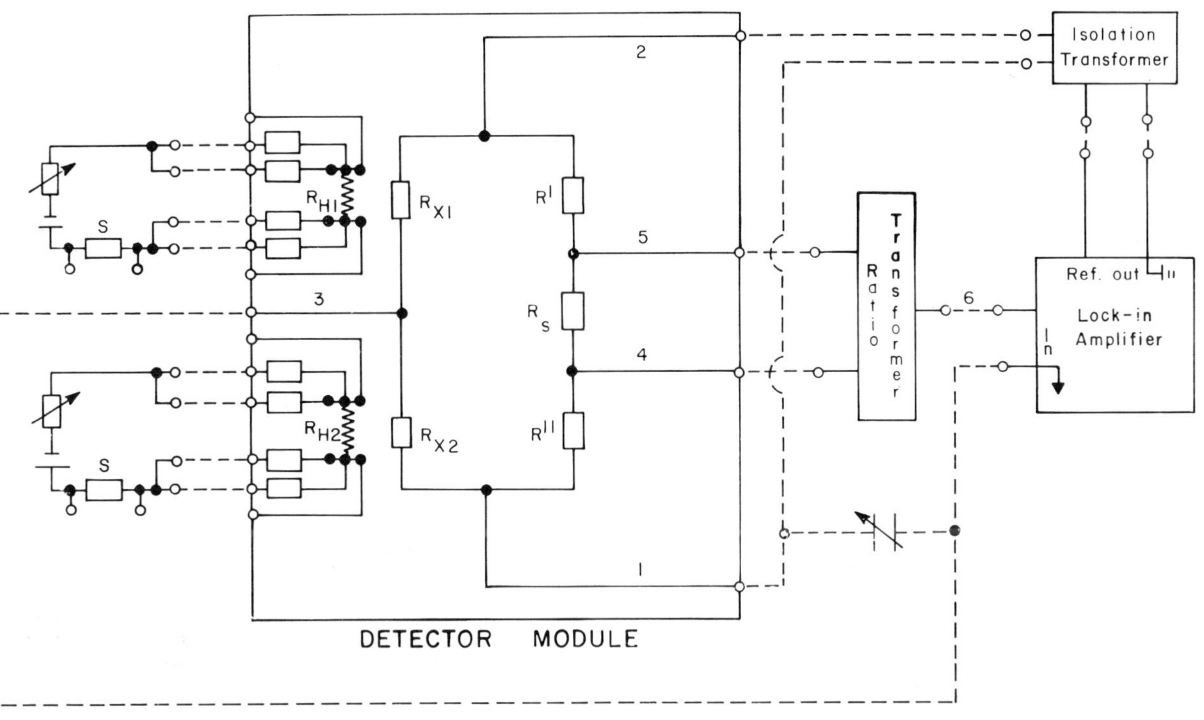

Fig. 1.32 Electric circuit diagram. (Hengstberger, 1975)

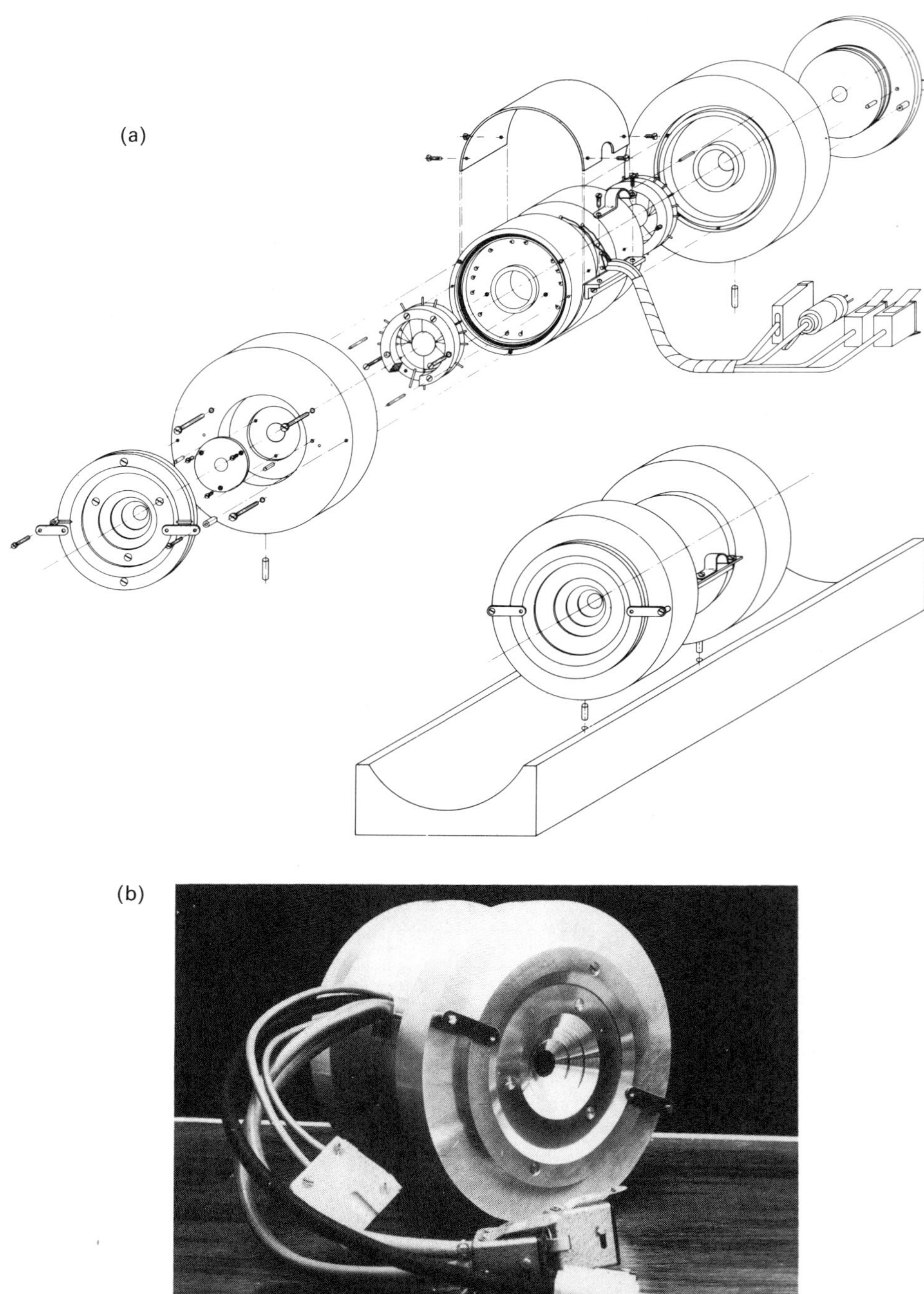

Fig. 1.33 (a) Exploded view of absolute radiometer. (b) Photograph of absolute radiometer. (Hengstberger, 1975)

In another significant event, the World Meteorological Organization (1977) approved the establishment of the so-called World Radiometric Reference (WRR), which was based on absolute radiometry and replaced the ill-fated International Pyrheliometric Scale of 1956. Thus, in a single eventful year, photometry was put on a radiometric basis and the radiation scale used in meteorology was switched to an absolute radiometric scale. Both events benefitted radiometry in general and removed a lot of confusion about the relationship of the units used in these different application areas of what is basically the same discipline.

A comparison of laser power scales at 633 nm and at a radiant power level of 3 mW was conducted between the ETL (Japan), NBS (United States), PTB (Federal Republic of Germany), and NPL (U.K.) in the period between 1972 and 1976 (Honda and Endo, 1978). The transfer standards were four calorimetric laser power meters, which were calibrated in all participating laboratories. The results showed a mutual scale consistency of better than 1%.

The PTB reported the development of absolute radiometers for laser-power measurements in the range from 1 mW to 10 W (Möstl, 1978). The instruments consisted of cone-shaped detector elements, covered with a specularly reflecting absorber on the inside. A heater wire was wound around the outside of the cone, which was surrounded by a gold-plated heat shield. A thermopile was used to measure the temperature drop across a thermal resistor between the cone and a heat-sink. The uncertainty quoted for this device originally (0.5%) was later reduced through careful characterization procedures to a level of about 0.1% (Stock, 1985). A schematic drawing of the instrument is shown in Fig. 1.34.

The power range was eventually extended using a water-cooled cavity device with an operating range from 3 W to 1 kW (Möstl, 1979; Stock, 1985). A compensated cone calorimeter with bolometric sensor windings between the heater windings on the outside of the cones served as the primary standard for laser-pulse energies up to 100 mJ (Stock, 1985).

The National Research Council of Canada developed absolute radiometers that incorporated thin-film thermopiles prepared by photoetching (Boivin and Smith, 1978). Figure 1.35a shows a schematic diagram of these absolute radiometers. The receiver element was composed of a 32-mm diameter, 0.25-mm-thick glass disk on which a nickel-chromium thermopile was fabricated, and a 24-mm diameter, 0.25-mm-thick copper-ceramic disk on which a 6-mm diameter thin-film chromium heating element was deposited. These two disks were bonded together to form the detector element (Fig. 1.35b). The absorber was a high-density goldblack.

The absolute radiometers were fully characterized in terms of the usual instrumental correction factors and one was eventually used to measure the

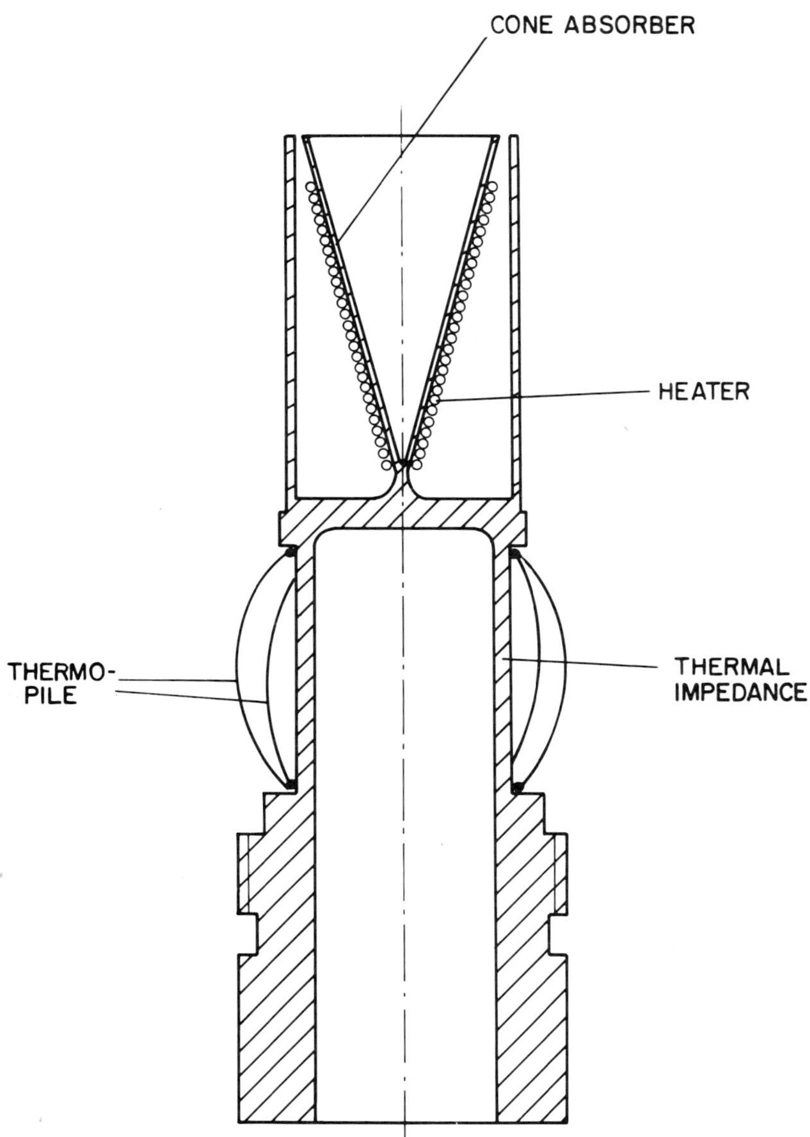

Fig. 1.34 Absolute radiometer. (Möstl, 1978)

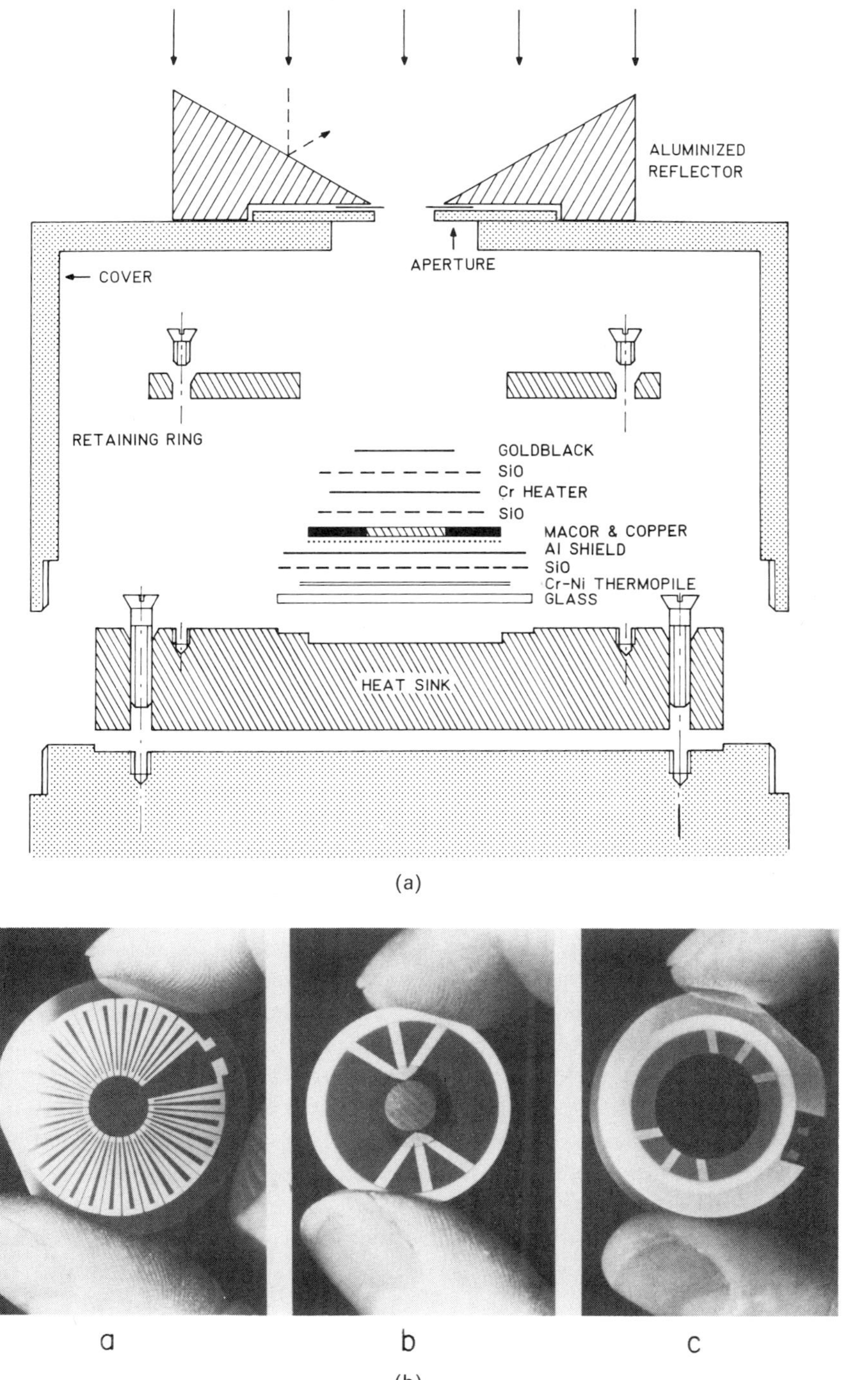

Fig. 1.35 (a) Schematic diagram of absolute radiometer. (b) Assembled detector element. (From Boivin and Smith, 1978; by permission of *Applied Optics*)

irradiance from five spectral irradiance standard lamps in combination with both a $V(\lambda)$ filter and an infrasil filter placed between the radiometer and the $V(\lambda)$ filter. The total irradiance from the lamps was calculated from their spectral irradiance values and the spectral transmittances of the filters as well as being measured by both the NRC radiometer and a Laser Precision pyroelectric substitution radiometer. The scatter of the measured values ranged from -0.58 to $+0.72\%$, consistent with an estimated error of 0.5% for power levels of 0.05 mW or greater.

The absolute radiometers were used for calibrating incandescent lamps for spectral irradiance over the wavelength range 400–700 nm (Boivin, 1980). A secondary silicon radiometer was calibrated against one of the absolute radiometers at nine discrete wavelengths using a krypton laser. Since its linearity had been determined in separate measurements, this detector could then be used in combination with both narrow-band interference filters (20–25-nm halfwidth) and wider-band absorption filters (100-nm halfwidth) to measure the radiation from the incandescent lamp within the filters' passbands. These values were then used to calculate the spectral irradiances of the lamp by assuming they obeyed a Planckian-type distribution. When comparing the obtained values with NBS spectral irradiance standards, the values of which had been adjusted after an international comparison, differences averaged about 0.5% with a maximum of 1%.

New models with a higher responsivity, smaller time-constant, lower noise-equivalent power, and greater durability followed (Boivin, 1985; Boivin and McNeely, 1986). They consisted of two aluminum rings with 0.006-mm–thick mylar membranes attached to one side. The first membrane was the substrate for a bismuth thin-film heating element covered with an insulating layer and goldblack on the front and a 0.013-mm–thick electroplated copper disk at the rear. The second substrate had a 30-junction thin-film silver-bismuth thermopile on its front surface and a reflective coating on the rear. A schematic diagram of the absolute radiometers, the configurations of the thermopile and of the heating element, and photographs of the two rings making up the detector element are shown in Fig. 1.36.

The new absolute radiometers were fully automated under the control of a desktop computer linked to auxiliary measuring instruments via an IEEE-488 interface. Their overall accuracy was estimated as plus or minus 0.1%. A comparison of three of the new absolute radiometers with two self-calibrated silicon photodiodes (see Chapter 7), using five Krypton laser lines between 476 and 676 nm at power levels around 1 mW, indirectly confirmed this figure.

A self-calibrating cavity radiometer developed by the Eppley Laboratory in the United States was incorporated in the instrumentation package of the

Nimbus 7 satellite for the Earth Radiation Budget experiment (Hickey et al., 1980; Hickey, 1985; Karoli et al., 1983). The sensor was a copper-on-constantan thermopile and the cavity receiver consisted of an inverted cone within a cylinder. It was used to monitor the solar constant. Results presented in 1985 (Hickey, 1985) indicated a decrease of this constant by about 0.015% per year over the period 1978–1985.

Improved versions of the active cavity radiometers were reported by Willson (1979) in the form of the ACR IV. The new instruments differed from the previous models (Willson, 1973) through the use of a specular absorber on the cavity walls, a compensated operating principle as well as improved mechanical construction techniques and electronic circuits. The instruments were used for measurements of the solar constant both in a sounding rocket experiment and in the Active Cavity Radiometer Irradiance Monitor experiment during NASA's Solar Maximum Mission (Willson, 1985). The three ACR IVs used for the latter mission detected temporary changes in the solar constant due to sun spots, plages, and periodic modes of solar oscillation as well as a long-term decrease in the solar constant that could be associated with the solar cycles. A comparison with some of the instruments used to define the World Radiometric Reference scale of 1977 indicated that the ACR IV series seemed to measure 0.4% higher than the WRR scale. A further version referred to as ACR V (Fig. 1.37) was reported as well (Willson, 1980). Its main feature was an improved cavity design, which avoided the detrimental effect on the absorptance of an imperfect meniscus of the cavity cone. A thin tube was soldered to the meniscus and excess paint was drawn through this tube during the painting process, resulting in a near perfect meniscus. This increased the cavity absorptance to 0.99988 or more and reduced the uncertainty from this correction to plus or minus 0.00002. (See also Section 2.3.4.) It was estimated that the radiation scale realized with ACR V would have an uncertainty relative to SI of plus or minus 0.1%.

Advances on several fronts were being made with radiometers at the NPRL. The theory of the instrumental corrections was extended by the inclusion of the correction for spatial responsivity variations in the theoretical framework (Hengstberger, 1979). This was followed by the development of small, portable radiometer heads and of a fully automated electronic operating system for the absolute radiometers (Hengstberger and Appenroth, 1983). They incorporated detector elements based on thin disks of beryllium oxide, with the heater and sensor layers deposited on opposite sides using thick-film technology. A thin coating of 3M Nextel brand, type 101-C10 velvet black coating was deposited over the heater layer to serve as absorber and the absorptance was again enhanced by placing the detector element in the center of a hemispherical aluminum reflector. The thick-film

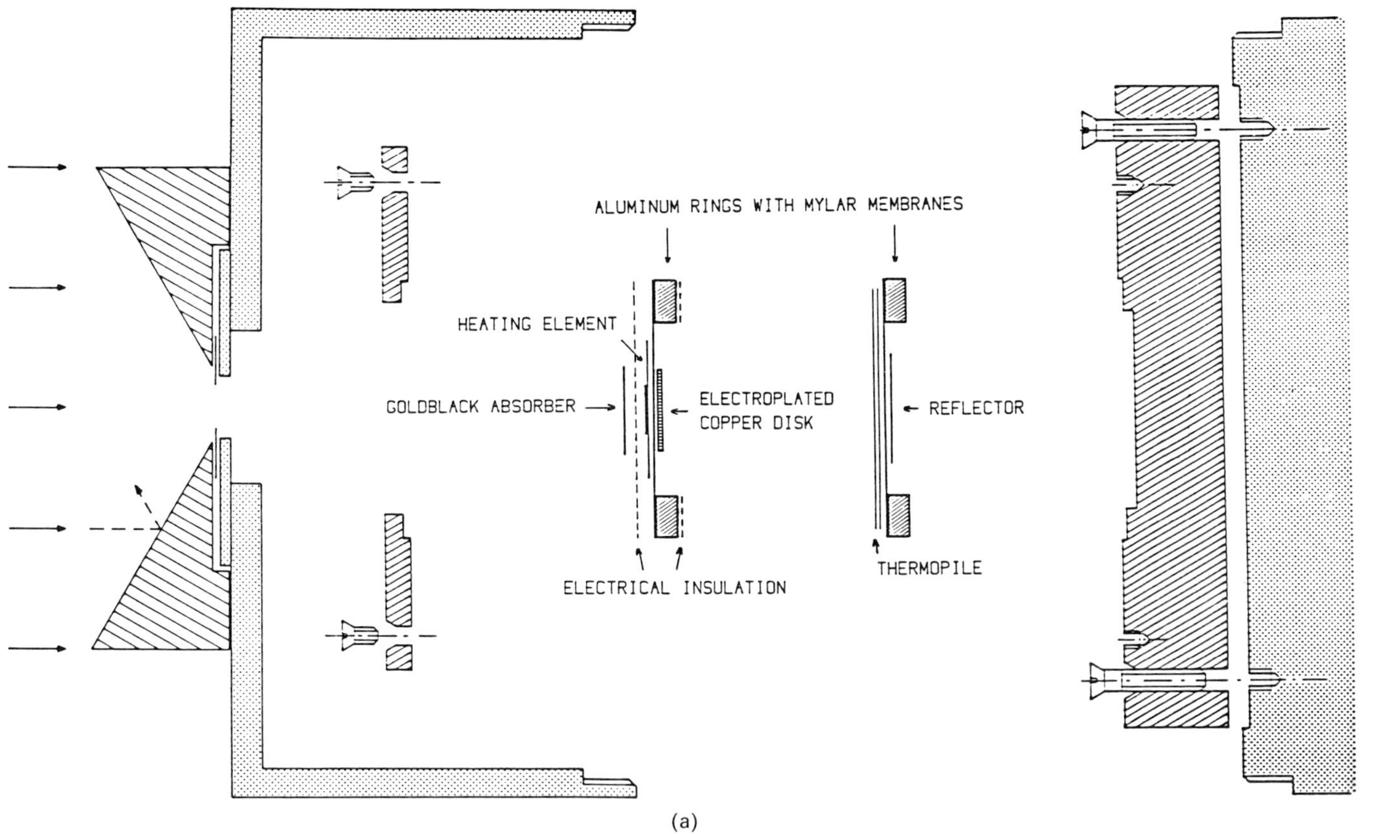

ALUMINUM RINGS WITH MYLAR MEMBRANES
HEATING ELEMENT
GOLDBLACK ABSORBER
ELECTROPLATED COPPER DISK
REFLECTOR
THERMOPILE
ELECTRICAL INSULATION

(a)

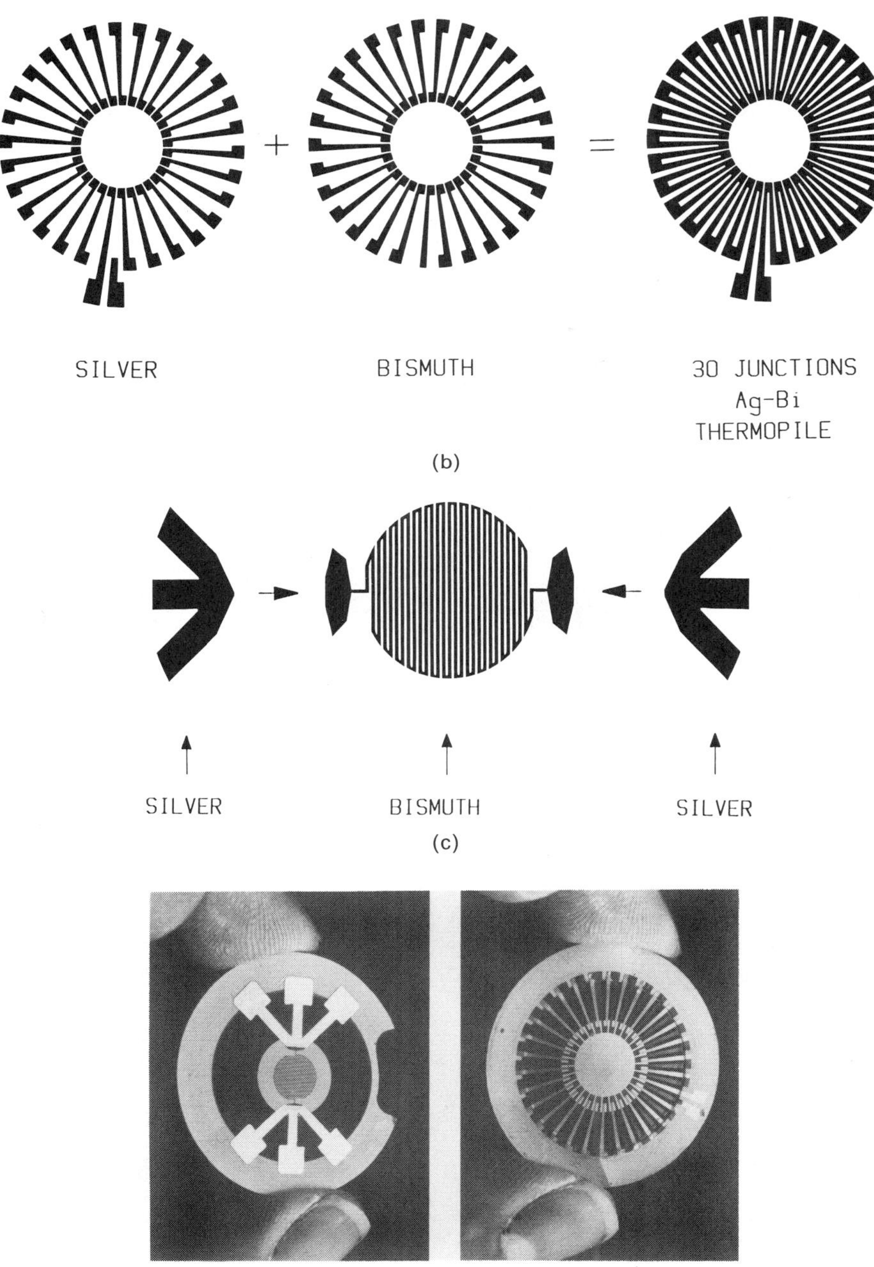

Fig. 1.36 (a) Schematic diagram of new absolute radiometer. (b) Thermopile configuration. (c) Heater element. (d) Photograph of detector element.

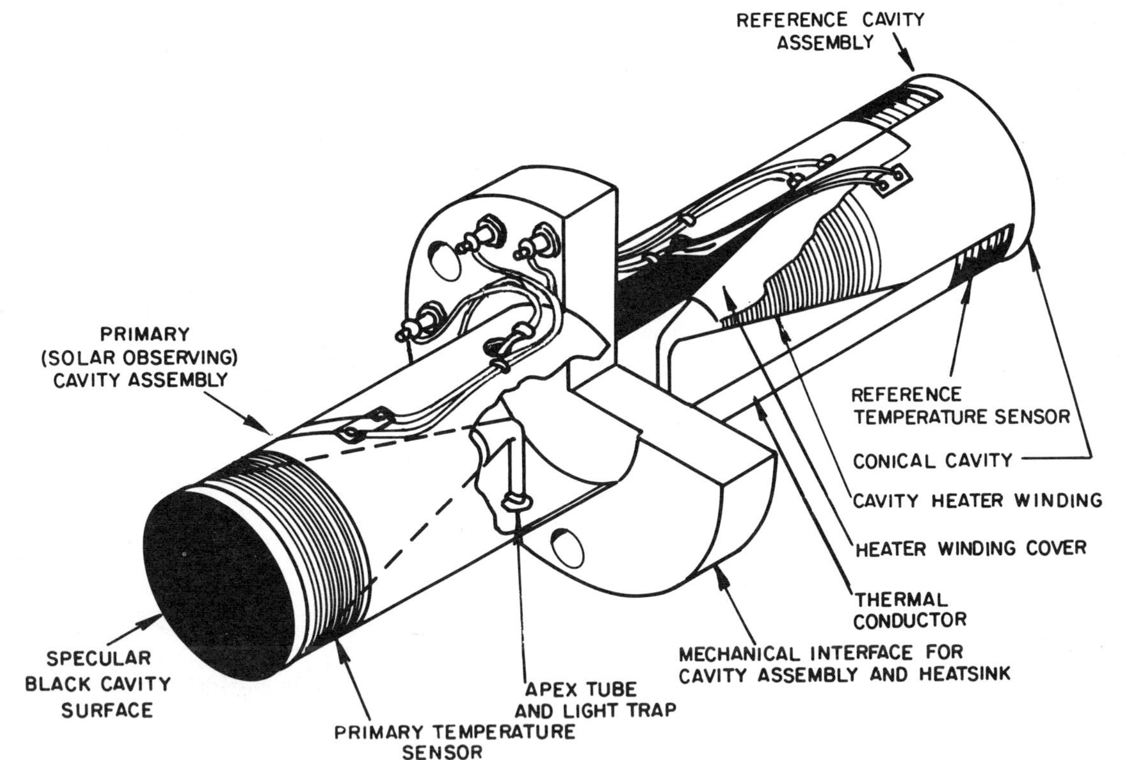

Fig. 1.37 ACR V. (From Willson, 1980; by permission of *Applied Optics*)

heaters and sensors were far superior to the previously used thin-film equivalents in terms of mechanical durability. At the same time, the high thermal conductivity of beryllium oxide combined with its high electrical resistance simplified the production of the detector elements by acting as an electrical insulator between the heater and the bolometric detector and as thermal diffusor at the same time. Responsivity variations from the center of the radiometer aperture (1 cm^2) to the edges decreased to less than 0.2% for the new radiometer heads at the cost of an increased thermal time-constant and a higher noise-equivalent irradiance. The increased thermal time-constant of the new detector elements was reduced by means of electronic compensation circuits (see Section 1.3.8) incorporated in the heater feedback of the radiometer electronics.

The radiometer instrumentation was under the control of a desktop computer linked to the individual instruments in the system via an IEEE-488 interface bus. Electronic feedback circuits kept the temperature difference between the active and the compensating detector elements constant at a selectable level by adjusting the electrical heater power in the active detector element. A second servo loop compensated for capacitive and inductive imbalances in the AC bridge (Fig. 1.32), which would have otherwise caused quadrature signals at the bridge output and interfered with the detection of small signals in phase with the bridge supply voltage. The in-phase signal components represented the resistive changes in the bolometric detector layers with temperature. A radiometer head of the new type together with the associated instrumentation is depicted in Fig. 1.38.

Instruments of this type were later acquired by metrology laboratories in a number of other countries. The latest version (Hengstberger et al., 1987) includes a thermostatted radiometer module in the standard radiometer head in close vicinity with a thermostatted preamplifier and achieves noise-equivalent irradiances in the order of 10 $nWcm^{-2}$ in air. Other studies involving the use of these instruments include measurements of the lead-heating correction (Gentile and Rastello, 1980; Etchechoury and Cogno, 1982) and mathematical procedures to calculate the absorbed radiant power from the electrical powers measured with the shutter in the open and closed positions in the presence of drift (Hengstberger and Dressler, 1982; Cogno and Ezpeleta, 1985).

At the Japanese National Research Laboratory for Metrology (NRLM), Ono (1979a) made an important contribution to absolute radiometry with his studies of designs that would minimize responsivity variations across the detector element. He developed useful guidelines and mathematical models for analyzing and predicting the thermal behavior of detector elements and followed his study with the description of a practical instrument designed along these lines (Ono, 1979b). It consisted of an aluminum disk of 0.02-mm

Fig. 1.38 Photograph of new NPRL radiometer head and control electronics.

thickness and 15-mm diameter. The front surface carried an evaporated heater layer covered by a black absorber coating, while the rear surface was coated with evaporated gold to reduce radiation losses. A radial thermopile of 24 pairs of copper-constantan thermocouples was used as thermal detector. A schematic diagram of the detector element is shown in Fig. 1.39a. A photograph of the assembled device appears in Fig. 1.39b.

A Laser Precision ECPR was used at the Instituto de Optica in Madrid as the basis for the calibration of incandescent lamps for spectral irradiance in the visible region (Carreras and Corrons, 1981). A set of narrow-band interference filters as well as an auxiliary silicon photodiode were used in the measurements. The uncertainty of the realization was estimated as plus or minus 1.5% and a comparison with an NBS spectral-irradiance standard lamp gave a maximum difference over the region between 403 and 800 nm of 0.8%. Using the realized spectroradiometric scale to calculate the luminous intensity of the lamps, the values agreed with measurements against four luminous-intensity standards from the BIPM within similar margins.

An absolute radiometer with an unusually short time-constant (0.1 s) was developed in the German Democratic Republic (Müller and Ratz, 1982). It consisted exclusively of thin-film components separated by 25-nm-thick foils of celluloseacetate mounted on a glass ring. The thermal detector was a radial thin-film thermopile consisting of 24 antimony and antimony-doped bismuth thermocouples. An absorber layer made of evaporated silverblack (because of its lower heat capacity compared with goldblack), a thin-film gold or silver heating element, and a thermal diffusing layer of 100-nm-thick silver were the other components of the detector element.

Absolute radiometers for realizing the candela were developed at the National Institute of Metrology (NIM) in Beijing, Peoples Republic of China (Gao zhizhong et al., 1983). They used cone-shaped detector elements in a compensated configuration. The copper cones were 25 mm long, with a half-vertex angle of 11.3° and an opening diameter of 10 mm. The heater was a coiled manganin wire attached to the inside of the cavity, which was then coated with 3M black paint. Next, 102 pairs of nichrome-constantan thermocouples were attached radially to the outside of the cone.

After a full characterization procedure, the seven instruments produced differed from the mean of the group by less than 0.1%. The radiometric realization of the candela (estimated uncertainty 0.28%) differed from an earlier realization with a platinum-point blackbody (estimated uncertainty 0.33%) also by less than 0.1%.

Work on ever-improved absolute radiometers was also continued at the Royal Meteorological Institute of Belgium (IRMB), resulting in a whole series of radiometers with detector elements in the form of flat-bottom cylindrical cavities (Crommelynck, 1985). These were coated on the inside

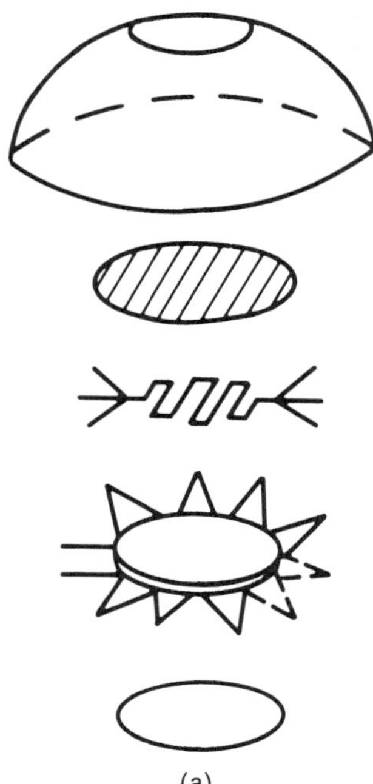

(a)

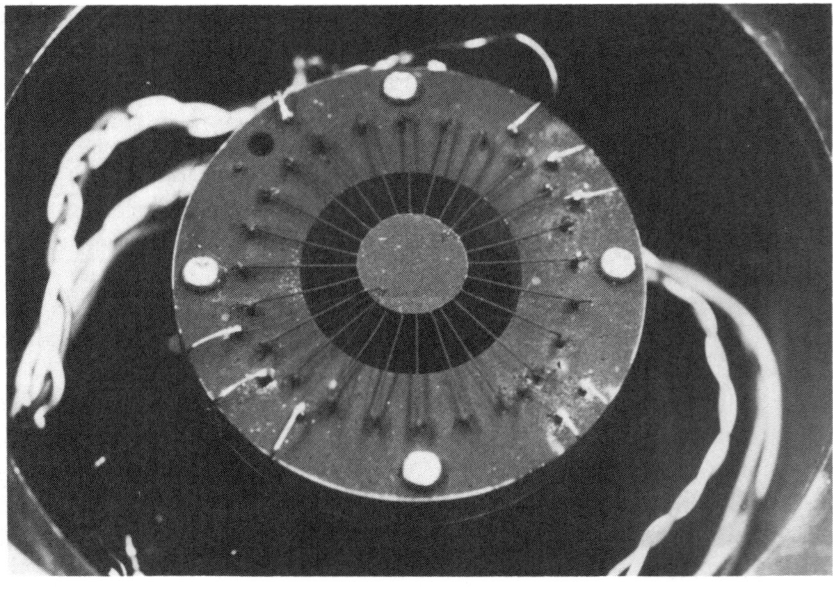

(b)

Fig. 1.39 (a) Schematic diagram of absolute radiometer. (From Ono, 1979b; by permission of the *Japanese Journal of Applied Physics*) (b) Photograph of assembled detector element.

with diffuse velvet black paint (3M) and their outer surfaces were gold-coated. Two identical, fully functional cavities were used in a dual compensated configuration. The electrical heating elements were located under the flat cylinder bottom and a differential thermopile between the two cavities served as thermal detector. The radiometric scale at the IRMB was based on the mean of 5 dual-channel instruments (in effect, 10 absolute radiometers). These were fully characterized experimentally (Crommelynck, 1982) both in air and in vacuum. Another of the instruments was integrated into an instrumentation package for a spacelab mission to perform a measurement of the solar constant (Crommelynck and Van de Velde, 1984). A schematic diagram of this absolute radiometer is shown in Fig. 1.40.

In spite of the failure elsewhere of an earlier attempt (Ginnings and Reilly, 1973) to use a cryogenic absolute radiometer for a measurement of the Stefan–Boltzmann constant and for high-accuracy measurements of the thermodynamic temperature scale, work on this subject had continued in the temperature section of the NPL since 1972. The problem with diffraction was resolved through an effective design philosophy and the project was successfully completed by 1983 (Quinn and Martin, 1984, 1985). The value found for the Stefan–Boltzmann constant differed by only 0.013% from the value calculated from fundamental constants, which was less than the combined standard deviation of the measured and calculated values. The overall configuration of the apparatus is depicted in Fig. 1.41.

The detector element was a cylindrical cavity made of 0.25-mm–thick electroformed copper coated on the inside with a 0.1-mm–thick layer of 3M-C101 paint. Its dimensions were 70 mm wide by 380 mm long with a reentrant cone attached to the closed end. The cavity was connected via a thermal resistor to a superfluid liquid helium reservoir at 2 K, with the temperature of this heat-sink controlled to about 0.04 mK over a period of 2 hours. The thermal detector was a germanium resistance thermometer (200–1000 Ω) read via a bridge circuit with a resolution of 2 parts in 10^6. The rise of the cavity temperature for a 1-mW power dissipation was 4.5 K and the corresponding power resolution was 2 nW.

The cryogenic absolute radiometer did not achieve its high accuracy because of its low noise-equivalent power (which is not much lower than for the best room-temperature instruments operated in a vacuum), but because of vanishingly small instrumental corrections. The most important of these gains was the perfect equivalence between radiant and electrical heating of the cavity, due to the extremely small radiation losses from and temperature gradients in the low-temperature cavity. This resulted in virtually all the generated heat flowing to the heat-sink by conduction via the thermal resistance, independent of the exact place where the two forms of energy were converted to heat. As superconducting current leads were used, the

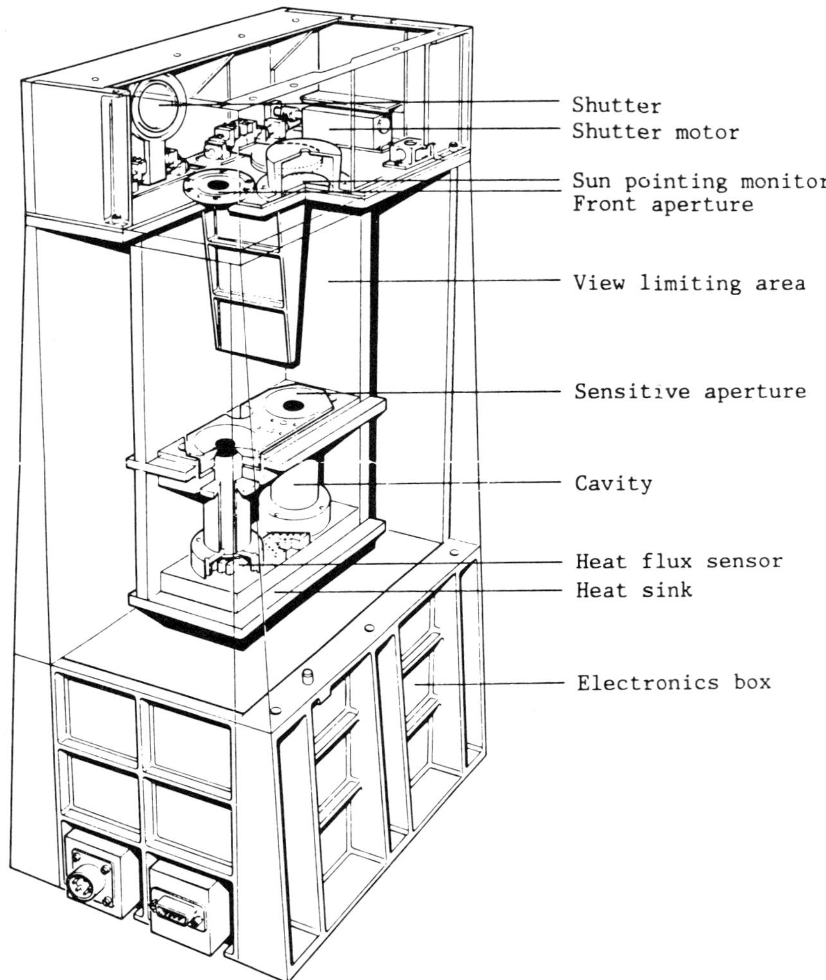

Fig. 1.40 Absolute radiometer flown on space lab mission. (From Crommelynck and Domingo, 1984; by permission of the American Association for the Advancement of Science. Copyright 1984 by the AAAS)

lead-heating effect also disappeared. A further advantage of the low-temperature operation was that the thermal diffusivity of copper is about 1000 times smaller at 4 K than at room temperature. Therefore, a relatively large and heavy (300-g) cavity such as the one used still had a reasonable time-constant (3 minutes) at that temperature.

The experience gained from this instrument was eventually also applied to the design of a modified version for radiometric purposes (Martin et al.,

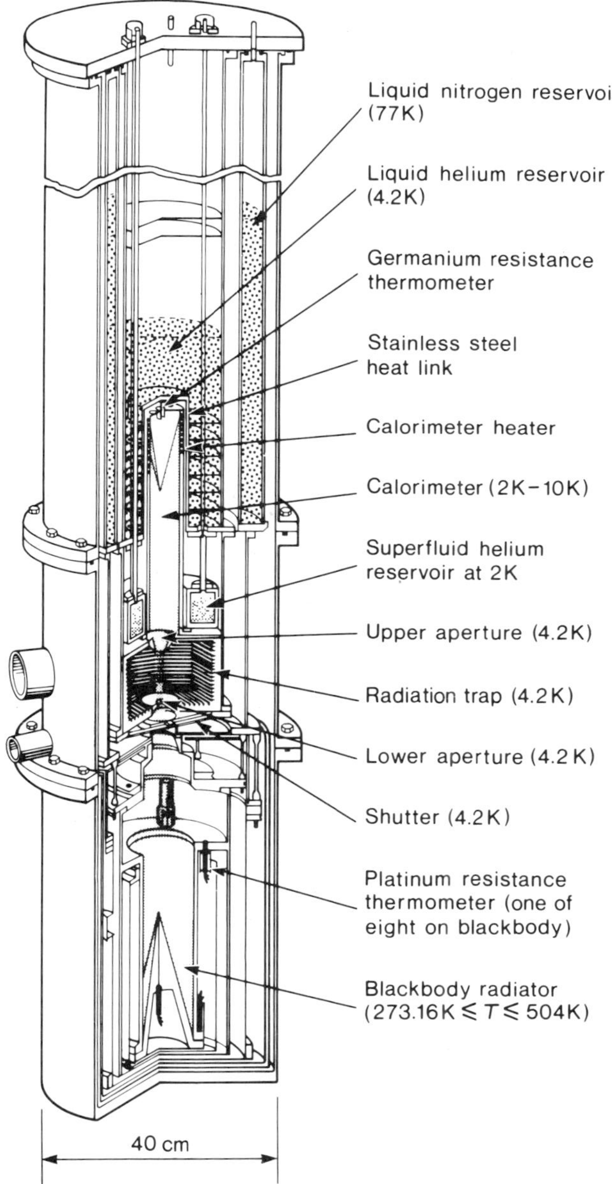

Fig. 1.41 Cryogenic absolute radiometer used for the measurement of the Stefan-Boltzmann constant and of thermodynamic temperatures. (From Quinn and Martin, 1985; by permission of the Royal Society, London)

1985). The detector element was again in the shape of a cylindrical cavity with 0.25-mm-thick walls of electroformed copper. The dimensions, however, were only 50 mm in diameter and 150 mm in length. The front end was closed by a plate with a radiometer aperture of 12-mm diameter and the back wall at the other end was inclined at 30° to the axis of the cylinder. It was coated internally with a specularly reflecting paint (Chemglaze Z 302), resulting in an effective cavity absorptance of 0.99998 over the entire visible spectrum. The cavity was connected via a thermal resistor to a heat-sink held at a constant temperature of about 5 K by a commercial helium cryostat. Bifilar constantan heater wires were wound around the outside of the cylindrical cavity and the connecting leads were made of superconducting niobium wire. The cavity temperature was measured with a germanium resistance thermometer. A schematic diagram of the instrument is shown in Fig. 1.42. It was mainly used for the calibration of transfer standards by means of a stabilized krypton ion laser. Such transfer detectors were then employed for a variety of radiometric purposes, including the realization of the candela according to its new radiometric definition (Goodman and Key, 1986).

The radiometric scale realized with the cryogenic absolute radiometers was found to differ from Gillham's (1962) scale by less than 0.2%. Agreement with self-calibrated silicon photodiodes (see Chapter 7) was found at the 0.02% level (Fox, 1985) and comparisons with synchrotron radiation standards (see Chapter 7) also produced agreement within the experimental uncertainty of the comparison of about 0.4% (Fox et al., 1986).

A prototype of another helium-cooled absolute radiometer has been constructed at Atmospheric and Environmental Research, Inc., in the United States and as of 1988 is being studied and evaluated (Foukal et al., 1985). Its target application areas are meteorology and space research.

Absolute radiometers for measuring the average power of laser radiation and the pulse energy of pulsed lasers were reported from Russia (Il'in, 1985). The instruments employed conical detector elements with a vertex angle of 15° and a base diameter of 20 mm. The cone material was copper blackened with AK-243 black enamel in the case of the power detector and vacuum-annealed tantalum for the energy detector. Heating coils of manganin wire were wound around the outside of the cones, which were connected via eight fin-shaped strips to larger copper cones on the outside. These fins contained 100 copper-constantan thermocouples each, resulting in detectivities of 0.14 to 0.2 V/W for the power probe and 5 to 8 mV/J for the energy probe. The operating ranges were 1 mW to 1 W and 0.01 to 10 J, respectively.

A comparison of radiant power scales at 488 and 633 nm was conducted by 10 metrology laboratories between 1985 and 1986 (Zalewski, 1986).

Five of the participants (NRC of Canada, NPL of Great Britain, PTB of West Germany, NML of Australia, and NPRL of South Africa) used scales

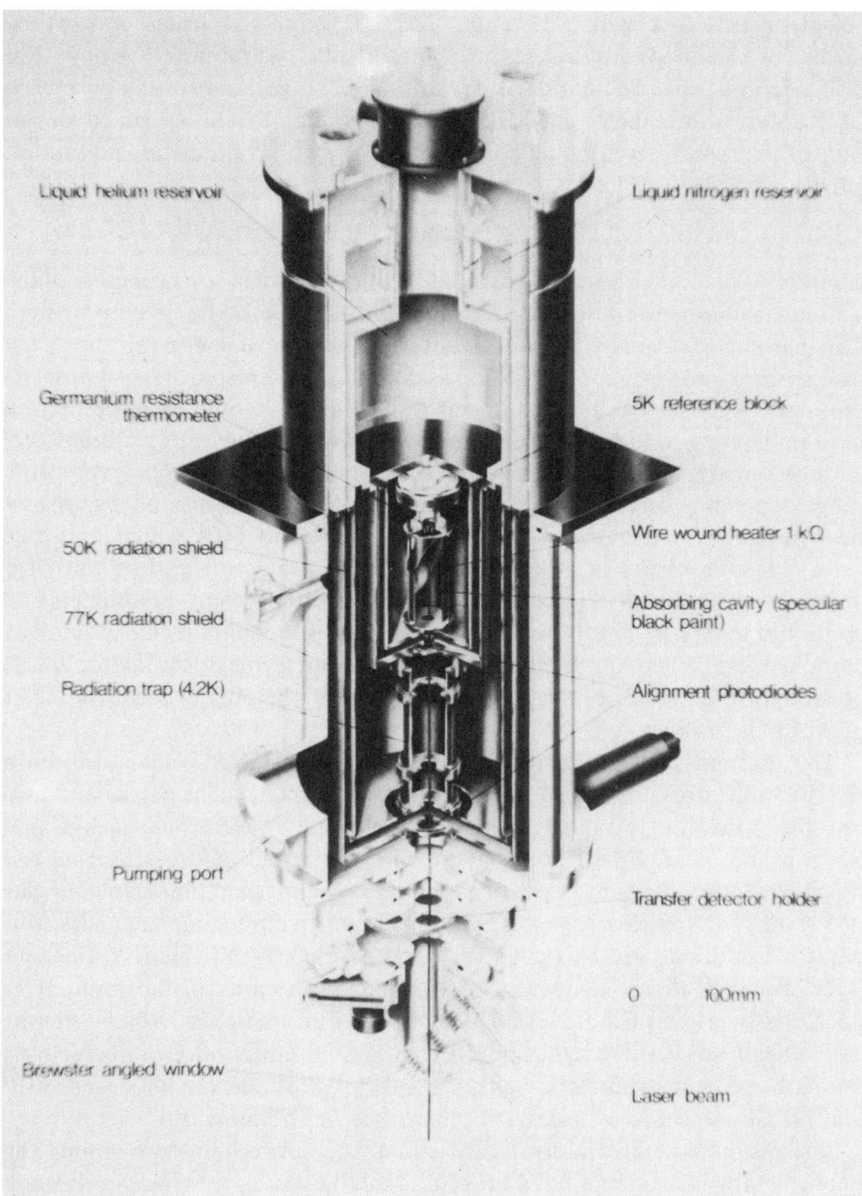

Fig. 1.42 Cryogenic absolute radiometer with radiometric application. (From Martin et al., 1985; by permission of *Metrologia*)

based on absolute radiometers and the rest employed scales realized by means of silicon photodiodes with a predictable quantum efficiency. (See also Section 7.5.) The standard deviation of the scales realized by means of absolute radiometers was about 0.7%, similar to the standard deviation of the scales based on silicon photodiodes with a predictable quantum efficiency.

1.3.4.7 Comparative Data on Absolute Radiometers

In order to facilitate the comparison of the most important parameters of the absolute radiometers described in Section 1.3.4, Table 1.5 has been prepared. The parameters chosen are not necessarily figures of merit for the listed instruments and their interpretation is complex. The responsivity figure (R) especially is rather treacherous in this respect and it should therefore not be used on its own to judge the performance of an instrument. For radiometers with bolometric detectors, for example, it could have any value in a very wide range depending on the bridge operating voltage, with increasing responsivities sometimes associated with increasing NEP and NEI values and vice versa. In spite of the fact that it can often be rather misleading (also for absolute radiometers with thermal detectors other than bolometers), it continues to be listed in comparative tables and is therefore included here for completeness. Where the responsivity was stated in terms or irradiance, it was converted to a power responsivity by dividing by the area of the radiometer aperture, if specified.

The thermal time-constant (τ) and the NEP and NEI values are more reliable indicators for the suitability of an instrument for a particular task, but the possibility of shortening time-constants by electronic means (see Sections 1.3.7 and 1.3.8) should also be kept in mind before attaching too much weight to this parameter. In cases where the thermal time-constant was not given, but the electronically shortened effective time-constant was, this value is listed followed by the letter "e." Where noise-equivalent irradiances were specified, these values were multiplied by the area of the radiometer aperture to obtain the noise-equivalent power and, where the NEP figure was available, it was divided by the aperture area to obtain a value of the NEI. As the measurement bandwidth was not stated by most authors, the listed NEP and NEI figures do not necessarily refer to a 1-Hz bandwidth.

The estimated uncertainty figure contains information both about the quality of the instrument and about the available means for characterizing it. An element of personal judgment is usually also included in this figure. In line with the most recent recommendations of the International Bureau of Weights and Measures concerning the statement of uncertainties (Giacomo, 1981), the listed value is, wherever possible, expressed at a confidence level of 66% (one standard deviation). In cases where authors have stated their

TABLE 1.5 Comparative Data on Absolute Radiometers

Author(s)	Year	Type	R (V/W)	τ (s)	NEP (nW)	NEI (nW/cm^2)	s (%)
Coblentz, Emerson	1916	T	—	—	—	—	1
Guild	1937	T	—	240	—	—	—
Rutgers	1951	B	—	15	—	—	0.5
Gillham	1962	T	0.0015	400	—	—	0.3
Gillham	1962	T	0.073	12	—	—	0.2
Gillham	1962	T	0.040	14	—	—	0.2
Ooba	1965	B	—	—	—	—	0.4
Blevin, Brown	1967	T	0.053	15	140	500	0.2
Blevin, Brown	1967	B	0.013	4	4000	500	0.2
Crommelynck	1967	T	—	30	—	—	—
Bischoff	1968	T	0.056	20	1800	700	0.5
Kendall, Berdahl	1970	B	—	1800	—	—	0.12
Kendall, Berdahl	1970	T	0.008	7	—	—	0.07
Blevin, Brown	1971	T	0.090	20	50	50	0.02
Geist	1972	T	0.034	—	—	—	0.18
Willson	1973	B	—	—	30000	30000	0.3
Ginnings, Reilly	1973	B	—	—	0.2	—	—
Brusa, Fröhlich	1975	—	—	—	—	—	0.22
Brusa, Fröhlich	1975	B	—	—	—	—	0.44
Doyle et al.	1976	P	—	—	—	—	0.35
Hengstberger	1977a	B	0.060	20	30	30	—
Möstl	1978	T	0.002	7	700	—	0.5
Boivin, Smith	1978	T	0.093	15	50	50	0.5
Boivin, Smith	1978	T	0.043	15	50	50	0.5
Boivin, Smith	1978	T	0.005	2.2	250	250	0.5
Ono	1979b	T	0.171	14	—	—	—
Ono	1979b	T	0.102	8.2	—	—	—
Müller, Ratz	1982	T	0.636	0.1	1000	1270	2.0
Hengstberger, Appenroth	1983	B	—	40	300	300	0.1
Gao zhizhong	1983	T	0.210	14	—	—	0.1
Laser Precision	1985	P	—	3	100	200	0.5
Laser Precision	1985	P	—	32	5	10	0.5
Martin et al.	1985	B	2	180	20	—	0.004
Stock	1985	T	0.050	20	—	—	0.1
Hengstberger et. al.	1985	B	—	40	100	100	0.1
Boivin, McNeely	1986	T	0.300	2.8	20	—	0.1
Brusa, Fröhlich	1986	B	—	2e	—	—	0.05

estimated accuracies without quoting a confidence level, a 66% level was assumed.

Where parameter values were not stated, this is indicated in the table with a dash. The symbols used for the detector types are T for thermopile or thermocouple, B for bolometer, and P for pyroelectric. In order to avoid listing wrong data, values have only been entered in the table if they could be established from the literature with a reasonable degree of confidence. Where it was not possible to extract any of the required numerical data, the instrument was omitted from the table.

1.3.5 General Design Considerations

The design of an absolute radiometer and specifically of its detector element depends on many factors. Some of them are identical to the general criteria applied to the design of thermal radiation detectors, while others are specific to the subject matter. Among the most important general considerations are the power and irradiance levels at which the absolute radiometer will operate. The irradiance level has a major influence on the choice of absorber material, while the power level determines the selection of the appropriate thermal resistance to the environment, which has to be high enough to guarantee a minimum detector output and low enough to prevent thermal damage to the detector element. For uniformly irradiated absolute radiometers with radiometer apertures of about 1 cm^2, a classification in terms of irradiance as given in Table 1.6 would seem reasonable (Hengstberger, 1978).

In the first category, the detector elements can be thermally and electrically optimized for the highest possible detectivity, taking into account such factors as the thermal capacity, thermal resistance to the environment, physical dimensions, and noise sources. (See Chapter 3.) Absolute radiometers used for photometric measurements fall into this category.

Optimum detectivity is of less concern in the second group and can therefore be traded off for a more rugged construction that is able to withstand a wider range of environmental conditions. Examples for instruments in this category can be found in absolute radiometers used for meteorological purposes.

TABLE 1.6 Classification in Terms of Irradiance

Irradiance Range	Design Class
<1 mW/cm^2	I: low irradiance
1 mW/cm^2 to 1 W/cm^2	II: medium irradiance
>1 W/cm^2	III: high irradiance

In the third category, there is such an abundance of radiant power that detectivity is no longer an important criterion. Instead, the dissipation and removal of the generated heat and the protection of the absorber surface from thermal damage is the major consideration. Efficient convection cooling of the radiometer element has to be provided at power levels exceeding 1–10 W, while water cooling is commonly used for instruments from 100 to 1000 W upwards. Most available surface absorbers can withstand CW irradiance levels of between 10 and a few hundred W/cm^2. For pulsed sources, surface damage in most absorbers only occurs at peak irradiance levels exceeding 10^5 to 10^6 W/cm^2 or at pulse energies of 0.1 to 10 J, with both limitations being applied simultaneously. Above these levels, volume-absorbers have to be used instead of surface-absorbers.

Apart from the ability to withstand the incident power or irradiance levels, the next most important characteristics of an absolute radiometer include accuracy, detectivity, and convenience of use. In order to achieve a high accuracy, it should be attempted to keep the various correction factors as small as possible and/or to make them easily measurable. For the absorber, this implies a low reflectance (which should also not vary much with wavelength, angle of incidence, or location) as well as low thermal resistance and (in most cases) high electrical resistance. The responsivity of the radiometer element should be as constant as possible across the surface and the detectivity of the absolute radiometer should be as high as possible subject to the limitations regarding the expected power and irradiance levels as well as the fundamental noise sources. However, the requirements for achieving a high accuracy are often contrary to the objective of obtaining the highest possible detectivity and convenience of use. Examples of this conflict are the extra wires attached to the heating element for the measurement of the lead-heating correction, the incorporation of a layer with a high thermal conductivity in the detector element to reduce response variations, and the use of detector elements larger than the radiometer aperture for both the achievement of a higher effective absorption factor for the detector element and the interception of diffracted radiation. All of these measures result in a higher potential accuracy for the instrument but may, and usually will, reduce the detectivity and increase the thermal time-constant.

The time-constant of an absolute radiometer is one of the most important factors affecting its convenience of use. When measuring steady-state quantities in photometric, radiometric, and meteorological applications, measuring periods for single measurements in the order of several minutes are usually acceptable, although there are certainly also requirements for faster measurements. Assuming the acceptability of a measuring period in the order of minutes and the need for signals to settle for several time-constants to achieve a given accuracy, it follows that the instruments should have time-constants

TABLE 1.7 Relative Difference from Steady-State Temperature as a Function of Time

Temperature Difference (%)	Time Expressed in Number of Time-Constants
1	4.6
0.1	6.9
0.01	9.2
0.001	11.5

in the order of seconds to tens of seconds. Table 1.7 lists the settling period in terms of the number of time-constants required to meet a given error limit under the assumption of an exponential thermal-detector response characteristic.

The total time-constant of an absolute radiometer depends both on the thermal time-constant of the detector element and the electrical time-constant of the amplifier circuits as well as on possible feedback loops involving the detector output and the heater supply. In absolute radiometers using thermopiles or bolometric detectors, the thermal time-constant of the detector element is usually the limiting factor. While pyroelectric absolute radiometers have appreciably shorter thermal time-constants, they generally require electronic time-constants of about the same order of magnitude to achieve a comparable detectivity. The time-constants of thermal detectors can be reduced by electronic response compensation techniques (see Section 1.3.7) and the response times can be further improved by operating with electronic feedback. (See Section 1.3.8.)

A further factor, which undoubtedly has a major effect on the convenience of use of an absolute radiometer, is the degree of automation of the measuring process. A full array of analog and digital control techniques as well as a wide range of microprocessors, desktop computers, programmable instruments, and interfaces can be employed for this purpose. Numerous examples of successful implementations of automated systems can be found in the literature. (See Section 1.3.4.)

It is generally necessary to incorporate some thermal compensation in the design of the radiometer head because of the thermal time-constant of the detector element, drifts in the temperature of the thermal environment (heat-sink) of the detector, temperature fluctuations caused by adiabatic air pressure changes (Stock, 1983), and component drifts. This measure is not required for absolute radiometers based on pyroelectric detectors. The compensation is accomplished by employing a differential technique based on two detector elements. Sometimes only one of these can be irradiated; in others cases, the two detector elements are completely interchangeable (i.e., they can both be used for the measurement of the incident radiant power). In both cases, the two elements should be as identical as possible, which implies that both should have identical absorber layers, heating elements, substrates, and sensors. When only one detector element can be irradiated, the absolute radiometer is described as compensated, while the second type is sometimes referred to as dual-compensated. Further aspects of compensation are discussed in Section 1.3.7.

1.3.6 Operating Modes of Absolute Radiometers

Even if an absolute radiometer is used passively (Fig. 1.7b)—i.e., without an electronic feedback system—there are a number of possible operating options. One can, for example, record the detector output signal for a sequence of exposures of the detector element to the unknown radiant power and then calibrate the obtained signal by heating the detector element with electrical power (*e* in Fig. 1.43a). The sequence of exposures is produced by opening (*o*) and closing (*c*) the shutter a number of times, in each case allowing sufficient time for the signal to settle within the required level of uncertainty. (See Table 1.7.) The recorder could be a simple chart recorder or some form of data-acquisition system. Alternatively, one can determine the approximate power level with this procedure and, in a second sequence of exposures, record the

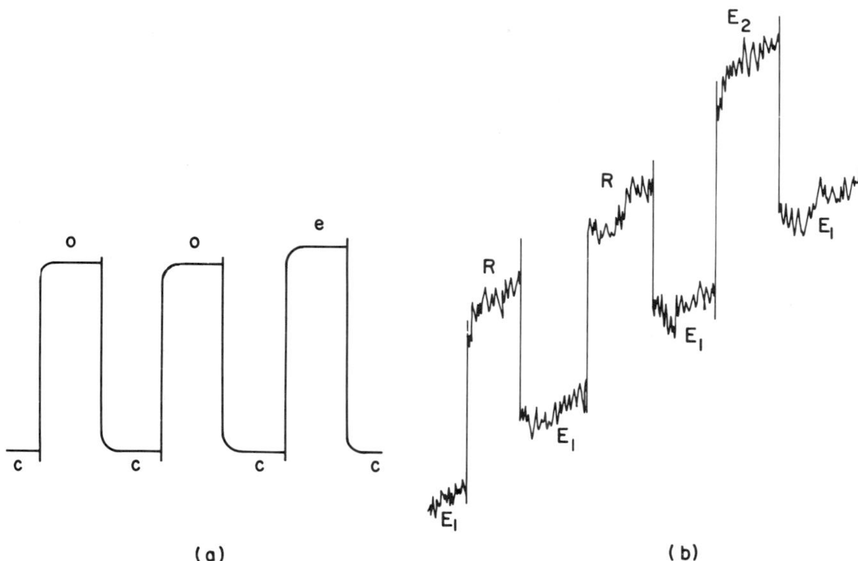

Fig. 1.43 Detector output of absolute radiometer in response to radiant or electrical heating. (a) Absolute. (b) Differential, with offset.

detector output signal while switching between the unknown radiant power (R) and electrical power (E_1) of the previously determined, approximate value. The advantage of this second measuring sequence is that it is a difference measurement between a known and an unknown power of approximately the same magnitude. For this reason, it is possible to offset the detector signal by a constant amount and use a more sensitive recorder range for the difference measurements, thus achieving a higher resolution. For the best results and the shortest measurement time, the operation of the shutter should be well synchronized with the electrical power switch and the switching times should be short. A typical chart recorder trace for a measuring sequence of this kind is depicted in Fig. 1.43b.

The traces can then be evaluated by manual interpolations and measurements with a ruler or they can be digitized and processed by computer. Suitable procedures for this purpose have been developed. (See Section 1.3.9.)

For detectors with very long thermal time-constants, where a direct comparison of the steady-state detector outputs is too time-consuming, use has been made instead of the rate of change of the detector signal to compare the electrical and radiant heating powers (Guild, 1937). Due to modern fabrication techniques for the detector elements, electronic time-constant compensation techniques, and the additional error potential of the "drift-mode" technique, it is seldom employed in modern absolute radiometers.

A measuring sequence in the active operating mode also consists of a series of measurements with the shutter alternately opened and closed. However, an electronic servo system is used to keep the detector output the same for both shutter positions. This is achieved by adjusting the amount of electrical power supplied to the detector element in such a way that the temperature of the detector element remains constant. The amount of incident radiant power is then inferred from the difference between the electrical power levels in the two shutter positions.

The active operating mode offers a number of possible variants, depending on whether the absolute radiometer is thermally compensated or uncompen-

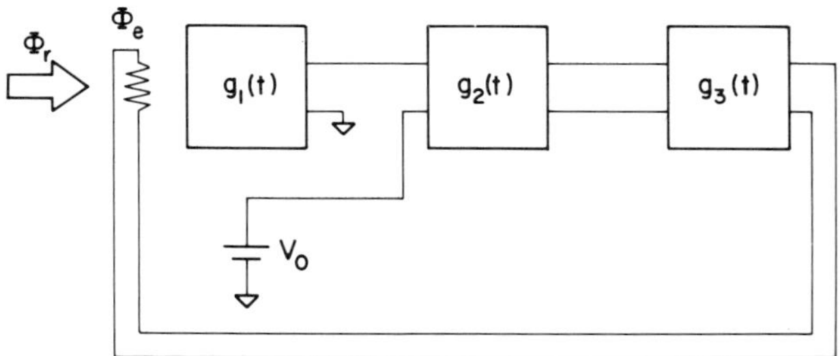

Fig. 1.44 Servo system for uncompensated absolute radiometer.

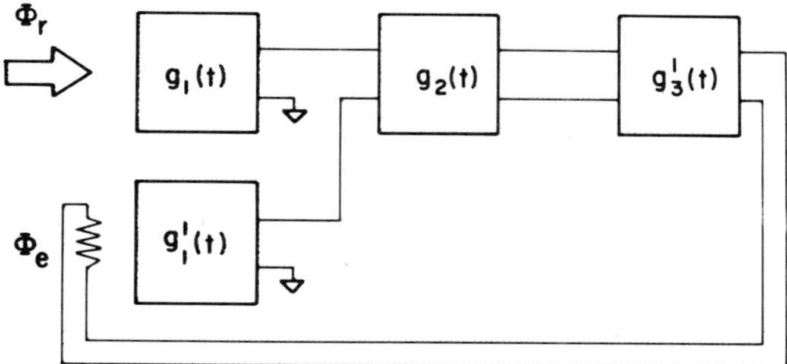

Fig. 1.45 Servo system for compensated absolute radiometer with compensating detector in feedback loop.

sated (Crommelynck, 1982). The servo system for an uncompensated instrument is depicted schematically in Fig. 1.44.

An example for this type of system would be an electrically calibrated pyroelectric radiometer. However, it is distinguished from similar systems incorporating other detectors in that the electrical power dissipated in the detector element in the radiant heating cycle is zero. For an explanation of the three gain stages—$g_1(t)$, $g_2(t)$, and $g_3(t)$—Section 1.3.8 should be consulted.

For a compensated instrument, there is a choice of using the feedback-controlled electrical power to heat either the nonirradiated detector element (Fig. 1.45) or the detector element exposed to the radiant power source.

In the second case, the detector signal can either be compared with the sum of the output of the compensating detector plus a constant reference voltage (Fig. 1.46) or with the output of the compensating detector heated by a constant reference voltage (Fig. 1.47).

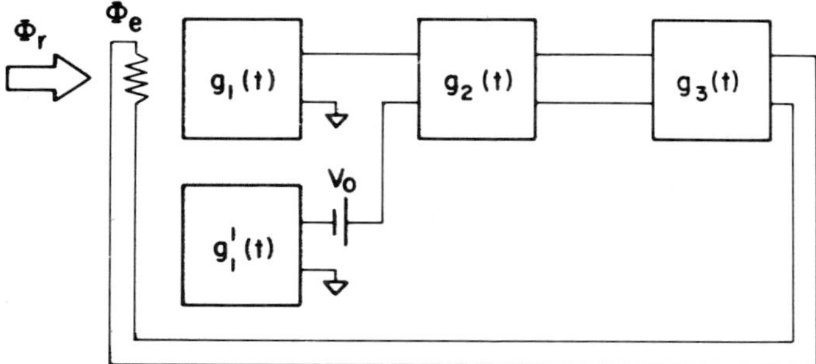

Fig. 1.46 Servo system for compensated absolute radiometer with irradiated detector in feedback loop. Its output is compared with the output of the compensating detector to which a constant offset is added.

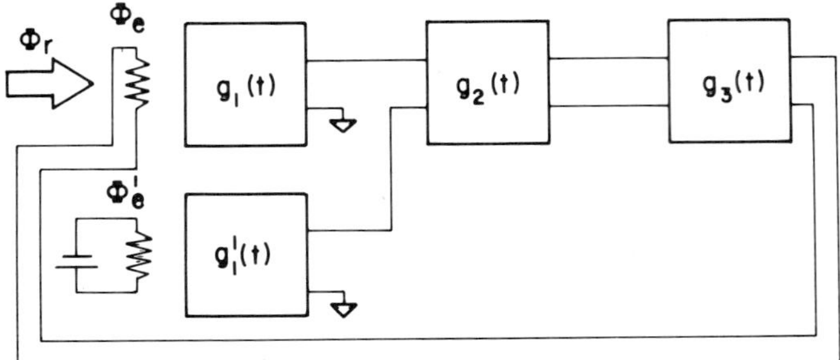

Fig. 1.47 Servo system for compensated absolute radiometer with irradiated detector in feedback loop. Its output is compared with the output of the compensating detector, which is heated with a fixed reference power.

1.3.7 Response-Accelerating Techniques

When the detector element of an absolute radiometer with a lumped thermal capacity C and a thermal resistance R to the environment is heated with a power input P, its temperature change is determined by a differential equation such as Eq. (1.20), with the constant b being replaced by $1/R$. For an initial temperature $T(t = 0) = T_0$, the solution has the well-known form

$$T(t) - T_0 = (T_\infty - T_0)[1 - \exp(-t/\tau)], \tag{1.32}$$

with the thermal time-constant τ equal to RC and the steady-state temperature T_∞ given by PR. The electrical circuit analog is depicted in Fig. 1.48a.

In order to examine the behavior of the circuit in the frequency domain, it

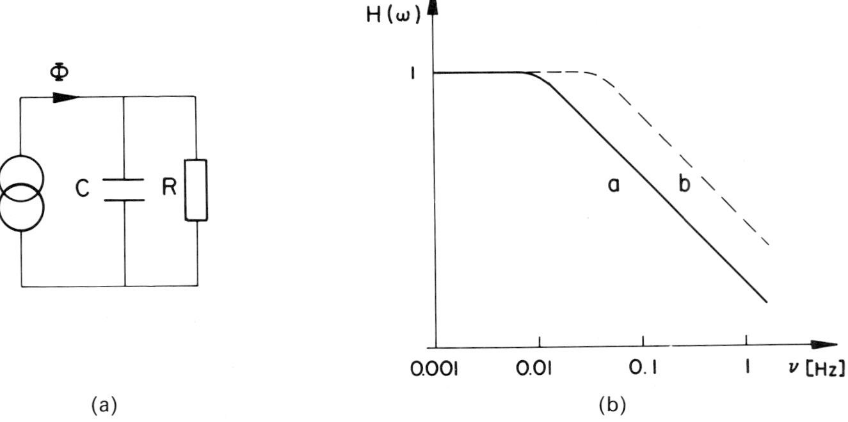

Fig. 1.48 (a) Equivalent electrical circuit representation of an absolute radiometer with lumped thermal resistance and capacity. (b) Frequency response of lumped parameter model.

is usual in electronic circuit analysis to solve the differential equation characterizing the circuit through Laplace transform techniques. If necessary, the solution can then also be transformed back into the time domain. Without going into any detail about Laplace transforms, it should just be noted that the circuit in Fig. 1.48a has a transfer function in the Laplace domain of the form $R/(1 + \tau s)$. Here s denotes the complex frequency variable $(\sigma + j\omega)$, with $\omega = 2\pi f$ and $j = \sqrt{-1}$, f being the frequency in hertz. The form of the transfer function and the values of the variables in it (e.g., τ) are characteristic for the frequency response of the circuit, which for the case of the circuit in Fig. 1.48a has the form depicted in Fig. 1.48b.

As the transfer function of two electronic circuits in series is the product of their individual transfer functions, it is obvious that one should be able to change the frequency response of a given circuit to a more desirable form by merely adding a series circuit with the necessary characteristics. Normally, the cascading of circuit elements is not ideal, because the added circuit will load the original circuit and thus disturb the characteristics of the transfer function. However, with the use of operational amplifiers, this problem can be solved very elegantly. The transfer function of a single operational amplifier with an input impedance Z_i and a feedback impedance Z_f has the form $A/[1 + Z_i/Z_f(1 - A)] \approx -Z_f/Z_i$, where A is the open-loop gain of the amplifier ($A \gg 1$) and Z_i and Z_f are the input and feedback impedances, respectively.

If one wishes to accelerate the response of the circuit in Fig. 1.48a, which represents an absolute radiometer with a given time-constant τ, one can therefore simply add an operational amplifier with a transfer function of the form $(1 + \tau s)$, which cancels the radiometer's transfer function term $1/(1 + \tau s)$. Several circuits with this property have been reported in the literature and they are depicted in Figs. 1.49a through 1.49c.

In the circuit of Fig. 1.49a, the values of R and C have to be chosen in such a way that RC matches the thermal time-constant of the absolute radiometer, while for circuit 1.49b, RC should equal twice the thermal time-constant of the radiometer (Lilley et al., 1975). The circuit in Fig. 1.49c includes an input RC-filter for noise reduction (R_3 and C_3). It has a constant gain of R_1/R_2 at low frequencies, and in this circuit R_2 and C_2 should be selected so that R_2C_2 matches the thermal time-constant of the radiometer (Brinkworth and Hughes, 1975). The function of the components R_1 and C_1 is to restrict the high-frequency gain of the circuit to avoid undue noise amplification. In all cases, the drop in the frequency response of the radiometers (Fig. 1.48b) is shifted to higher frequencies (curve b) just as if the time-constant of the absolute radiometer had been decreased by a corresponding amount. Responses accelerated by up to 10 times have been reported with these techniques.

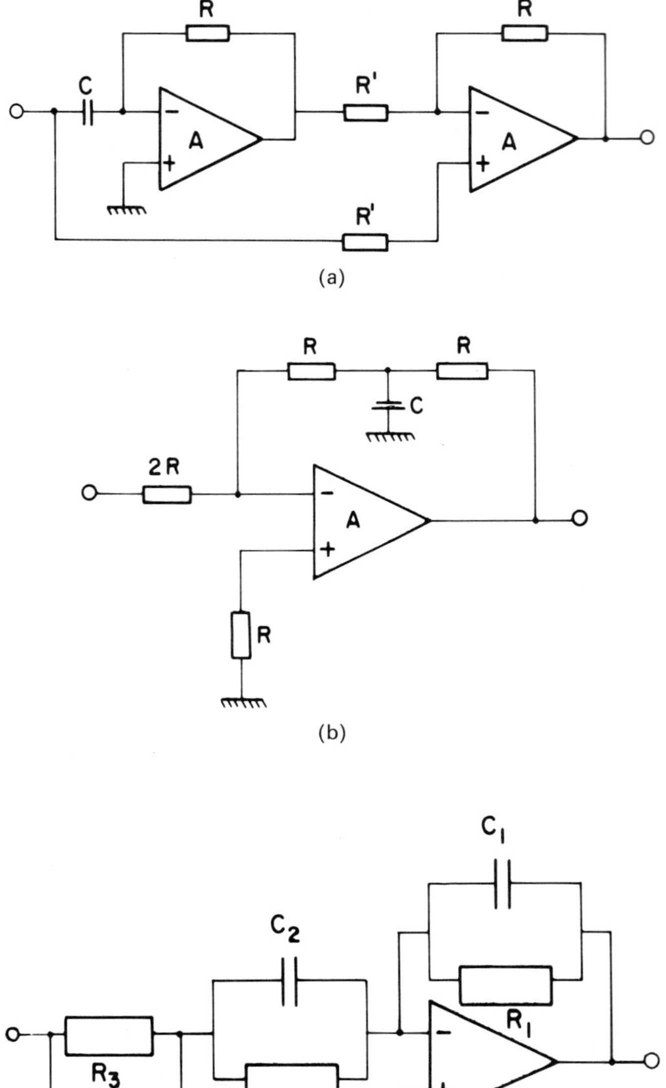

Fig. 1.49 Three examples of response-accelerating circuits. (Brinkworth and Hughes, 1975; Lilley et al., 1975)

Many detector elements, however, do not show the same frequency response as the simple *RC* filter with lumped parameters, which has as its signature a high-frequency asymptotic slope of −20 dB/decade (Fig. 1.48b). It is more common to find an asymptotic slope of only −10 dB/decade, which is characteristic for a network where the thermal resistance and capacitance are distributed rather than lumped parameters. In that case, the electronic compensation of the frequency response is somewhat more difficult, but the problem has also been solved. Malcorps (1983) has shown that this type of frequency response is equivalent to that of an infinite number of filters connected in series, each of them with one pole and one zero. The circuit for realizing one of these cascade filters is identical to the circuit in Fig. 1.49c, but Malcorps suggested the addition of a resistance in series with C_2 to add a second pole for ensuring the stability of the circuit. He also showed that a cascade of six such filters, each with different *RC* values, is a good approximation for at least two decades above the kink (break frequency) in the frequency response curve. In order to invert this filter for the compensation of the radiometer's frequency response, it is merely necessary to put the realized cascade filter into the feedback of an operational amplifier. The realization of this circuit by Malcorps (1983) is shown in Fig. 1.50. The block *H* is the six-element filter to be inverted, amplifier A_1 is introduced for stability reasons and to invert the signal, A_2 adds the input and feedback signals, and A_3 realizes the high gain required for the proper inversion.

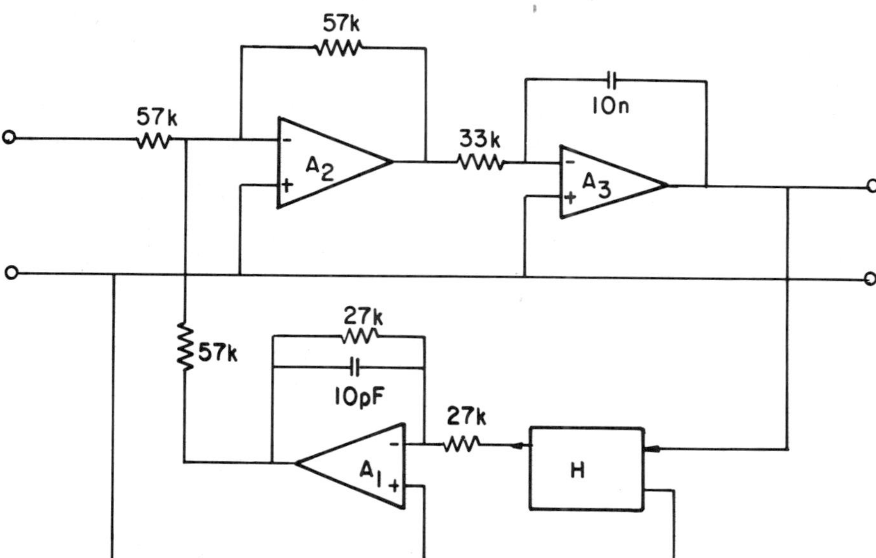

Fig. 1.50 Response-accelerating circuit for distributed parameter model. (Malcorps, 1983)

1.3.8 Feedback Control

The reasons why automatic control systems are employed in modern absolute radiometers are the same as when such systems are used in other application areas. They include:

(a) The system output (electrical heating) can be made to track a specified input function (radiant power) in an automatic fashion.
(b) The system performance depends less on variations in parameter values and unwanted disturbances.
(c) The system offers a means to achieve a desired transient or steady-state response.

Against these advantages, one has to weigh certain disadvantages:

(a) There is a potential for instability, which must be addressed and overcome by proper design philosophies.
(b) Additional amplification stages may be required to compensate for a decrease in system gain.
(c) Additional components and circuits are required for the feedback loop.

In order to analyze the behavior of a feedback control system, it is usual in classical control theory to represent a single-input, single-output system by means of a diagram such as the one in Fig. 1.51a.

The simplifying assumptions, useful for this model, are linearity and constant coefficients. Although few real systems comply strictly with these assumptions, they are often valid for a sufficiently narrow operating range. Such systems are conveniently analyzed through the use of Laplace transforms and frequency domain techniques and they can also be simulated with computers. The system depicted in Fig. 1.51a has a single input or reference

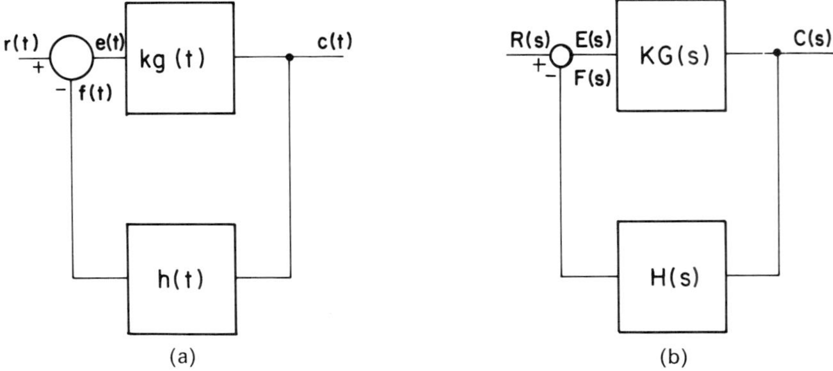

Fig. 1.51 Schematic feedback control system. (a) Time domain. (b) Frequency domain.

signal r(t) and a single output or controlled signal c(t). The input and output are related by a product of the system's impulse response function g(t) and an adjustable gain K. The error signal is e(t), the feedback signal is f(t), and the impulse response function of the feedback path is h(t).

If the system is to be analyzed with frequency domain techniques, one first has to perform a Laplace transform on the various system components, the result of which is depicted in Fig. 1.51b. The Laplace transform of the system's impulse response is known as the forward transfer function KG(s), and the transform of the feedback impulse response is the feedback transfer function H(s). When the feedback signal is not connected to the summing junction, the so-called open-loop transfer function of the system (around the loop) is KG(s)H(s). When the feedback loop is closed, the system output (controlled signal) can be expressed as

$$\mathrm{C}(s) = \frac{\mathrm{KG}(s)\mathrm{R}(s)}{1 + \mathrm{KG}(s)\mathrm{H}(s)}. \tag{1.33}$$

In the time-domain, the solution for the system output c(t) could be obtained by an inverse Laplace transform of C(s), which is, however, a much more complex operation in most cases than the simple symbolic expression suggests. For this purpose, a numerical solution of the differential equations on a computer may well be preferable except in simple cases where the inverse Laplace transform in analytical form is relatively straightforward. For most real problems, the system performance is evaluated in terms of certain desirable features of C(s) and the complete solution in the time-domain is not obtained for this purpose. The features that can be analyzed in this way are stability, steady-state accuracy, and satisfactory transient and frequency response.

In order to show in principle how feedback can reduce the effective thermal time-constant of an absolute radiometer, one can take the simple lumped parameter equivalent of the detector element (Fig. 1.48a) as a starting point. As has been discussed in Section 1.3.7, the transfer function of such an instrument is $R/(1 + \tau s)$ and its temperature increase in response to a sudden (step change) supply of power P is given by Eq. 1.32. If a gain K is added at the output of the system, the right-hand side of Eq. 1.32 has to be multiplied by K. This means that the temperature rise would be K times larger but the time-constant would remain unchanged. If, however, a gain K is added to the system transfer function and a feedback loop with H(s) = 1 is connected around it, the closed-loop response C(s) becomes (Schivell et al., 1982)

$$\mathrm{C}(s) = \frac{KR}{1 + KR} \frac{1}{1 + \left(\dfrac{\tau}{1 + KR}\right)s}. \tag{1.34}$$

It can immediately be seen from this equation that the effective time-constant (expression in brackets) is lower than the open-loop value by the fraction $1/(1 + KR)$. The gain K is usually the product of three separate gains (see Section 1.3.6): namely, G_1, the detector output per unit temperature change (e.g., in V/°C); G_2, the amplifier gain; and G_3, the power gain of the detector per unit detector output (e.g., in W/V). Therefore, K has the dimension [W/°C] and, since the thermal resistance R has the dimension [°C/W], KR is a dimensionless quantity. The time-constant will only be significantly affected if KR is in the order of one or larger.

Elements that are often, but not always included in the second gain stage in the feedback loop include full PDI (proportional-derivative-integral) regulators or a subset thereof (i.e., PI, I, or P, etc.), square-root elements, and time-constant compensating elements. A schematic circuit diagram for a PDI regulator is depicted in Fig. 1.52. As can be seen, the signal is fed to three operational amplifiers in parallel, which are operated as a proportional amplifier (A_1), as a differentiator (A_2), and as an integrator (A_3), respectively. The outputs of the three PDI elements are summed via amplifier A_4. The advantage of including an integrator in parallel with or instead of a proportional amplifier is that the error signal in the control loop tends to zero in this case. The derivative arm is often added for extra immunity to fast disturbances and for a smooth approach to the steady-state value without overshoot.

In order to find the optimum setting for the three individual PDI gains, a number of tuning techniques have been developed empirically (Smith, 1976;

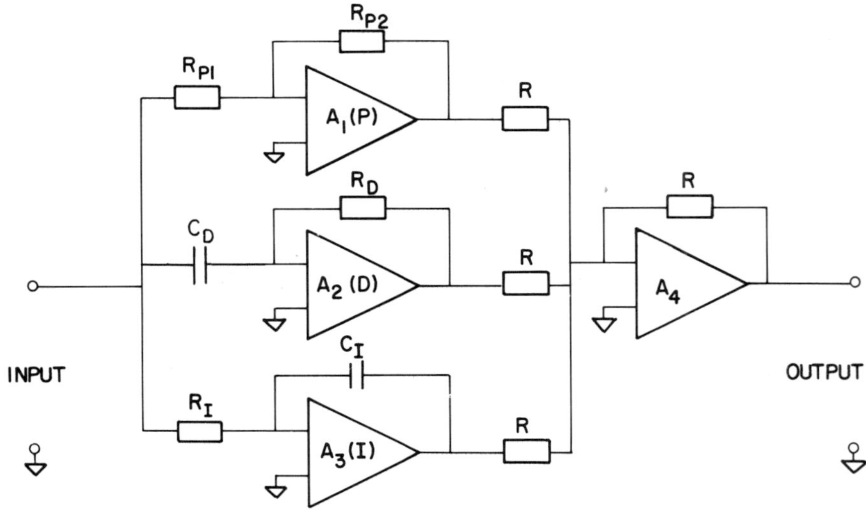

Fig. 1.52 PDI regulator circuit.

Stephanopoulos, 1984). One of these is the so-called Ziegler-Nichols closed-loop method, which works as follows:

(a) Disconnect the integral and derivative arms from the summing amplifier and adjust the proportional gain to a minimum or maximum.
(b) Bring the system to an equilibrium at the desired power level and change the proportional gain until the system reaches a state of sustained oscillation (constant oscillation amplitude).
(c) Calculate the gain and period of oscillation (in seconds) as reached in step (b). Then reduce the proportional gain R_{P2}/R_{P1} by a factor of 1.7, and set the integrator time-constant $R_I C_I$ equal to half and the differentiator time-constant $R_D C_D$ to one-eighth of the measured period of oscillation.

The reason why most feedback loops for absolute radiometers include a square-root element is that the feedback signal is usually the voltage applied to the heater of the detector in the feedback loop. Since this voltage determines the power dissipated in the heating element in a quadratic form (i.e., as $P = V^2/R$), it introduces a nonlinearity into the system response and this is often avoided by the introduction of the square-rooter. This measure has two main benefits. First, it allows one to make a single adjustment of the feedback gain to obtain optimum operating conditions, which are valid for all operating power levels. Second, it greatly simplifies the theoretical analysis of the feedback system. However, if a system is tuned for a particular power level and the measured radiant powers are quite small relative to that level, a practically linear operation of the feedback system around that level can be achieved even without the square-rooter.

The advantage gained by adding the various feedback components should always be carefully weighed against potential losses in accuracy or signal-to-noise levels. Cross-checks should always be made with results obtained in the open-loop operating mode to ensure that the feedback system does not introduce significant systematic errors. (See Section 6.5.9.) While there is no reason why a properly designed feedback system should cause a significant systematic error in measurements with an absolute radiometer, a cross-check with open-loop results would seem to be mandatory to ensure full metrological confidence in the results. Such a test should be an integral part of the instrumental characterization process.

1.3.9 Data Processing

The processing of the data acquired with a particular absolute radiometer generally depends on the specific application for which it is employed. Since the observed signals always contain drift and noise components, the main purpose of the data-processing stage is to reduce the influences of these components on the measurement results.

For absolute radiometers with pyroelectric detectors, the chopping frequency is high enough in most cases to reduce drift effects to a negligible magnitude. Consequently, the noise-reduction process can be reduced to simple analog or digital averaging (filtering).

In the case of absolute radiometers with larger time-constants (as, for example, those employing thermopile or bolometric detectors), both drift and noise have to be considered in the data processing. The standard technique for reducing drift-related effects is the acquisition of data with the detector element alternately exposed to (shutter open) or shielded from (shutter closed) the radiant power source. This generally results in two sets of curves, corresponding to the periods with the shutter open or closed, respectively. These can either be detector outputs recorded in a passive operating mode or electrical powers measured in an active operating mode. (See Section 1.3.6 and Fig. 1.43.) The acquired data in this case will represent a time-series of electrical powers $P_{ei}(t_i)$, with odd indices (i) corresponding to shutter-closed positions and even indices to shutter-open positions.

If the radiant power from the source changes in an unpredictable fashion (as, for example, in meteorological applications), the processing could consist of the fitting of a polynomial to two consecutive (or all) shutter-closed data and the subtraction of the fitted curve from the individual data in the shutter-open position. This would provide reasonable drift compensation but only partial noise reduction since only the set of shutter-closed data is smoothed in the process.

If one measures a stable source on the other hand, polynomials can be fitted to both sets of data with improved noise rejection. Hengstberger and Dressler (1982) analyzed the fitting of individual second-degree polynomials of the form

$$a + bt + ct^2 \tag{1.35}$$

to the two data sets. The difference between the two data sets could then be expressed by a polynomial of the same type. Figure 1.53 shows the result of this procedure for the absolute radiometer at the NPRL at a radiant power level of approximately 1 μW.

Cogno and Ezpeleta (1985) investigated the use of a joint least-squares model, which ensures the same slope for the curves through both sets of data and completely eliminates the drift effect from the measurements. The use of this model is of course only applicable if the source is sufficiently constant. They also suggested the use of the autocorrelation function of the noise as a check on the degree of statistical correlation of the noise function. In order to study the effect of the residual correlation, which they observed in this way, the authors investigated an autoregressive model of order k fitted to the two time-series of shutter-open and shutter-closed data. The main advantage of this model was that it produced a natural distinction between drift and noise.

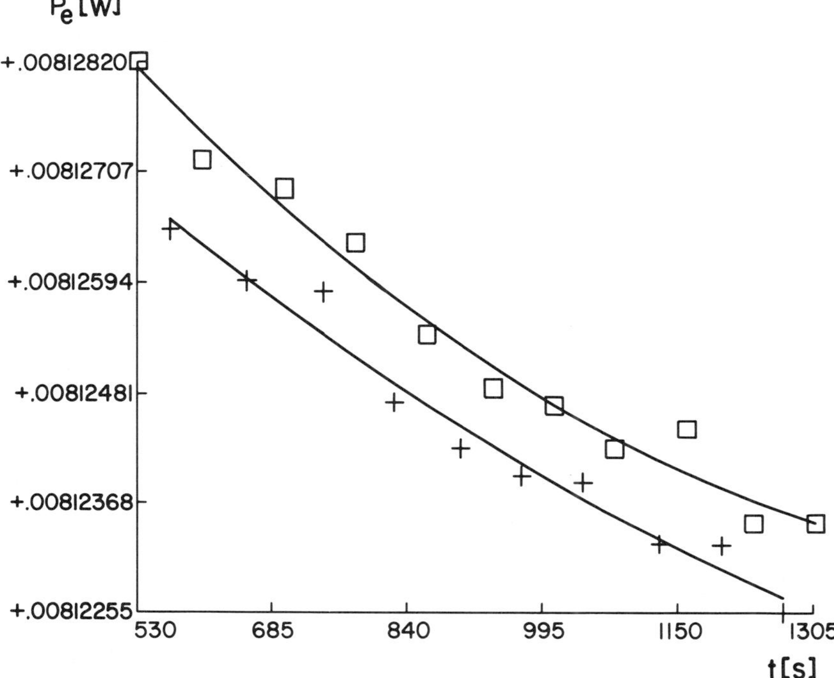

Fig. 1.53 Measured shutter-closed (□) and shutter-open (+) powers in watts as a function of time for a series of measurements with a servo-controlled absolute radiometer. Fitted polynomials are also shown.

The results indicated that the difference in the calculated power values for the two models amounted to less than 0.1 % for sets of more than 150 data points.

Premoli and Rastello (1986) included the dynamic response of the thermal detector in the mathematical model of the process. They developed an algorithm based on alternate linear and nonlinear optimization stages and proved the suitability of their sophisticated method on a large set of experimental data.

The emergence of powerful data-processing techniques specifically for use with absolute radiometers complements the advances being made regarding materials, design, electronics, theory, and device characterization and augurs well for continued progress in this area of metrology.

1.3.10 Applications

In the first few decades after the invention of the absolute radiometer, the principal applications for these instruments were solar-radiation measurements and the study of the blackbody radiation laws, including the determination of the constants in these laws. The field saw a considerable revival after the start of the space programs, the invention of the laser, and the formulation of proposals for a radiometric method to realize the candela in the early 1960s. Today, absolute radiometry is flourishing in the areas of traditional radiometry, photometry, and meteorology. Applications have multiplied, with absolute radiometers being used for the calibration of broadband power (including laser power and fiber-optic power) and irradiance meters, the calibration of the spectral response of detectors, the realization of the candela, the realization of the thermodynamic temperature scale, more and more accurate measurements of the Stefan–Boltzmann constant, the realization of spectral irradiance scales, measurements of the solar constant and of the earth's radiation budget, and measurements of the thermal performance of space craft and satellites. The availability of commercial absolute radiometers has enabled many smaller radiometry laboratories to eliminate their dependance on calibrated secondary standards. They can now calibrate their equipment against their own purchased absolute standards, which have specified uncertainties of about 0.5% (one sigma confidence level), or they can use these absolute instruments directly for critical measurements. Both of these measures ensure a higher accuracy for the end user. The accuracy of the most sophisticated absolute radiometers at metrology laboratories is now in the order of 0.03 to 0.3% (one sigma confidence level), and the best absolute radiometers used for meteorological purposes have similar uncertainty specifications.

APPENDIX 1.1

Acronyms of Associations, Government Agencies, Equipment, and Scientific Terms

ACR	active cavity radiometer
ANSI	American National Standards Institute
BIPM	Bureau International des Poids et Mesures
BSI	British Standards Institution
CCP	Consultative Committee on Photometry of the International Committee on Weights and Measures
CCPR	Consultative Committee on Photometry and Radiometry of the International Committee on Weights and Measures
CGPM	General Conference on Weights and Measures
CIE	International Commission on Illumination
CIPM	International Committee on Weights and Measures
CSIR	Council for Scientific and Industrial Research (South Africa)
DAMW	Deutsches Amt für Messwesen und Warenprüfung der DDR (East Germany)

ECPR	electrically calibrated pyroelectric radiometer
ECR	electrically calibrated radiometer
ESR	electrical substitution radiometer
ETL	Electrotechnical Laboratory (Japan)
IEC	International Electrotechnical Commission
IGY	International Geophysical Year
IMEKO	International Measurement Confederation
INM	Institut National de Metrologie (France)
IPS	International Pyrheliometric Scale
IRMB	Royal Meteorological Institute of Belgium
ISO	International Organization for Standardization
JPL	Jet Propulsion Laboratory (United States)
NASA	National Aeronautics and Space Administration (United States)
NBS	National Bureau of Standards (United States)
NEI	noise-equivalent irradiance
NEP	noise-equivalent power
NIM	National Institute of Metrology (Peoples Republic of China)
NML	National Measurement Laboratory (Australia)
NPL	National Physical Laboratory (United Kingdom)
NPRL	National Physical Research Laboratory (South Africa)
NRC	National Research Council (Canada)
NRLM	National Research Laboratory for Metrology (Japan)
NSL	National Standards Laboratory (Australia)
PDI	proportional-derivative-integral
PEL	Physics and Engineering Laboratory (New Zealand)
PMO	Physikalisch Meteorologisches Observatorium (Switzerland)
PTB	Physikalisch-Technische Bundesanstalt (West Germany)
PVF	polyvinylfluoride
SI	International System of Units
VNIIM	Mendeleev Institute of Metrology (Soviet Union)
WMO	World Meteorological Organization
WRR	World Radiometric Reference

REFERENCES

Abbot, C. G. (1911). The silver disk pyrheliometer. *Smithsonian Misc. Coll.* **56**, 1.

Ångstrom, K. (1886). Sur une nouvelle méthode de faire des mesures absolues de la chaleur rayonnante, ainsi qu'un instrument pour enregistrer la radiation solaire. *Nova Acta Reg. Soc. Sci. Ups.* **9**, 1.

Ångstrom, K. (1890). Beobachtungen über die Strahlung der Sonne. *Wied. Ann.* **39**, 294.

Ångstrom, K. (1893). Eine elektrische Kompensationsmethode zur quantitativen Bestimmung strahlender Wärme. *Nova Acta Reg. Soc. Sci. Ups. Ser.* **III**, 1 (English translation in *Phys. Rev.* **1**, 365 of 1893).

Ångstrom, K. (1899). The absolute determination of the radiation of heat with the electrical compensation pyrheliometer, with examples of the application of this instrument. *Astrophys. J.* **9**, 332.

Annals of the International Geophysical Year, Instruction Manual, Part IV: Radiation Instruments and Measurements. Vol. 5 (1958). Pergamon Press, New York.

Betts, D. B., and Gillham, E. J. (1968). An international comparison of radiometric scales. *Metrologia* **4**, 101.

Bischoff, K. (1968). Ein einfacher Absolutempfänger hoher Genauigkeit. *Optik* **28**, 183.

Blevin, W. R. (1970). Diffraction losses in radiometry and photometry. *Metrologia* **6**, 39.

Blevin, W. R., and Brown, W. J. (1966). Black coatings for absolute radiometers. *Metrologia* **2**, 139.

Blevin, W. R., and Brown, W. J. (1967). Development of a scale of optical radiation. *Austr. J. Phys.* **20**, 567.

Blevin, W. R., and Brown, W. J. (1969). Corrections in radiometry for absorption by atmospheric water vapour. *Metrologia* **5**, 28.

Blevin, W. R., and Brown, W. J. (1971). A precise measurement of the Stefan–Boltzmann constant. *Metrologia* **7**, 15.

Blevin, W. R., and Geist, J. (1974). Influence of black coatings on pyroelectric detectors. *Appl. Opt.* **13**, 1171.

Blevin, W. R., Collins, B. G., and Brown, W. J. (1969). A comparison of two radiometric scales. *Appl. Geophys.* **77**, 189.

Bocquet, P. (1979). "Contribution à la réalisation d'un radiomètre absolu à calibration par faisceau électronique. Construction d'un prototype." Dipl. Ing. thesis. Conservatoire National des Arts et Métiers, Paris.

Boivin, L. P. (1980). Calibration of incandescent lamps for spectral irradiance by means of absolute radiometers. *Appl. Opt.* **19**, 2771.

Boivin, L. P. (1985). Current developments in ESR's at NRC. *Proc. of an International Meeting on Advances in Absolute Radiometry, Cambridge, Massachusetts*, P. V. Foukal, ed., pp. 38–41. Atmospheric and Environmental Research, Cambridge, Mass.

Boivin, L. P., and McNeely, F. T. (1986). Electrically calibrated absolute radiometer suitable for measurement automation. *Appl. Opt.* **25**, 554.

Boivin, L. P., and Smith, T. C. (1978). Electrically calibrated radiometer using a thin film thermopile. *Appl. Opt.* **17**, 3067.

Bonhoure, J. (1969). Comparaison internationale des échelles radiométriques. *Rev. Gén. de l'Electr.* **78**, 1001.

Born, M., and Wolf, E. (1964). *Principles of Optics.* 2nd ed. Pergamon Press, Oxford, England.

Brinkworth, B. J., and Hughes, T. D. R. (1975). A method of obtaining fast response in solar pyranometers. *J. Phys. E: Sci. Instr.* **8**, 902.

Brown, W. J. (1975). A radiometric realization of the photometric units. *Metrologia* **11**, 111.

Brown, W. J., and Blevin, W. R. (1971). Usage direct d'un radiomètre absolu pour maintenir le candela. *Proc. CCP.* Bureau International des Poids et Mesures, Paris.

Brusa, R. W. (1983). *Solar Radiometry.* Publication no. 598. Physikalisch-Meteorologisches Observatorium and World Radiation Centre, Davos, Switzerland.

Brusa, R. W., and Fröhlich, C. (1975). Realization of the absolute scale of total irradiance. *Scientific discussions of the IV. International Pyrheliometer Comparisons.* World Radiation Centre, Davos, Switzerland.

Brusa, R. W., and Fröhlich, C. (1986). Absolute radiometers (PMO6) and their experimental characterization. *Appl. Opt.* **25**, 4173.

Bureau International des Poids et Mesures (1973). *Le Système International d'Unités*. 2e edition. BIPM, Paris.

Callendar, H. L. (1911). The radio-balance. A thermoelectric balance for the absolute measurement of radiation, with applications to radium and its emanation. *Proc. Phys. Soc.* (London) **23**, 1.

Carreras, C., and Corrons, A. (1981). Absolute spectroradiometric and photometric scales based on an electrically calibrated pyroelectric radiometer. *Appl. Opt.* **20**, 1174.

Coblentz, W. W. (1914). Various modifications of thermopiles having a continuous absorbing surface. *Bull. Bur. Std.* **11**, 131.

Coblentz, W. W., and Emerson, W. B. (1916). Studies of instruments for measuring radiant energy in absolute value: an absolute thermopile. *Bull. Bur. Std.* **12**, 503.

Coblentz, W. W., and Stair, R. (1933). The present status of the standards of thermal radiation maintained by the Bureau of Standards. *U.S. Bur. Stds. J. Res.* **11**, 79.

Cogno, J. A., and Ezpeleta, M. A. (1985). Mathematical models for the calculation of power absorbed by absolute radiometers. *Metrologia* **21**, 157.

Collet, E., and Wolf, E. (1978). Is complete spatial coherence necessary for the generation of highly directional light beams? *Opt. Lett.* **2**, 27.

Comité Consultatif de Photométrie et Radiométrie (*CCPR*), 9e Session (1977). Bureau International des Poids et Mesures, Paris.

Commission Internationale d'Eclairage (1970). *International Lighting Vocabulary*. CIE Publ. no. 17. CIE, Vienna.

Commission Internationale d'Eclairage (1983). *The Basis of Physical Photometry*. 2nd ed. CIE Publ. no. 18.2. CIE, Vienna.

Commission Internationale d'Eclairage (1985). *Electrically Calibrated Thermal Detectors of Optical Radiation* (*Absolute Radiometers*). CIE Publ. no. 65. CIE, Vienna.

Commission Internationale d'Eclairage (1986). *Colorimetric Illuminants*. First ed. CIE Publ. no. S001. CIE, Vienna.

Conference Générale des Poids et Mesures, 16e Session (1979). Bureau International des Poids et Mesures, Paris.

Crommelynck, D. A. (1967). *Contribution à l'Etude d'un Radiomètre Absolu: Caracteristiques d'un Prototype Experimental*. Technical Note 85. (113) World Meteorological Organization.

Crommelynck, D. A. (1982). *Fundamentals of Absolute Pyrheliometry and Objective Characterization*. NASA Conference Publication 2239. National Aeronautics and Space Administration, Scientific and Technical Information Branch, United States.

Crommelynck, D. A. (1985). The IRMB/SSD absolute radiometric base—objectives and developments. *Proc. of an International Meeting on Advances in Absolute Radiometry, Cambridge, Massachusetts*. P. V. Foukal, ed., pp. 12–15. Atmospheric and Environmental Research, Cambridge, Mass.

Crommelynck, D. A., and Domingo, V. (1984). Solar irradiance observations. *Science* **225**, 180.

Crommelynck, D. A., and Van de Velde, H. (1972). *Asservissement de la Compensation Electrique de Radiomètres Absolus*. Technical Note 10. Institut Royal Météorologique de Belgique, Brussels, Belgium.

Crommelynck, D. A., and Vandenborre, W. (1967). *Compensation Automatique de Chauffage des Lamelles des Pyrhéliomètres du Type Ångstrom*. Technical Note 85. World Meteorological Organization.

Doyle, W. M., McIntosh, B. C., and Geist, J. (1976). Implementation of a system of optical calibration based on pyroelectric radiometry. *Opt. Eng.* **15**, 541.

Drummond, A. J. (1970). In *Advances in Geophysics*, H. E. Landsberg and J. van Mieghem, eds., vol. 14, pp. 1–52. Academic Press, New York and London.

Eppley, M., and Karoli, A. R. (1957). Absolute radiometry based on a change in electrical resistance. *J. Opt. Soc. Am.* **47**, 748.

Etchechoury, E. M., and Cogno, J. A. (1982). Correccion por calentamiento de conductores en radiometros absolutos. *Carta Metrologica* **5**, 23. Instituto Nacional de Tecnologia Industrial, Buenos Aires, Argentina.

Foley, J. T., and Wolf, E. (1985). Radiometry as a short wavelength limit of statistical wave theory with globally incoherent sources. *Opt. Commun.* **55**, 236.

Foukal, P., Hoyt, C., and Miller, P. (1985). A helium-cooled absolute radiometer for irradiance measurements in the laboratory and from space platforms. *Proc. of an International Meeting on Advances in Absolute Radiometry, Cambridge, Massachusetts*, P. V. Foukal, ed., pp. 52–55. Atmospheric and Environmental Research, Cambridge, Mass.

Fox, N. P. (1985). Uses of a cryogenic radiometer in absolute radiometry. *Proc. of an International Meeting on Advances in Absolute Radiometry, Cambridge, Massachusetts*, P. V. Foukal, ed., pp. 56–59. Atmospheric and Environmental Research, Cambridge, Mass.

Fox, N. P., Key, P. J., Riehle, F., and Wende, B. (1986). Intercomparison between two independent primary radiometric standards in the visible and near infrared: a cryogenic radiometer and the electron storage ring BESSY. *Appl. Opt.* **25**, 2409.

Franzen, D. L., and Schmidt, L. B. (1976). Absolute reference calorimeter for measuring high power laser pulses. *Appl. Opt.* **15**, 3115.

Friberg, A. T. (1979). On the existence of a radiance function for finite planar sources of arbitrary states of coherence. *J. Opt. Soc. Am.* **69**, 192.

Fröhlich, C. (1973). The relation between the IPS now in use and Smithsonian scale 1913, Ångstrom scale and absolute scale. *Proc. Symp. Solar Radiation.* Smithsonian Institution, Washington, D.C., p. 61.

Fröhlich, C., Geist, J., Kendall, J., and Marchgraber, R. M. (1973). The Third International Comparisons of Pyrheliometers and a comparison of radiometric scales. *Solar Energy* **14**, 157.

Gao zhizhong, Wang zhenchang, Piao dazhi, Mao shihua, and Yang chiuhong (1983). Realization of the candela by electrically calibrated radiometers. *Metrologia* **19**, 85.

Geist, J. (1971). *Fundamental principles of absolute radiometry and the philosophy of this NBS program (1968 to 1971)*. NBS Techn. Note 594-1. U.S. Government Printing Office, Washington, D.C.

Geist, J. (1972). Waveform-independent lock-in detection. *Rev. Sci. Instr.* **43**, 1704.

Geist, J. (1976). Trends in the development of radiometry. *Opt. Eng.* **15**, 537.

Geist, J., and Blevin, W. R. (1973). Chopper-stabilized radiometer based upon an electrically calibrated pyroelectric detector. *Appl. Opt.* **12**, 2532.

Geist, J., Schmidt, L. B., and Case, W. E. (1973). Comparison of the laser power and total irradiance scales maintained by the National Bureau of Standards. *Appl. Opt.* **12**, 2773.

Gentile, C., and Rastello, M. L. (1980). *Misure de sensibilita e dell'effetto di riscaldamento dei conduttori in un radiometro assoluto.* Technical Report. Istituto Elettrotecnico, Torino, Italy.

Giacomo, P. (1981). News from the BIPM. *Metrologia* **17**, 69.

Gillham, E. J. (1962). Recent investigations in absolute radiometry. *Proc. Roy. Soc.* (London) *Ser. A* **269**, 249.

Gillham, E. J. (1964). Further work on a radiometric method of perpetuating the unit of light. *Proc. Roy. Soc. (London) Ser. A* **278**, 137.

Ginnings, D. C., and Reilly, M. L. (1973). Calorimetric measurement of thermodynamic temperatures above 0°C using total blackbody radiation. In *Temperature: Its Measurement and Control in Science and Industry*, vol. 4, pp. 339. Instrument Society of America, Research Triangle Park, North Carolina, United States.

Goodman, T. M., and Key, P. J. (1986). *A radiometric realization of the candela.* NPL Report QU 75. National Physical Laboratory, Teddington, U.K.

Grum, F., and Becherer, R. J. (1979). *Optical Radiation Measurements*, vol. 1. Academic Press, New York.

Guild, J. (1937). Investigations in absolute radiometry. *Proc. Roy. Soc. Ser. A* **161**, 1.

Gunn, S. R. (1973). Calorimetric measurements of laser energy and power. *J. Phys. E.* **6**, 105.

Hamilton, C. A., Phelan, R. J., and Day, G. W. (1975). Pyroelectric radiometers: off the drawing board. *Opt. Spectra*, 37.

Hengstberger, F. (1975). *Entwurf eines Meßsystems zur absoluten Messung spektraler Verteilungen und Bau eines Absolutempfängers für elektromagnetische Strahlung im optischen Bereich.* D.Sc. thesis. Technical University Vienna.

Hengstberger, F. (1977a). *The absolute radiometer at the National Physical Research Laboratory.* CSIR Research Report 331. Council for Scientific and Industrial Research, Pretoria, South Africa.

Hengstberger, F. (1977b). *A fresh look at the lead-heating effect in absolute radiometry.* CSIR Research Report 333. Council for Scientific and Industrial Research, Pretoria, South Africa.

Hengstberger, F. (1977c). An improved theory of the instrumental corrections for absolute radiometers. *Metrologia* **13**, 69.

Hengstberger, F. (1978). The present state of the art in absolute radiometry. *Proc. IMEKO Symp. Photon-Detectors, Prague.* International Measurement Confederation (IMEKO) TC-2, Budapest.

Hengstberger, F. (1979). A correction for spatial responsivity variations in absolute radiometers. *S.-Afr. J. Phys.* **2**, 41.

Hengstberger, F., and Appenroth, T. (1983). The development of a fully automated absolute radiometer. *Proc. 20th Session CIE, Amsterdam.* International Commission on Illumination (CIE), Vienna.

Hengstberger, F., and Dressler, R. E. (1982). Numerical treatment of measurement data captured with absolute radiometers. *Proc. 10th Int. Symp. IMEKO Techn. Com. Photon-Detectors, West Berlin*, pp. 236. International Measurement Confederation (IMEKO) TC-2, Budapest.

Hengstberger, F., Belsey, D. C., Stevens, D. P., Bittar, A., and Hamlin, J. D. (1987). *The electronic operating system for the absolute radiometer.* PEL Report no. 956. Physics and Engineering Laboratory (PEL), Lower Hutt, New Zealand.

Hengstberger, F., Dressler, R. E., Monard, L. A. G., Kok, C. J., and Turner, R. (1985). Further advances with the fully automated absolute radiometer developed at the NPRL. *Proc. of an International Meeting on Advances in Absolute Radiometry, Cambridge, Massachusetts*, P. V. Foukal, ed., pp. 34–37. Atmospheric and Environmental Research, Cambridge, Mass.

Hengstberger, F., Thain, E., and Turner, R. (1977). *A new measuring system for realizing photometric and radiometric scales.* CSIR Research Report 332. Council for Scientific and Industrial Research, Pretoria, South Africa.

Hickey, J. R., Stowe, L. L., Jacobowitz, H., Pellegrino, P., Maschhoff, R. H., House, F., and vonder Haar, T. H. (1980). Initial solar irradiance determinations from Nimbus 7 cavity radiometer measurements. *Science* **208**, 281.

Hickey, J. R. (1985). Analysis of calibration of Nimbus 7 radiometry. *Proc. of an International Meeting on Advances in Absolute Radiometry, Cambridge, Massachusetts*, P. V. Foukal, ed., pp. 30–33. Atmospheric and Environmental Research, Cambridge, Mass.

Honda, T., and Endo, M. (1978). International intercomparison of laser power at 633 nm. *IEEE J. Quantum Electronics* **QE-14**, 213.

IGY (1958). *Annals of the International Geophysical Year*, Vol. V, part VI. Pergamon Press, Oxford.

Il'in, A. S. (1985). PI-4 and PI-5 standard primary instrument transducers for laser radiation. *Meas. Techn.* **28**, 331.

International Organization for Standardization (1973). *Quantities, Units, Symbols, Conversion Factors and Conversion Tables.* ISO 1000-1973. ISO, Geneva.

Karoli, A. R., Hickey, J. R., and Frieden, R. G. (1983). Self-calibrating cavity radiometers at the Eppley Laboratory. *Proc. SPIE Conf.: Applications of Optical Metrology Techniques and*

Measurements, Arlington, Virginia, vol. II, pp. 43–50. Society of Photo-Optical Instrumentation Engineers, Bellingham, Washington, United States.

Kartachevskaia, V. E. (1961). Experimental determination of the luminous efficacy of radiation. *Vses. Nauch. no. Issled. Inst. Metrol. Trudy. Inst. Kom.* **56**, 36.

Kartachevskaia, V. E. (1962). Travaux dans le domaine d'application des methodes de mesures énergétiques aux mesures photométriques. *Proc. CCP*, 5e Session. Bureau International des Poids et Mesures, Paris.

Kendall, J. M., and Berdahl, C. M. (1970). Two blackbody radiometers of high accuracy. *Appl. Opt.* **9**, 1082.

Kok, C. J. (1972). Absolute radiometry and the unit of light. *S. A. Engineer & Electrical Review*, 17.

Kostkowski, H. J., Erminy, D. E., and Hattenburg, A. T. (1970). In *Advances in Geophysics*, H. E. Landsberg and J. van Mieghem, eds., vol. 14, pp. 111–127. Academic Press, New York and London.

Kurlbaum, F. (1892). In *Die Thaetigkeit der Physikalisch-Technischen Reichsanstalt in den Jahren 1891 und 1892*, 6. Physikalisch-Technische Reichsanstalt (PTR), Braunschweig, Federal Republic of Germany.

Kurlbaum, F. (1893). Die Thaetigkeit der Physikalisch-Technischen Reichsanstalt in den Jahren 1891 und 1892. *Z. Instrumentenkunde*, March 1893, 122.

Kurlbaum, F. (1894). Notiz über eine Methode zur quantitativen Bestimmung strahlender Wärme. *Wied. Ann.* **51**, 591.

Kurlbaum, F. (1898). Über eine Methode zur Bestimmung der Strahlung in absolutem Maass und die Strahlung des schwarzen Körpers zwischen 0 und 100 Grad. *Wied. Ann.* **65**, 746.

Kurlbaum, F. (1899). Änderung der Emission und Absorption von Platinschwarz und Russ mit zunehmender Schichtdicke. *Ann. der Phys.* **1**, 846.

Kurlbaum, F. (1900). Temperaturdifferenz zwischen der Oberfläche und dem Inneren eines strahlenden Körpers. *Ann. der Phys.* **2**, 546.

Laser Precision Corporation (1985). *Rs-5900 electrically calibrated pyroelectric radiometer (ECPR)*. Datasheet. Laser Precision Corporation, Utica, New York.

Lilley, P., Warner, D. H., Mordecai, B. A., and Spencer, A. J. (1975). The use of electronic prediction to achieve fast response from a simple thermal mass flow meter. *J. Phys. E: Sci. Instr.* **8**, 3.

Lummer, O. and Kurlbaum, F. (1892). Ueber die Herstellung eines Flächenbolometers. *Z. Instrumentenkunde* **12**, 81.

Malcorps, H. (1983). Method to increase the bandwidth of heat flux meters. *Rev. Sci. Instr.* **54**, 381.

Martin, J. E., Fox, N. P., and Key, P. J. (1985). A cryogenic radiometer for absolute radiometric measurements. *Metrologia* **21**, 147.

Martinez-Herrero, R., and Mejias, P. M. (1981). Characterization and reconstruction of planar sources that generate identical intensity distributions in the Frauenhofer zone. *Opt. Lett.* **6**, 607.

Martinez-Herrero, R., and Mejias, P. M. (1982). Relation among planar sources that generate the same radiant intensity at the output of a general optical system. *J. Opt. Soc.* **72**, 765.

Martinez-Herrero, R., and Mejias, P. M. (1984). Radiometric definitions for partially coherent sources. *J. Opt. Soc. Am. A* **1**, 556.

Martinez-Herrero, R., and Mejias, P. M. (1986). Radiometric definition from second-order coherence characteristics of planar sources. *J. Opt. Soc. Am. A* **3**, 1055.

Möstl, K. (1978). Empfängernormale zur Messung der Strahlungsleistung von Dauerstrich-Lasern. *Feinwerktechnik & Messtechnik* **86**, 72.

Möstl, K. (1979). *Empfängernormale für Laserstrahlung.* PTB-Bericht PTB-Opt-8, pp. 58–75. Physikalisch-Technische Bundesanstalt, Braunschweig, Federal Republic of Germany.

Müller, J. E., and Ratz, P. (1982). Strahlungsthermosäule für Absolutmessungen. *Feingerätetechnik* **31**, 175.

Nicodemus, F. E. (1963). Radiance. *Am. J. Phys.* **31**, 368.

Nicodemus, F. E. (1976). *Self-Study Manual on Optical Radiation Measurements: Part I—Concepts*, chapters 1–3. NBS Techn. Note 910-1. U.S. Government Printing Office, Washington, D.C.

Nicodemus, F. E. (1978). *Self-Study Manual on Optical Radiation Measurements: Part 1—Concepts*, chapters 4–5. NBS Techn. Note 910-2. U.S. Government Printing Office, Washington, D.C.

Ono, A. (1979a). A design of thermal detectors of radiation with uniform response over the surfaces of the receivers. *Metrologia* **15**, 127.

Ono, A. (1979b). An absolute radiometer with uniform response; fabrication, characterization and analysis. *Japanese J. Appl. Phys.* **18**, 1995.

Ooba, N. (1965). Radiomètre absolu. *Proc. Comité Consultatif de Photométrie, 6e Session*, pp. 23–27. Bureau International des Poids et Mesures, Paris.

Phelan, R. J., and Cook, A. R. (1973). Electrically calibrated pyroelectric optical-radiation detector. *Appl. Opt.* **12**, 2494.

Premoli, A., and Rastello, A. P. (1986). An analysis algorithm for thermal detectors of radiation. *IEEE Trans. Instr. Meas.* **IM-35**, 612.

Preston, J. S. (1963). A radiometric method of perpetuating the unit of light. *Proc. Roy. Soc. Ser. A*, **272**, 133.

Putley, E. H. (1982). History of infrared detection—Part 1. The first detectors of thermal radiation. *Infrared Phys.* **22**, 125.

Quinn, T. J., and Martin, J. E. (1984). Radiometric measurements of the Stefan-Boltzmann constant and thermodynamic termperature between −40°C and +100°C. *Metrologia* **20**, 163.

Quinn, T. J., and Martin, J. E. (1985). A radiometric determination of the Stefan–Boltzmann constant and thermodynamic temperatures between −40°C and 100°C. *Phil. Trans. R. Soc. Lond. A* **316**, 85.

Ready, J. F. (1971). *Effects of high-power laser radiation.* Academic Press, New York.

Rozenberg, G. V. (1973). Coherence, observability and the photometric aspect of beam optics. *Appl. Opt.* **12**, 2855.

Rutgers, G. A. W. (1951). A new absolute radiometer. *Physica* **17**, 129.

Schivell, J., Renda, G., Lowrance, J., and Hsuan, H. (1982). Bolometer for measurements on high-temperature plasmas. *Rev. Sci. Instr.* **53**, 1527.

Shumaker, J. B. (1983). *Self-Study Manual on Optical Radiation Measurements: Part 1—Concepts*, chapter 10, "Introduction of coherence in radiometry." NBS Techn. Note 910-6. U.S. Government Printing Office, Washington, D.C.

Siegel, R., and Howell, J. (1972). *Thermal Radiation Heat Transfer.* McGraw-Hill, New York.

Smith, C. L. (1976). Controller tuning that works. *Instr. & Control Syst.*, September 1976, 43.

Sparrow, E. M., and Cess, R. D. (1966). *Radiation Heat Transfer.* Brooks/Cole, Belmont, California.

Stair, R. (1970). In *Advances in Geophysics*, H. E. Landsberg and J. van Mieghem, eds., vol. 14, pp. 83–109. Academic Press, New York and London.

Stephanopoulos, G. (1984). *Chemical Process Control.* Prentice Hall, London.

Stock, K. D. (1983). Effect of gusts of wind on a thermal detector of radiation. *Rev. Sci. Inst.* **54**, 1708.

Stock, K. D. (1985). Electrically calibrated radiometers at the PTB—Advances and applications. *Proc. of an International Meeting on Advances in Absolute Radiometry, Cambridge, Massachusetts*, P. V. Foukal, ed., pp. 42–45. Atmospheric and Environmental Research, Cambridge Mass.

Walther, A. (1968). Radiometry and coherence. *J. Opt. Soc. Am.* **58**, 1256.

Walther, A. (1978). Propagation of the generalized radiance through lenses. *J. Opt. Soc. Am.* **68**, 1606.

Welford, W., and Winston, R. (1987). Generalized radiance and practical radiometry. *J. Opt. Soc. Am. A* **4**, 545.

West, E. D., Case, W. E., Rasmussen, A. L., and Schmidt, L. B. (1972). A reference calorimeter for laser energy measurements. *NBS J. Res.* **76A**, 13.

Willson, R. C. (1973). Active cavity radiometer. *Appl. Opt.* **12**, 810.

Willson, R. C. (1979). Active cavity radiometer type IV. *Appl. Opt.* **18**, 179.

Willson, R. C. (1980). Active cavity radiometer type V. *Appl. Opt.* **19**, 3256.

Willson, R. C. (1985). Solar irradiance variability monitoring by active cavity radiometers. *Proc. of an International Meeting on Advances in Absolute Radiometry, Cambridge, Massachusetts*, P. V. Foukal, ed., pp. 6–9. Atmospheric and Environmental Research, Cambridge, Mass.

Wolf, E. (1978). Coherence and radiometry. *J. Opt. Soc. Am.* **68**, 6.

World Meteorological Organization (1977). *World Radiometric Reference.* Resolution 10 (Executive Committee XXX), pp. 136–137. WMO, Geneva.

Zalewski, E. F. (1986). Addendum to the report on the preliminary round of the comparison of the national standards of absolute spectral responsivity. *Proc. CCPR.* document CCPR/86-10. Bureau International des Poids et Mesures, Paris.

Additional References

Balderston, J., and King, P. W. (1980). Development of radiometers for high power thermal radiation. SPIE vol. 234: *New Developments and Applications in Optical Radiation Measurement.* Society for Photo-Optical Instrumentation Engineers, Bellingham, Washington.

Baskin, I. M., and Paderin, L. Ya. (1982). An absolute detector for radiative heat fluxes. *Izmeritel'naya Tekhnika* **11**, 44. Cover-to-cover translation: *Meas. Tech.* (U.S.) **25**, 929.

Byrne, P. O., and Farmer, F. T. (1972). A self-calibrating blackbody radiometer for use in the UV, IR and visible spectrum. *J. Phys. E: Sci. Instr.* **5**, 590.

Carson, R. S., and Pietrzyk, Z. A. (1978). An absolute calorimeter for a high-power high-energy CO_2 laser. *J. Phys. E: Sci. Instr.* **11**, 998.

Edwards, J. G. (1967). An accurate carbon cone calorimeter for pulsed lasers. *J. Sci. Instr.* **44**, 1967.

Fisk, G. A., and Gusinow, M. A. (1977). Circulated-liquid calorimeter for the detection of high-power and high-energy pulsed laser signals. *Rev. Sci. Instr.* **48**, 118.

Gerashchenko, O. A., Sazhina, S. A., and Paniashvili, M. S. (1973). An absolute two sided thermal radiometer. *Teplofizika i Teplotekhnika* **25**, 52. Translated in *Heat Transfer—Soviet Research*, **7**, 56 (1975).

Gerashchenko, O. A., Sazhina, S. A., Grishchenko, T. G., and Lebedev, A. D. (1975). Dynamic characteristics of an absolute radiometer. *Heat Transfer—Soviet Research* **7**, 116.

Greenfield, E. (1986). Measuring laser power accurately. *Laser Focus/Electro-Optics*, March 1986, 134.

Hengstberger, F. (1977). Mesure des caractéristiques de sensibilité d'un radiomètre absolu: étude d'un cas concret. *Proc. CCPR*, 9e Session, pp. 115–131. Bureau International des Poids et Mesures, Paris.

Hesser, R. J., and Potter, W. D. (1983). Use of active cavity radiometers as absolute radiometers and transfer standards. *Proc. SPIE Conf.: Applications of optical metrology, techniques and measurements, Arlington,* Vol. II, 128. Society of Photo-Optical Instrumentation Engineers, Bellingham, Washington, United States.

Kandpal, H. C., and Joshi, K. C. (1986). Directional heat losses in radiometers. *Appl. Opt.* **25**, 1693.

Kmito, A. A. (1977). Analysis of the errors in pyrheliometers that are caused by the coating of the receiver. In *Radiation Research in the Atmosphere,* vol. 393, pp. 78–82 (in Russian). Gidrometeoizdat, Leningrad.

Kmito, A. A., Anspok, E. S., Agapova, V. P., and Razin'kov, V. V. (1977). Pyrheliometer with thermoelectric cooling of the receiver surface. In *Actinometry, Atmospheric Optics, Ozonometry,* vol. 384, pp. 74–79. Gidrometeoizdat, Leningrad.

Lobo, P. C. (1986). An electrically compensated radiometer. *Solar Energy* **36**, 207.

Malcorps, H. (1981). Frequency-response of heat fluxmeters. *J. Phys. E.: Sci. Instr.* **14**, 1054.

Malcorps, H. (1982). Influence of convection, conduction and radiation on the frequency response of heat fluxmeters. *Rev. Sci. Instr.* **53**, 362.

Masaki, M. (1978). Development of an absolute radiometer. *J. Mech. Eng. Lab.* (Japan) **32**, 43 (in Japanese).

Niimura, M., Dooling, J., Zich, R. N., Brock, R. N., and York, T. M. (1985). Large aperture, high-speed calorimeter for high-energy optical pulses. *Rev. Sci. Instr.* **56**, 2253.

Nikolsky, G. A. (1976). Increase in the accuracy of the reproduction of the absolute pyrheliometric scale. In *Radiation in the Atmosphere,* pp. 138–154. Leningrad University Press, Leningrad.

Pradhan, M. M., and Garg, R. K. (1982). Pyroelectric null detector for absolute radiometry. *Appl. Opt.* **21**, 4456.

Predtechenskij, A. V., and Skljarov, Jv. A. (1976). Bolometric pyrheliometer with automatic compensation. In *Radiation Research in the Atmosphere,* vol. 370, pp. 3–71 (in Russian). Gidrometeoizdat, Leningrad.

Preston, J. S. (1972). Spectral response of laser cone calorimeters. *J. Phys. E: Sci. Instr.* **5**, 1014.

Shcherbina, D. M., Profatilova, N. I., Orlov, V. A., and Mavasnev, Yu. Z. (1979). Instruments for measuring the parameters related to intensive radiation fluxes. *Proc. VIII IMEKO Congress, Moscow, pp.* S14–5–S14–10. International Measurement Confederation.

Simpson, P. A., Johnson, E. G., and Etzel, S. M. (1984). *A Calorimeter for Measuring High-Energy Optical Pulses.* NBSIR 84-3008, NBS, Boulder, Colorado. U.S. Government Printing Office, Washington, D.C.

Smith, R. L., Russel, T. W., Case, W. E., and Rasmussen, A. L. (1972). A calorimeter for high-power CW lasers. *IEEE Trans. Instr. Meas.* **IM-21**, 434.

Tietz, G. E. (1977). Spatial variation in the response of laser calorimeters. *Appl. Opt.* **16**, 1136.

Wiedner, S., and Jaoudi, E. (1978). Autobalancing radiometer. *Am. J. Phys.* **46**, 935.

Zatkovic, J. (1978). Symetrický system ako referenčny prijimač v optickej rádiometrii. *Jemná mechanika a Optika* (Czechoslovakia) **1978/1**, 7.

2

Absorbers of Optical Radiation

K. MÖSTL

Physikalisch-Technische Bundesanstalt
Braunschweig, Federal Republic of Germany

2.1 INTRODUCTION

When optical radiation is incident on matter, a number of various physical or chemical processes can occur: absorption, reflection, transmission, luminescence, Raman scattering, photoelectric phenomena, structural and chemical changes in the matter, and so on. For application in an absolute radiometer, the interaction desired is the conversion of radiant energy into heat. Any conversion into other than heat energy is a loss and may give rise to a measurement error. Therefore, the first objective when selecting an absorber is to choose materials in which the other phenomena mentioned are negligible or calculable. Particular care is necessary with absorbers for vacuum-ultraviolet radiation because of the emission of photoelectrons and chemical changes. With absorbers for high-power radiation, special attention must be paid to structural and chemical changes as a result of high absorber temperatures.

The heat produced in the absorber is usually not sensed directly, but, rather, via heat conduction that leads to an increase in temperature in a sensor attached close to the absorber. The second selection criterion is therefore that the absorber should have a high and uniform heat conductivity that should be fairly well known in order to calculate the heat-flow distribution for correction purposes. These calculations will be easiest when the absorber geometry is simple.

If the absorber is porous, an important portion of the heat conduction is caused by air contained in the pores. This must be taken into account when

ISBN 0-12-340810-5

performing experiments in vacuum. Absorbers to be exposed to high irradiance levels must have an extremely high heat conductivity to avoid impermissible excess temperatures. In the particular case of the absorption of very short laser pulses, the problem of high absorber temperature cannot be avoided by good heat conductivity, as the duration of the absorption process is too short to enable sufficient heat dissipation. In this case, the absorber must permit a high temperature without causing damage such as in a metal with a high melting point, or the energy must be optically dissipated. This is achieved when the radiation is absorbed in a volume instead of on a surface or when it is spread over a large surface area by means of multiple reflections. When applying chopped radiation, the responsivity of the radiometer may depend on the depth within the absorber layer where the radiation is mainly absorbed, because the transit time of the heat waves and, therefore, the phase will depend on this parameter (Stair et al., 1965). This effect, which may be wavelength-dependent, will decrease as the heat conductivity of the absorber increases.

In general, the absorptance of the absorber is measured before assembling the radiometer, and a remeasurement at a later date is difficult. This leads to the third criterion for selection: the absorptance should exhibit an excellent temporal stability. This stability is subject to thermal and mechanical stress; therefore, mechanical durability and good adhesion to the substrate are other desirable features.

A fourth selection criterion is the specific heat of the absorber. In order to attain a small time-constant for the radiometer, a low heat capacity is desirable. Detectors to be operated with modulated radiation demand a low heat capacity to avoid serious reduction in responsivity (Blevin and Geist, 1974). This also applies to calorimeters which measure the radiant energy of pulses and for which the responsivity is proportional to the heat capacity.

Criteria associated with reflectance and transmittance, which are obviously of great importance, are discussed in the next section.

2.2 ABSORPTANCE AND REFLECTANCE

When any energy conversion other than conversion to heat is avoided, the directional spectral absorptance $\alpha(\lambda, \upsilon, \theta)$ can be interpreted as heat-conversion efficiency of the radiation with wavelength λ incident from the direction (υ, θ). As the absorber is usually irradiated normally, it is common practice to omit the angle variables. However, when absorptance values are taken from the literature, one should be aware that α may depend on the geometric measurement conditions. It is easiest to define the absorptance indirectly. It is the relative part of the incident radiation (from a defined direction) that is not transmitted and not reflected. If a collimated monochromatic beam of radiant power $\Phi_0(\lambda)$ is directed onto a piece of matter, the integral over all radiation components with wavelength λ leaving the matter with angles $\upsilon < 90°$ to the normal is called the reflected power $\Phi_r(\lambda)$. Accordingly, the integral over the components of radiant intensity with $\upsilon > 90°$ is the transmitted power $\Phi_t(\lambda)$. The spectral reflectance is therefore

$$\rho(\lambda) = \Phi_r(\lambda)/\Phi_0(\lambda), \tag{2.1}$$

the spectral transmittance is

$$\tau(\lambda) = \Phi_t(\lambda)/\Phi_0(\lambda), \tag{2.2}$$

and, finally, the spectral absorptance is

$$\alpha(\lambda) = 1 - \tau(\lambda) - \rho(\lambda). \tag{2.3}$$

It is easy to design the absorber in such a way that $\tau(\lambda)$ is zero. High energy conversion efficiency is then equivalent to a low spectral reflectance:

$$\alpha(\lambda) = 1 - \rho(\lambda) \qquad (\tau(\lambda) = 0). \tag{2.4}$$

Equation (2.4) is the formula commonly used when dealing with thermal detectors. The aim of the preceding discussion was to clarify the number of restrictions that must be considered when this simple expression is used.

According to Eq. (2.4), the spectral absorptance can be derived from a measurement of the spectral reflectance, more exactly, from a measurement of the directional-hemispherical reflectance. That requires a directed irradiation (at normal incidence if possible) and a hemispherical integration of the reflected radiant intensity. Due to the Helmholtz reciprocity principle, an equivalent configuration is to irradiate hemispherically with uniform radiance and to measure the radiance of the surface in normal direction. (For measurement techniques, see Section 6.5.2.) A low spectral reflectance is desirable for three reasons. First, the absorptance becomes almost independent of wavelength; second, a comparatively high relative uncertainty for the reflectance measurement can be tolerated; and third, small temporal changes of the reflectance will not significantly affect the absorptance.

Another method of determining the spectral absorptance is to make use of Kirchoff's law

$$\alpha(\lambda) = \varepsilon(\lambda) \tag{2.5}$$

and to measure the spectral emittance $\varepsilon(\lambda)$ (Stierwalt, 1966; Hawks and Cottingham, 1970) or to calculate it (Bedford and Ma, 1974; Bedford et al., 1985). This method is of particular interest in the middle and far infrared ranges, where measurements of the hemispherical spectral reflectances are difficult.

If a radiometer is to be used for the measurement of polychromatic radiation of unknown spectral composition, the absorber must be selected so that it has an absorptance as constant as possible in the wavelength range of interest.

2.3 ABSORBER TYPES

In general, the criteria discussed in the previous two sections cannot all be realized at the same time. The absorber types discussed in this section are different approaches to the ideal absorber. For example, the relative importance of a high absorptance or a low time-constant depends on the application for which the radiometer is intended.

2.3.1 Cavity Absorbers

This section deals with cavities whose inner walls are coated with a matt black paint. In this case, there is no advantage in designing the shape of the cavity differently from a rectangular prism or a cylinder. But if specular reflectance is predominant, different shapes are superior, as discussed in Sections 2.3.4 and 2.3.5. One exception from this procedure is made: the cavities of cryogenic radiometers are summarized only in this section although one of them makes use of specular reflections.

A cavity with matt blackened walls and a small entrance aperture enables an absorptance approaching unity within 0.001 (Gillham, 1962; Bischoff 1964; Quinn and Martin, 1985).

The effective reflectance ρ_{eff} can be calculated if the reflectance ρ of the black coating and the geometry are known (Bauer and Bischoff, 1971; Bedford and Ma, 1974; Bedford et al., 1985). For a tubular cavity of length L and radius r, this is given in good approximation by

$$\rho_{\text{eff}} = \frac{\rho}{1-\rho}\frac{1}{1+(L/r)^2} \tag{2.6}$$

if one end face forms the entrance aperture and if only the opposite end face is irradiated directly by the incident beam. If the black paint is not ideally matt but also specularly reflecting, this can cause serious deviations from the predicted value of the effective reflectance (Geist and Richmond, 1971). The accuracy can also be affected when different parts of the cavity wall evaluate the absorbed power slightly differently due to different heat-flux distributions and due to a nonuniform responsivity of the (distributed) temperature sensing system.

Besides the high absorptance, another advantage of the cavity absorber is that a high thermal conductivity of the black paint is of minor importance, because most of the heat lost at the front surface of the paint is transferred to other parts of the wall and is therefore not lost for the sensor. An increase in the cavity length increases the effective absorption, but it also increases the heat capacity, which already is disadvantageously high. The response time is further enlarged by the lateral heat flow within the cavity walls and by the slow heat transfer to the air contained in the cavity. The response time will therefore be seriously underestimated when the product of the heat capacity of the cavity and its thermal resistance (reciprocal heat-flow rate) to the heat sink is taken as the time-constant. Moreover, the time dependence of the response cannot be described as a simple exponential function.

If the cavity is operated without a window, another disadvantage is added to this: adiabatic fluctuations of air pressure produce temperature fluctuations and thus increase the noise of the output (Stock, 1983). The time required to establish thermal equilibrium can be reduced when the cavity wall is subdivided into thermally insulated sections (Bischoff, 1964; Gillham, 1962) because the lateral heat flow is then reduced. At cryogenic temperatures, the disadvantages of the cavity absorber are greatly reduced: the heat capacity is much lower and the thermal conductivity is higher than at room temperature. The positive features remain unchanged. A cavity absorber is therefore the best choice for cryogenic radiometers (Ginnings and Reilly, 1972; Quinn and Martin, 1985). Martin et al. (1985) improved the absorptance of their cryogenic radiometer by replacing the diffuse black paint by a specular reflecting one. Of course, the primary reflection of the bottom plate (some percent) had to be prevented from directly escaping from the cavity. Therefore, the bottom was inclined by 30°. The resulting effective absorptance was 0.99998.

2.3.2 Disk Absorbers

The disk absorber is the simplest type of absorber. It consists of a disk-shaped substrate covered by an absorbing layer. The substrate should have a high thermal conductivity and is therefore usually made of silver or copper. The absorptance is mainly determined by the absorbing coating if this is thick enough. In contrast to the cavity absorber, the heat losses at the front surface of the coating (mainly air convection and thermal reradiation) are of importance. They are the main reason for the nonequivalence of radiative and electrical heating. The heat current flowing to the heat sink has to pass the thermal resistance formed by the black coating. The result is a temperature drop across this coating. In order to minimize these effects, the absorbing layer should be homogeneous in thickness and only just thick enough to be opaque.

The main advantage of the disk absorber is that it can be made with low heat capacity, thus achieving a small time-constant. If the disk is irradiated homogeneously, the lateral heat flow within the disk is negligible and the response-time-constant of the detector is given in a good approximation by the product of heat capacity of the absorber unit and its thermal resistance to the heat sink.

A special form of the disk absorber is the glass-disk absorber, where the absorbing element is a grey or colored glass, absorbing the radiation in volume instead of on a surface. The absorptance exhibits an excellent stability under chemical and mechanical stress, and the surface is not sensitive to dust. The reflectance is specular and therefore easy to measure. This type of absorber is mainly used for calorimeters measuring the energy of high-irradiance laser pulses (Edwards, 1970; Gunn, 1974; Xu, 1979). The radiative load can be reduced by using glass with a lower absorption coefficient, thus increasing the thickness of the absorption path. On the other hand, the heat capacity increases linearly with the thickness of the glass plate. An added disadvantage is the poor thermal conductivity of the glass, which, in conjunction with the large heat capacity, is responsible for the long response and cooling time of these detectors. It must also be mentioned that a determination of the relative spectral responsivity from a measurement of the spectral absorptance of the glass can be erroneous. Because of the poor thermal conductivity, the ratio between the front surface losses and the desired flow to the heat sink depends on where the radiation is absorbed within the glass disk. For example, the uv responsivity is reduced because the short wave radiation is absorbed within a thin layer behind the front surface of the glass and is thus subject to higher convection losses than radiation that is mainly absorbed closer to the temperature sensor at the rear surface.

2.3.3 Disk Absorbers with Reflectors

The effective absorptance of a disk absorber can be significantly increased by placing the disk center in the focus of a hemispherical mirror, thus returning most of the disk-reflected radiation (Müller, 1933; Hengstberger, 1975; Ono, 1979a). As the mirror requires a hole through which the radiation can pass to the absorber and because of the absorption losses of the mirror, the absorptance remains below 1. From the formulas given by Brandenburg (1964), Blevin and Brown (1971) deduced that the radius of the disk r_d must be significantly larger than the radius r_i of the area of the disk directly irradiated, so that its image, after reflection, falls entirely within the receiver (r_m being the mirror radius):

$$r_d = r_m r_i / (r_m - 2r_i) \tag{2.7}$$

The effective spectral reflectance of the receiver ρ_r is then given by

$$\frac{\rho_r}{\rho_b} = \frac{1 - \rho_m}{1 - \rho_m \rho_b} \tag{2.8}$$

where ρ_b is the spectral reflectance of the diffusely reflecting black coating without mirror and ρ_m is the spectral reflectance of the mirror. If the reflectance is more specular in character, Day et al. (1976) suggest a geometric design in which the beam impinges on the detector five times before exiting.

As Gillham (1953) showed, a hemispherical mirror can be applied for an in situ determination of the reflectance of a disk-type thermopile. However, when applying this method, the fact that the thermal radiation emitted by the absorber is also reflected by the mirror must be taken into consideration.

2.3.4 Cone Absorbers

Inner cones are frequently used as absorbers. The idea is to absorb the radiation by multiple specular reflections. For this purpose, the inner surface of the cone must reflect specularly to a high degree and the diffuse reflectance must be as small as possible. Then a beam, incident parallel to the cone axis, undergoes $n = 180°/\theta$ reflections before exiting. (Here, θ is the full cone angle in degrees.) The effective absorptance can then be roughly estimated from the reflectance ρ of the flat surface by

$$\alpha_{\text{eff}} = 1 - \rho^n. \tag{2.9}$$

As the angle of incidence with reference to the cone surface is not 90° and changes from reflection to reflection, a more accurate calculation of the effective absorptance must be performed using Fresnel's formulas and the optical constants of the absorber material. (See, e.g., Driscoll and Vaughan, 1978). Such computations show that the effective reflectance for parallel-polarized radiation is always smaller than that for the perpendicularly polarized one. To obtain an effective reflectance that does not depend on the status of polarization, the cone absorber must be designed in such a way that the specular reflectance for both planes of polarization is negligible. The remaining reflectance is due to unavoidable diffuse reflectance. Möstl (1978, 1986) found an absorptance of about 0.9985 for a 45° cone (diameter 12 mm) electroplated with black nickel (Schering Nickellyt-Schwarzglanz, current density 1 mA/cm^2, applied for 10 min.).

Figure 2.1 shows that the absorptance is unfortunately not completely homogeneous over the cone diameter. The dip in the center is due to the nonperfect apex giving rise to some direct specular reflectance. Although a calculation predicted that the fourfold specularly reflected residue would be

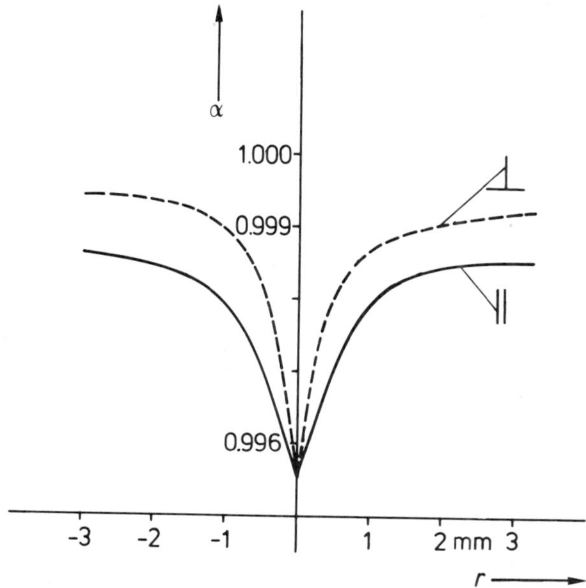

Fig. 2.1 Hemispherical spectral absorptance $\alpha(\lambda)$ of a 45° cone at $\lambda = 515$ nm (measured by the author at room temperature with an integrating sphere) as a function of the radius r (apex: $r = 0$) for perpendicular ($\perp$) and parallel ($\|$) polarized radiation.

negligible, there is still a small polarization dependence of the reflectance that obviously originates in the diffuse reflection. Not shown in the figure is the smooth decrease in absorptance for larger radii that is due to the increase of the solid angle of exit for the reflected radiation. This dependence of the angle of exit on the location on the cone surface is the reason why it is not advisable to coat the cone with a diffuse black paint. The result would be a rather inhomogeneous absorber. Zalewski et al. (1979) demonstrated this with a cone coated with 3M Nextel Black Paint exhibiting a mean effective absorptance of 0.9978.

They also prepared other cones in two steps. First, a thin coating of low-gloss black paint was applied, which was followed by a clear, high-gloss coating. For these cones, they found an effective absorptance of 0.9994 with a decrease in the surrounding of the apex. This inhomogeneity could be avoided by tubular termination of the cone center, improving the effective absorptance to 0.99988. Willson (1979) applied this experience to his absolute solar radiometer ACR IV, which was equipped—in contrast to the preceding types—with a specular reflecting black epoxy paint on its cone. The succeeding model ACR V also uses the tubular termination in order to avoid the formation of a meniscus at the apex. Willson (1980) reports an effective absorptance of 0.99988 for this cone absorber. Unfortunately, this improvement in absorptance results in an increase in heat capacity and temperature transit time. Some authors applied a diffuse black paint to their cones (Kendall and Berdahl, 1970; Gao et al., 1983). This restricts the application of the radiometers to that type of irradiation for which the effective absorption was determined.

An advantage of the just-mentioned black-nickel cone is that an absorbing layer 1 μm thick with a good thermal conductivity is sufficient to absorb almost the entire radiation, and the temperature drop across the absorbing layer is therefore negligible. This is especially advantageous for the measurement of cw laser power. A disadvantage is the longer time constant of the cone detectors. As Möstl (1978) showed, it depends approximately quadratically on the cone length. To minimize this length, the cone angle chosen should be as large as possible without sacrificing the high absorptance.

If the energy of laser pulses is to be absorbed, cones with a metallic nickel (Edwards, 1975) or chromium (Möstl et al., 1985) surface electroplated on copper cones are used. As these metals are highly reflecting, the cone angle must be small enough to make 12 to 15 reflections possible. In this case, particular attention must be paid to the polarization dependence of the absorptance.

2.3.5 Cylindro-Cone Absorbers

Cylindro-inner-cones have found application in radiometers measuring direct solar radiation (Kendall and Berdahl, 1970; Willson, 1973). The theory of the emittance of this special type of cavity (also applicable to absorptance) was developed by Bedford et al. (1985). It should be considered that a more complicated absorber geometry increases not only the absorptance but also the thermal mass. Geist et al. (1973) and later Brusa and Fröhlich (1978) used a cavity with an inverted-cone-shaped bottom painted with a specular reflecting black enamel. This design certainly has a higher thermal mass than a simple (inner) cone absorber, but the apex of the cone can be made more perfectly. The achieved absorptance of 0.9998 for solar radiation was accordingly high.

2.4 COMMON ABSORBER MATERIALS

2.4.1 3M Nextel Black Paint

This black paint (manufactured by the Minnesota Mining and Manufacturing Company) also goes by the name of Black Velvet Coating 2010 (if purchased in two-liter canisters) or 101 Velvet Coating C 10 (if in aerosol spray cans). Quinn and Martin (1985) and Smith (1984) report that Nextel Black consists of solid spheres of silicate material with diameters of between 5 and 100 μm (80% by mass) coated with carbon black (about 20% by mass) and a small amount of hydrocarbon binder, whereas the manufacturer's brochure only mentions spherical grounded glass pigments. Quinn and Martin (1985) published a scanning electron microscope photo showing these spheres. From the same manufacturer, a similar paint, but with two components, has the code no. 401. Its binder is an epoxy resin.

Nextel Black has an absorptance value of 0.973 in the visible range, which is not remarkably high. Nevertheless, it is very popular because it exhibits other favorable features: the mechanical stability is excellent, the thermal conductivity is reasonably high, and the paint is electrically nonconducting, thus making an insulating layer against the electrical heater unnecessary. Furthermore, Hsia and Richmond (1976) found that if sprayed onto the substrate, this paint exhibits a non-angle-dependent bidirectional reflectance in a wide range of angles. In the visible and near infrared range, the paint is an almost perfect diffuse reflector and the retroreflection is negligible. (Quinn and Martin (1985) estimated it to be in the order of 5×10^{-5} at a wavelength of 633 nm.) Compton et al. (1974) found that the reflectance at 28-μm wavelength is still diffuse, but becomes more and more specular with a corresponding reduction in the diffuse reflectance at longer wavelengths. Smith (1984) attributed the low specular reflectance that he found between 18 and 40 μm to the principal absorption bands of amorphous silicate. Stierwalt (1966) measured the spectral emittance of Nextel Black in the range from 3 to 40 μm. From his description of the apparatus used, it can be concluded that he determined the directional spectral emittance in a direction perpendicular to the surface. This means his results chiefly represent the retroreflection with maxima at about 9 and 21 μm. These are due to the reststrahlen bands in silica (Heaney et al., 1983). In another work, Stierwalt (1979) measured the spectral emittance of Nextel Black at cryogenic temperatures applying the same method.

Another result of the previously mentioned studies of Compton et al. was the realization that Nextel paint is suitable for vacuum applications where the temperature does not exceed 250°C. After a first phase of copious desorption of greasy matter, low outgassing rates were obtained and the coatings retained their cohesion.

Houck (1982) mixed Nextel Black with a SiC lens-grinding compound as an additional absorptive, which resulted in a better far-infrared absorptance. He applied this paint successfully in vacuum at liquid-helium temperatures. Although values are not given, it must be assumed that the heat capacity of this coating is higher than for ordinary Nextel Black.

TABLE 2.1 Some Properties of Just Opaque Coatings of Common Absorber Materials

Material	Typical Thickness (μm)	Areal Density (g/m^2)	Heat Capacity (Jm^{-2}K^{-1})	Thermal Resistance (μK W^{-1}m^2)	Outgassing Rate (Pa 1 s^{-1})	Remarks
3M Nextel Black	50[a]	50[a]	60[b]	70[b]	10^{-8}[c]	at 25°C
Parson's Black	10[d]	100[f]	100[g]	250[d]		
2-torr gold black	5[d]	0.2[d]	0.02[d]	700 in air[d]		density 0.2% of solid[d]
1-torr gold black	1[d]	0.6[d]	0.08[d]	70 in air[d]		density 2% of solid[d]
Martin Black	60[e]				nondetectable[e]	
Smoked carbon black	20[d]	25[f]	2[g]	600 in air[d]		

[a] Manufacturer's brochure.
[b] Blevin and Geist (1974).
[c] Compton et al. (1974).
[d] Blevin and Brown (1966).
[e] Wade and Willson (1975) and Smith (1984).
[f] Harris (1967).
[g] Estimated.

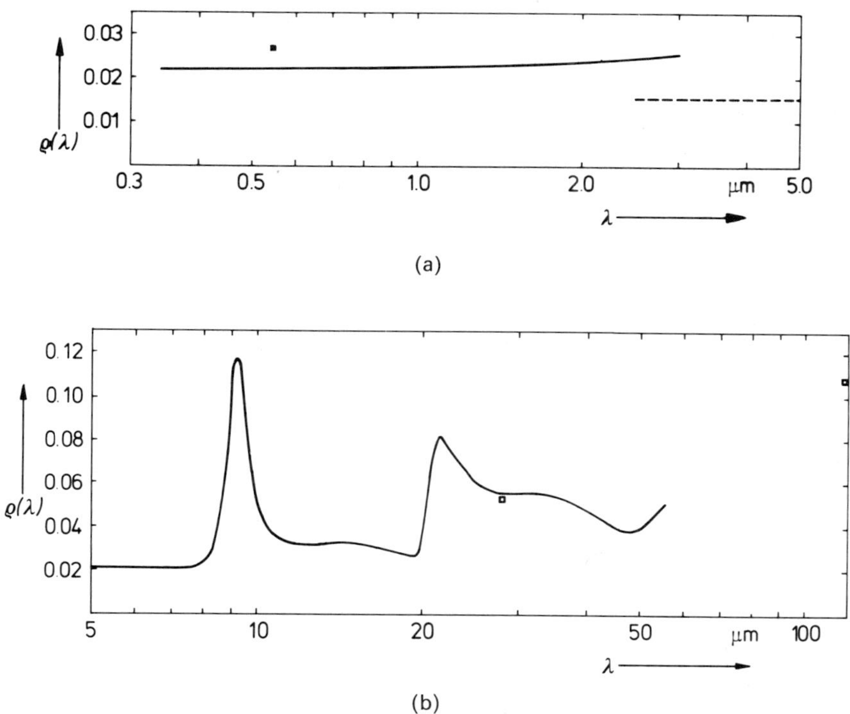

Fig. 2.2 Hemispherical spectral reflectance $\rho(\lambda)$ of Nextel Black Paint as a function of the wavelength λ at room temperature. (a) Spectral range from 340 nm to 5 μm. (——) after Wightman and Grum (1981) (method not specified); (□) after Bischoff (1964) by an integrating sphere method; (- - - -) after Clarke and Larkin (1985) by a hemispherical mirror method. (b) Spectral range from 5 to 118 μm. Curve continued from (a) after Clarke and Larkin (1985); points (□) calculated from measurements of diffuse and specular reflectance (angle of incidence 15°) performed by Compton et al. (1974).

As of 1988, Nextel Black is only available from stock because 3M Company has discontinued production and replaced it with ECP-2200. This paint is composed of jagged silica particles held together by an ethyl silicate binder containing a proprietary black dye that is, in contrast to Nextel Black, not a carbon black. Smith (1983) assumes this last-mentioned change to be responsible for the higher far-infrared specular reflectance of the new paint.

More data on Nextel Black and on other blacks are compiled in Table 2.1. Values reported on the hemispherical spectral reflectance by various authors for Nextel Black are put together in Fig. 2.2. Curves of the specular infrared reflectance that are of little use for the evaluation of the absorptance are not included but can be found in Smith (1982, 1983) and Smith and Wolfe (1982).

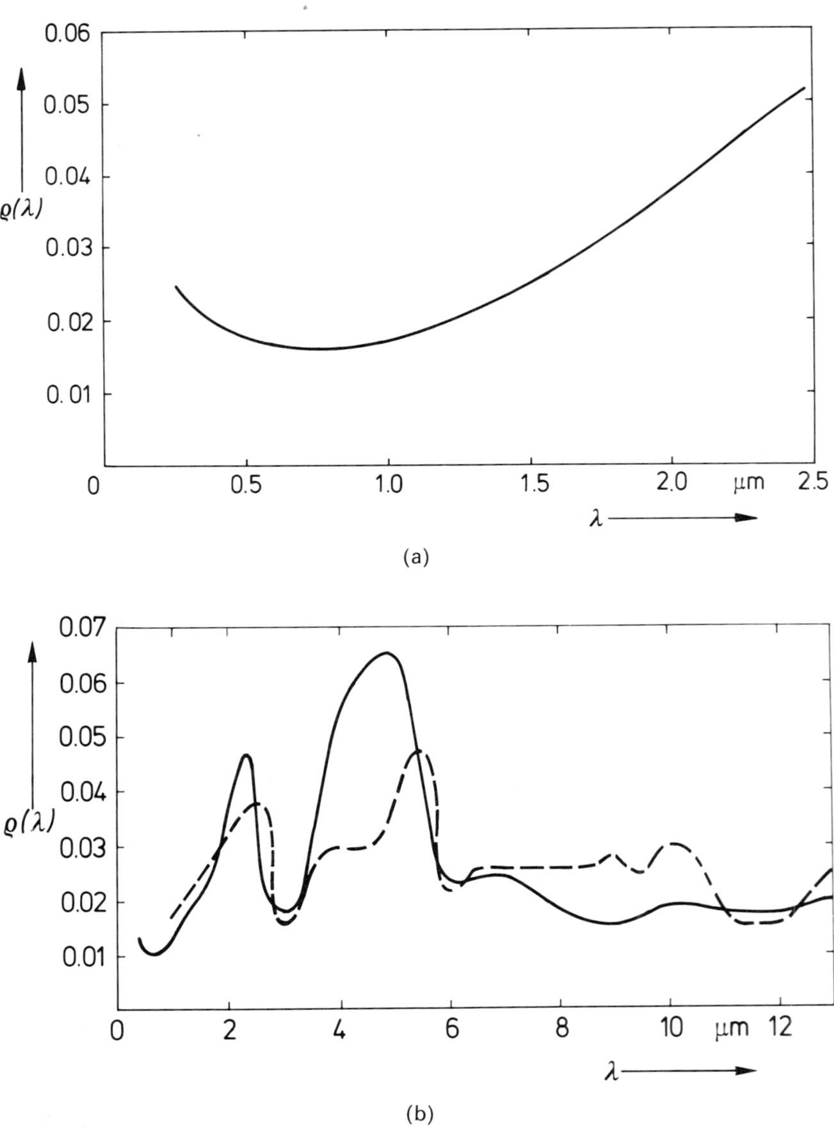

Fig. 2.3 Hemispherical spectral reflectance $\rho(\lambda)$ of Parson's Optical Black Lacquer as a function of the wavelength λ at room temperature. (a) After Blevin and Brown (1965), measured by a spheroidal mirror reflectometer. (b) (——) after Gillham (1953); (----) after Blevin and Brown (1965). Method the same as in (a).

2.4.2 Parson's Optical Black Lacquer

This paint is manufactured by Thom. Parson and Sons and must be applied in two steps. First, the substrate is sprayed (or brushed) with the Optical Black Lacquer undercoat, containing mostly carbon black dispersed in nitrocellulose (Harris, 1967), and then with a top coat of the same paint, until it appears that continued spraying is not increasing the absorptance of the coating (Geist and Richmond, 1971). The top coat containing a mixture of aniline black and ethyl cellulose is responsible for the low reflectance. When examined with a microscope, the surface appears ragged and uneven; no definite pattern is apparent, the irregularities being in the order of several micrometers (Harris, 1967). In the visible range, the absorptance is higher than for Nextel Black, but is more selective in the infrared, especially in the range from 1.5 to 6.5 μm (Gillham, 1953; Blevin and Brown, 1965), as Fig. 2.3 shows. Studies by Hsia and Richmond (1976) of the bidirectional reflectance showed that, if applied by brushing, this coating is an excellent diffuse reflector in the visible range of the spectrum at angles of incidence within about 30° of normal, but becomes more and more specular at increasing angles of incidence.

An attempt to improve the far-infrared absorptance was made by Philer and Houck (1971). They first sprayed the substrate with Parson's Optical Black Lacquer undercoat. Small short nylon fibers were dusted onto the still tacky paint. After the paint had thoroughly dried, a second coat of Parson's Black was lightly sprayed over the fibers. This coating is obviously less suited for applications requiring a low thermal resistance.

2.4.3 Eppley–Parson's Optical Black Lacquer

Sold by the Eppley Laboratory, this paint must be applied in the same way as Parson's Optical Black. Geist and Richmond (1971) assume that these lacquers are identical in composition although they found different reflectance values. Hsia and Richmond (1976) mention that the Eppley Laboratory purchased the rights to this coating from Thom. Parson and Sons. They found that Eppley–Parson's Optical Black shows a slight tendency toward retroreflectance when applied by brushing, but in general appears quite similar to the Parson's black.

2.4.4 Kodak Black Baking Lacquer

This lacquer is produced by Eastman Kodak Co. It is applied by spraying onto the substrate and subsequent baking at about 175°C. If the proportion of paint to air is adjusted during spraying, the surface can be made either

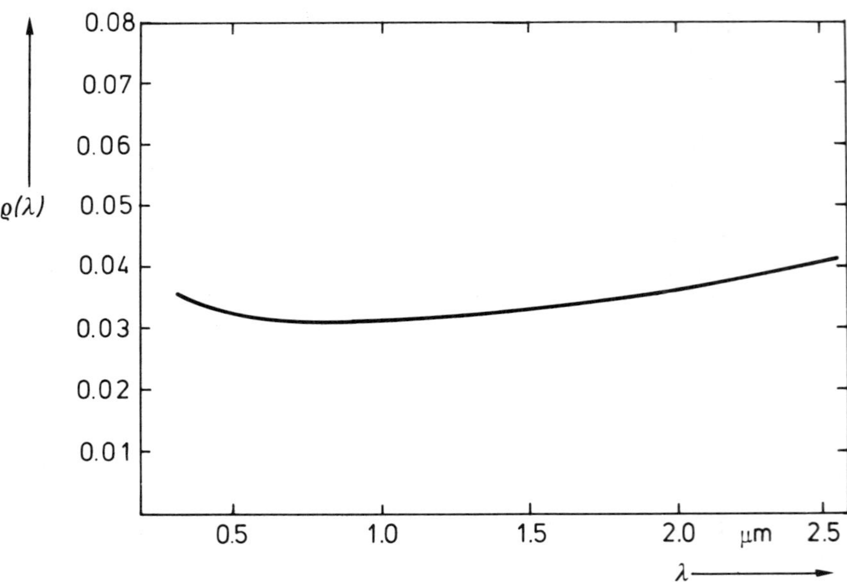

Fig. 2.4 Spectral reflectance $\rho(\lambda)$ of Kodak Black Baking Lacquer as a function of the wavelength λ. [After Wightman and Grum, 1981]

glossy or matt. An advantage of this coating is its wear resistance, but at about 3.5% (Wightman and Grum, 1981), the reflectance is unfavorably high (see Fig. 2.4) unless the lacquer is applied to cones and cavities where a specularly reflecting absorber is advantageous.

2.4.5 Krylon Ultraflat Black Enamel No. 1602

Wightman and Grum (1981) report that the reflectance of this spray enamel (manufactured by Borden Inc.) depends to some extent on the substrate. Their values for the reflectance together with those evaluated from absorptance measurements presented by Stierwalt et al. (1963) are summarized in Fig. 2.5.

2.4.6 Chemglaze Z-306 and Z-302

These specularly reflecting paints manufactured by Hugson Chemical Co. are carbon-black pigmented polyurethane coatings (Smith, 1984). They are suitable for absorbers when the multiple-reflection method is applied. Using Chemglaze Z-302 for their cryogenic radiometer, Martin et al. (1985) report a specular reflectance of 5% and a diffuse reflectance of about 1% in the visible

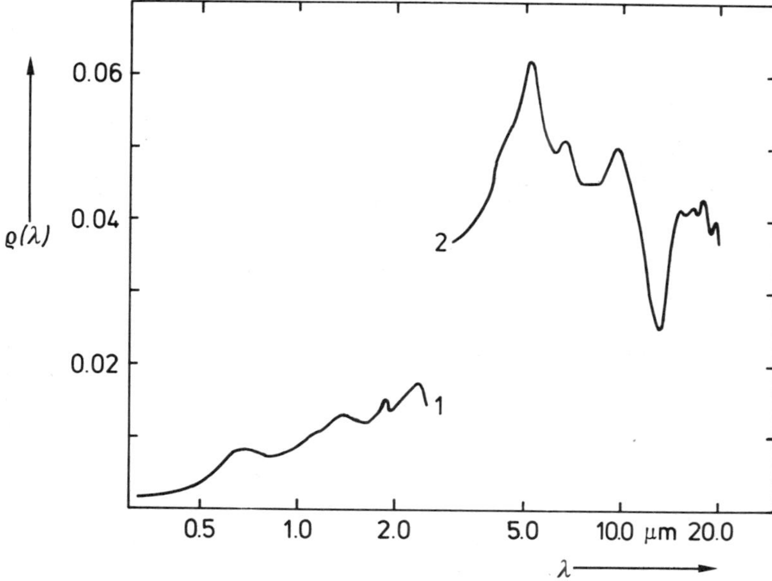

Fig. 2.5 Spectral reflectance $\rho(\lambda)$ of Krylon Ultraflat Black Enamel No. 1602 as a function of the wavelength λ. Curve 1 after Wightman and Grum (1981); curve 2 calculated from directional spectral emittance measurements (valid for a temperature range of 50 to 200°C) performed by Stierwalt et al. (1963).

range for the flat coating. The effective absorptance of their cavity was 0.99998. Booker (1982) measured the specular and the diffuse spectral reflectance of Chemglaze Z-302 deposited on an aluminum substrate in the uv range. His values and those taken from Stierwalt (1979) and Smith (1984) for Chemglaze Z-306 are compiled in Fig. 2.6.

2.4.7 Metal Blacks

When metals are vaporized in the presence of a low-pressure inert atmosphere, small crystallites of metal about 10 nm in diameter are formed in the gas. They are deposited in chains on the substrate in such a way that most of them are in contact with neighboring crystallites. This causes the small dc conductivity of these coatings. Their mass density is two to three orders of magnitude less than that of the solid metal. The interaction of the radiation-induced fields in the particles, which are much smaller than the wavelength, is very strong and must be considered in a theory explaining the optical properties of these deposits. Zaeschmar and Nedoluha (1972) used a model that, in principle, simulates the behavior of a complex impedance. It describes

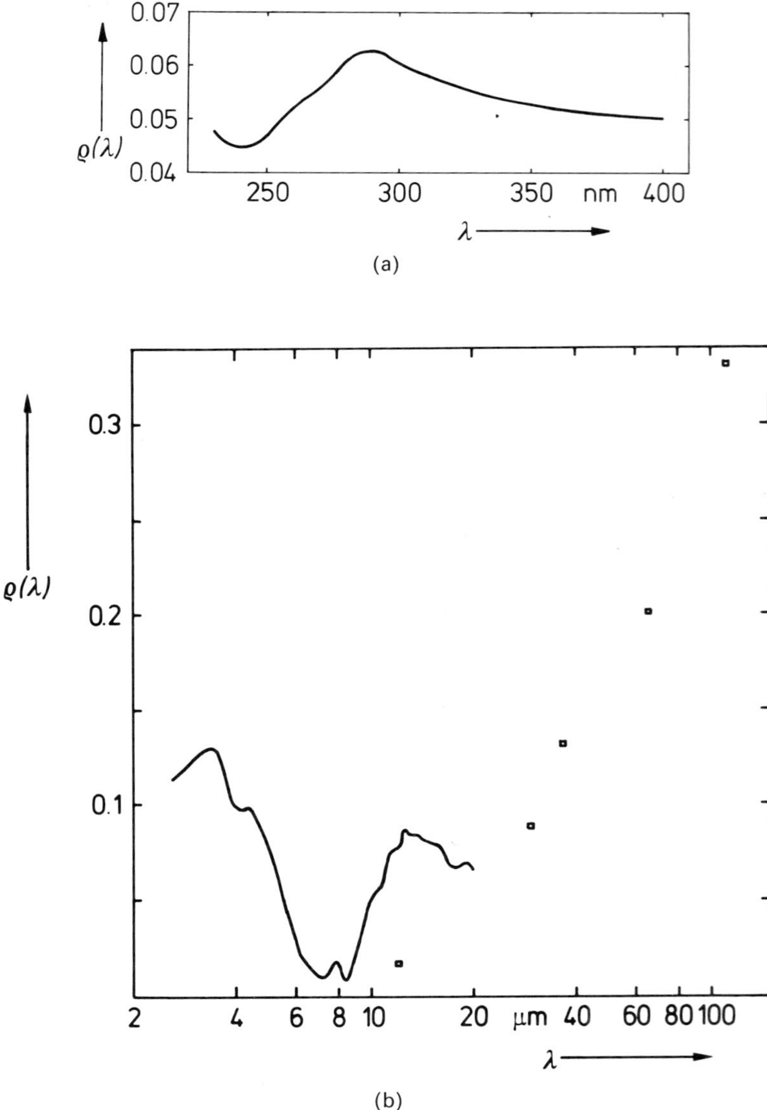

Fig. 2.6 Spectral reflectance $\rho(\lambda)$ of two Chemglaze glossy black paints in dependence of the wavelength λ. (a) Total spectral reflectance of Chemglaze Z-302, calculated from the specular (6° incidence) and diffuse spectral reflectance measured by Booker (1982) at room temperature. (b) Spectral reflectance of Chemglaze Z-306. The curve is calculated from the directional spectral emittance (observation perpendicular to the surface) measured by Stierwalt (1979) at 77 K. The points are measurements of the specular reflectance (6° incidence) at room temperature after Smith (1982).

quite well the infrared absorptance but is too simple to yield the reflectance quantitatively.

The deposits look black when they are evaporated at a gas pressure of the order of 100 Pa (1 torr). It is not surprising that their infrared absorptance depends on the kind of gas and its pressure, the rate of evaporation, and the substrate temperature. The scattering gas in which the cloud of crystallites is formed must be free of water vapor and oxygen.

The most common of these metal blacks is gold black. Compared with most paint coatings, it exhibits a higher absorptance and lower heat capacity. Its directional reflectance is almost Lambertian (Harris and Cuff, 1956). Unfortunately, the deposits are mechanically very delicate and may develop more and more metallic properties by sintering when stored at elevated temperatures. However, a treatment at 70°C for several hours stabilizes the deposit, so that sintering proceeds appreciably slower (Harris et al., 1950). Exposure to wetness, produced by different kinds of liquids, may also cause a structural collapse of the deposit (Harris and Beasley, 1952), recognizable by its brownish appearance and its strong increase in electrical conductivity.

Harris (1967) gave a review on the preparation technique and some physical properties of the deposits. He found an improved thermal stability for coating containing several percent copper. Depending on the exact evaporation conditions, it is possible to obtain a high-density (about 1/50 of solid gold) or a low-density (about 1/500 of solid gold) variant (Blevin and Brown, 1965, 1966). The high-density variant exhibits a higher electrical conductivity and a higher infrared reflectance than the low-density variant, but the latter is not suitable for vacuum applications (Blevin and Brown, 1966) because its thermal conductivity is mainly due to incorporated air. On the other hand, its low heat capacity is an interesting feature. Blevin and Brown (1971) overcame the disadvantages of low-density gold black by using an undercoat of Nextel Black Paint that was top-coated with a low-density deposit just thick enough for maximum absorptance but thin enough for a favorable low thermal resistance.

Hemispherical spectral reflectance curves of four samples of gold black are given in Fig. 2.7. Two of them were prepared in a nitrogen atmosphere of 1 torr and the other two at 2 torr. The reflectance measurements were all performed with the same spheroidal mirror method. Nevertheless, there are significant differences between samples prepared under equal conditions. This indicates the restricted reproducibility of the absorptance of gold black. Values from the literature can therefore only serve as an indication but cannot replace one's own measurements of the spectral reflectance. Sometimes an additional measurement of the transmittance will be necessary to ensure that the deposit is opaque.

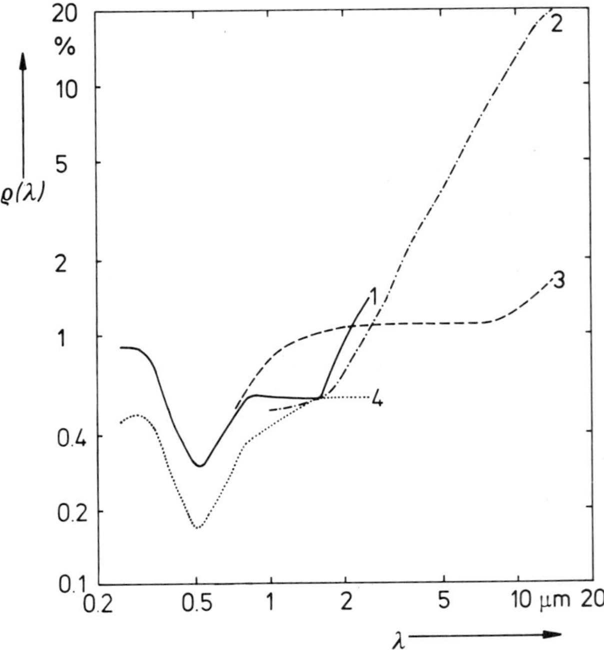

Fig. 2.7 Typical hemispherical spectral reflectance $\rho(\lambda)$ of gold blacks as a function of the wavelength λ after Blevin and Brown (1965) and (1966). Curves 1 and 2 belong to two different samples of 1-torr gold black and curves 3 and 4 to different samples of 2-torr gold black.

Other metal blacks have rarely been applied because they are less stable. A review can be found in Harris's monograph (1967). Platinum black deposited electrolytically is mentioned in Section 2.4.9.

2.4.8 Metal Films

The optical behavior of metal films with a thickness d is independent of wavelength λ if, first, λ is larger than about 4 μm, so that the static electrical conductivity σ determines the optical constants n and k, and, second, d is small compared with λ/n (Woltersdorff, 1934). The absorptance α is then given by (Hettner, 1946):

$$\alpha = \frac{2g}{(1+g)^2}, \qquad \text{with} \quad g = \frac{2\pi\sigma d}{\mathrm{c}} \tag{2.10}$$

(c = speed of light). Choosing d, it is possible to make $g = 1$, thus obtaining the maximum achievable value $\alpha = 0.5$ for the absorptance. In this case, the reflectance and the transmittance both exhibit the value 0.25. Such films can be used for absorbers in the infrared with extremely low heat capacity such as thin film bolometers (Liddiard, 1983; Dragovan and Moseley, 1984). Brett and Sullivan (1965) measured the spectral transmittance over a wavelength range from 2 to 153 μm for aluminum films on a collodion film substrate. They found it possible to prepare films with a nonselective transmittance in their range of measurement.

A modification of this technique is described by Gillham (1962), who sputtered about 50 alternate layers of (dielectric) tin oxide and platinum to a total thickness of about 1 μm on a substrate and achieved a smoothly decreasing absorptance from about 0.935 at 0.5 μm to about 0.915 at 1.4 μm and finally 0.87 at 2 μm.

2.4.9 Black Coatings by Electrochemical Processes

Martin Marietta Aerospace Co. developed an electrochemical process that, when applied to aluminum, produces a porous aluminum oxide layer containing a black dye in the upper 60 μm of the substrate. This black coating is called Martin Marietta's Optical Black. Wade and Willson (1975) published photographs of the microstructure of the surface and the spectral emittance in the range from 3 to 125 μm. The surface is sealed to make it extremely inert, whereas carbon-black-based paints react rapidly with atomic oxygen forming carbon monoxide (Leger, 1982), thus making them unsuitable for space experiments. The low outgassing rate, even at elevated temperatures, may be another advantage of Martin Marietta's Optical Black. Further measurements of its spectral emittance were performed by Stierwalt (1979). He found such samples to be much better infrared absorbers than samples of common black anodized aluminum. Measurements that he performed on another group of samples, treated by the identical anodizing process, except for the elimination of the dye, showed that at wavelengths longer than 8 μm, the aluminum oxide alone absorbs as well as the black anodized surface.

An improvement of the Martin Marietta process producing a somewhat better near-infrared absorptance and a strong decrease in far-infrared specular reflectance was introduced by Pompea et al. (1983, 1984) as Infrablack. Data on the hemispherical spectral reflectance of both coatings, developed at Martin Marietta Aerospace, are compiled in Fig. 2.8. The curves 1 and 3 are not exactly comparable because Stierwalt measured the directional emittance perpendicular to the surface instead of the hemispherical emittance; moreover, these measurements were performed at lower temperatures. They do, however, give an impression of the far-infrared behavior of this coating. Measurements of the specular infrared reflectance can be found in the papers

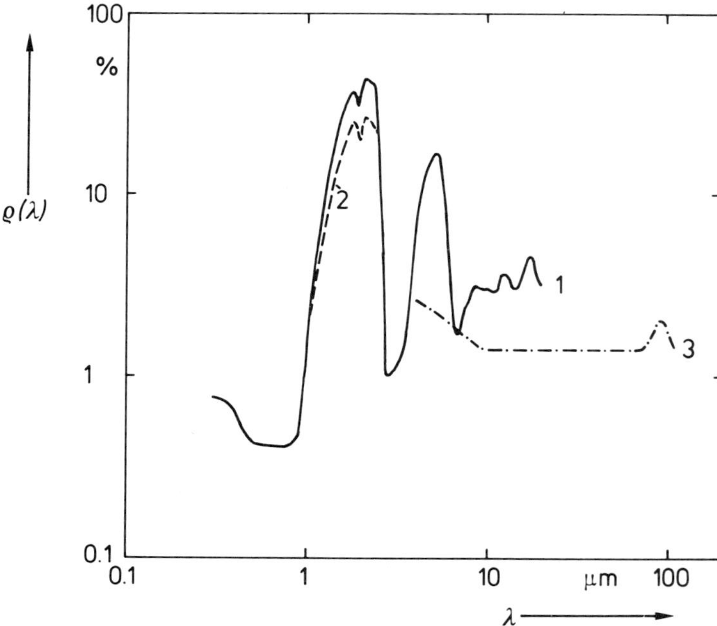

Fig. 2.8 Typical hemispherical spectral reflectance $\rho(\lambda)$ of Martin Black (curve 1) and Infrablack (curve 2) coatings as a function of the wavelength λ. Curves are composed from data given by Pompea et al. (1983, 1984). Curve 3 is evaluated from measurements of the directional spectral emittance at 77 K (Stierwalt, 1979).

of Smith (1982, 1984) and Smith and Wolfe (1982) and in the literature cited above.

A simple chemical-immersion technique producing an ultrablack nickel-phosphorus coating was described by Johnson (1979). Properly cleaned substrates were placed in a commercial, electrodeless nickel bath for 15 min. to 2 h. and afterwards rinsed in distilled water and dried. They were subsequently immersed in a strong oxidizing acid such as nitric acid for 5 to 15 s., then rinsed in distilled water and ethanol, and finally dried. Electron microscope photographs showed that the acid had selectively edged the surface, leaving a honeycomb structure with microscopic conical pores that function as light traps. Figure 2.9 shows a typical spectral reflectance curve for this black, ranging from 320 to 2140 nm. Measurements on electrolytically deposited platinum black, optimized for sensitive pyroelectric detectors, were published by Clarke and Larkin (1985). They found great discrepancies for different samples, especially beyond a wavelength of 15 μm. The infrared reflectance of a chrome black specimen was measured by Stierwalt (1979) and

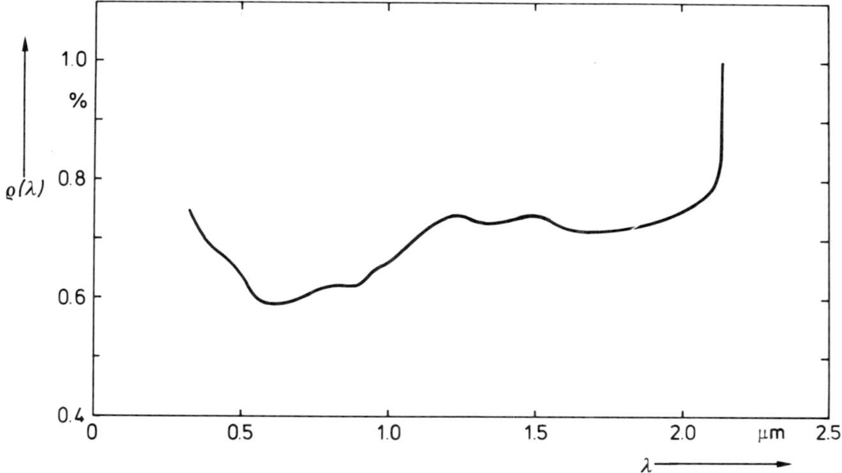

Fig. 2.9 Spectral reflectance $\rho(\lambda)$ of the nickel-phosphorous coating described by Johnson (1979) as a function of the wavelength λ. Measurements were performed with an integrating sphere.

showed an increase from 0.1 at 3 μm to about 0.5 at 23 μm. This coating can be prepared in a similar way to that described for nickel black in the section on cone absorbers so that it reflects almost perfectly specularly, thus making it suitable for cone absorbers.

2.4.10 Carbon Blacks

The term *carbon blacks* comprises all absorber layers that use the absorbing properties of amorphous carbon. Two main preparation techniques are employed. One is the deposition of soot from a strongly sooting flame such as burning camphor (Pfund, 1937; Blevin and Brown, 1965, 1966) or acetylene (Harris and Cuff, 1956; Harris, 1967; Batten, 1985). The other one is the brushing or spraying of a suspension of finely ground charcoal, bone coal, or vegetable coal (Betts, 1965) in a suitable binder. The soot blacks are very fragile and exhibit a high thermal resistance (Gillham, 1962) and are therefore rarely used. Simple carbon suspensions are usually replaced by more sophisticated black paints. Extinction coefficients of amorphous carbon grains were measured by Koike et al. (1980) and Borghesi et al. (1983).

2.4.11 Other Blacks

Table 2.2 summarizes other blacks in use and the literature describing these.

TABLE 2.2 Some Less Common Absorber Materials for Radiometers and the Literature Describing Them

Material	Literature
Barnes Standard Black	Stierwalt et al. (1963)
Bostic Black Paint	Willey et al. (1983)
Cat-a-Lac Black	Hsia and Richmond (1976), Stierwalt (1966, 1979)
Cornell Black	Smith (1980, 1983, 1984)
Denver Black Paint	Willey et al. (1983)
Dupont Black	Harris and Cuff (1956)
Lampblack	Harris and Cuff (1956)
Mautz Black	Hsia and Richmond (1976)
Sicon Black Paint	Eisenman and Bates (1964), Stierwalt et al. (1963)
Silver black	Vogel (1981)

2.5 SUMMARY

Reflection losses depend on the material used for absorption, but often these losses can be strongly reduced by giving the absorber a cavity-shaped form. This can be a macroscopic cavity with blackened walls or a porous surface, which can be understood as an assembly of microcavities. Attention must be paid to the question of whether a specular or a diffuse reflecting coating is the better choice for the particular application. Operation at liquid-helium temperatures allows much larger cavities than room temperature operation because of the strongly increased thermal conductivity (by a factor of about 10^4) and the strongly decreased heat capacity (about a thousandth of the room temperature value).

In any case, the following error sources should be considered:

(1) Does the absorber convert the absorbed radiant energy exclusively into heat?

(2) Is the absorber opaque?

(3) How large are the effective reflection losses? For experiments to measure them, refer to Section 6.5.2. Is one of these methods (or a new one) applicable to the absorber in its final assembly?

(4) Is the absorptance uniform across the sensitive area or input port, respectively? For the consequences of a non-uniformity refer to Section 6.5.7.

(5) How much does the thermal resistivity of the black coating contribute to the non-equivalence of radiant and electrical heating? The qualitative answer given in this chapter is: the smaller the thermal resistivity, the smaller

the correction. A quantitative treatment of this problem is given in Section 6.5.6.

REFERENCES

Batten, C. E. (1985). Spectral optical constants of soot from polarized angular reflectance measurements. *Appl. Opt.* **24**, 1193.

Bauer, G., and Bischoff, K. (1971). Evaluation of the emissivity of a cavity source by reflection measurements. *Appl. Opt.* **10**, 2639.

Bedford, R. E., and Ma, C. K. (1974). Emissivities of diffuse cavities: Isothermal and nonisothermal cones and cylinders. *J. Opt. Soc. Am.* **64**, 339.

Bedford, R. E., Ma, C. K., Chu, Z., Sun, Y., and Chen, S. (1985). Emissivities of diffuse cavities. 4: Isothermal and nonisothermal cylindro-inner-cones. *Appl. Opt.* **24**, 2971.

Betts, D. B. (1965). The spectral response of radiation thermopiles. *J. Sci. Instrum.* **42**, 243.

Bischoff, K. (1964). Die spektrale Empfindlichkeit thermischer Strahlungsemfänger. *Optik* **21**, 521.

Blevin, W. R., and Brown, W. J. (1965). An infra-red reflectometer with a spheroidal mirror. *J. Sci. Instrum.* **42**, 385.

Blevin, W. R., and Brown, W. J. (1966). Black coatings for absolute radiometers. *Metrologia* **2**, 139.

Blevin, W. R., and Brown, W. J. (1971). A precise measurement of the Stefan–Boltzmann constant. *Metrologia* **7**, 15.

Blevin, W. R., and Geist, J. (1974). Influence of black coatings on pyroelectric detectors. *Appl. Opt.* **13**, 1171.

Booker, R. L. (1982). Specular UV reflectance measurements for cavity radiometer design. *Appl. Opt.* **21**, 153.

Borghesi, A., Bussoletti, E., Colangeli, L., Minafra, A., and Rubini, F. (1983). The absorption efficiency of submicron amorphous carbon particles between 2.5 and 40 μm. *Infrared Phys.* **23**, 85.

Brandenburg, W. M. (1964). Focusing properties of hemispherical and ellipsoidal mirror reflectometers. *J. Opt. Soc. Am.* **54**, 1235.

Brett, D. A., and Sullivan, E. J. (1965). Infrared transmittance of thin aluminum films on collodion substrates. *J. Opt. Soc. Am.* **55**, 1556.

Brusa, R. W., and Fröhlich, C. (1978). *The PMO-6 Radiometer and its Characterization.* Publ. No. 563 of the World Radiation Center, Davos, Switzerland.

Clarke, F. J. J., and Larkin, J. A. (1985). Measurement of total reflectance, transmittance and emissivities over the thermal IR spectrum. *Infrared Phys.* **25**, 359.

Compton, J. P., Martin, J. E., and Quinn, T. J. (1974). Some measurements of outgassing properties and far-infrared reflectivities of two optical blacks. *J. Phys. D: Appl. Phys.* **7**, 2501.

Day, G. W., Hamilton, C. A., and Pyatt, K. W. (1976). Spectral reference detector for the visible to 12-μm region; convenient, spectrally flat. *Appl. Opt.* **15**, 1865.

Dragovan, M., and Moseley, S. H. (1984). Gold absorbing film for a composite bolometer. *Appl. Opt.* **23**, 654.

Driscoll, W. G., ed., and Vaughn, W., asst. ed. (1978). *Handbook of Optics.* McGraw-Hill, p. 10–10.

Edwards, J. G. (1970). A glass disk calorimeter for pulsed lasers. *J. Phys. E: Sci. Instrum.* **3**, 452.

Edwards, J. G. (1975). A standard calorimeter for pulsed lasers. *J. Phys. E: Sci. Instrum.* **8**, 663.

Eisenman, W. L., and Bates, R. L. (1964). Improved black radiation detector. *J. Opt. Soc. Am.* **54**, 1280.

Gao, Z., Wang, Z., Piao, D., Mao, S., and Yang, C. (1983). Realization of the candela by electrically calibrated radiometers. *Metrologia* **19**, 85.

Geist, J. C., and Richmond, J. H. (1971). *On the Absorptance of Cavity-Type Receivers.* NBS Technical Note 575. U.S. Government Printing Office, Washington, D.C.

Geist, J., Schmidt, L. B., and Case, W. E. (1973). Comparison of the laser power and total irradiance scales maintained by the National Bureau of Standards. *Appl. Opt.* **12**, 2773.

Gillham, E. J. (1953). A method for measuring the spectral reflectivity of a thermopile. *Brit. J. Appl. Phys.* **4**, 151.

Gillham, E. J. (1962). Recent investigations in absolute radiometry. *Proc. Roy. Soc.* **A269**, 249.

Ginnings, D. C., and Reilly, M. L. (1972). Calorimetric measurement of thermodynamic temperatures above 0°C using total blackbody radiation. In *Temperature: Its Measurement and Control in Science and Industry*, T. P. Preston-Thomas and R. L. Shepard, eds., Vol. 4, Part 1. Pittsburgh: Instrum. Soc. Am., p. 339.

Gunn, S. R. (1974). Volume-absorbing calorimeters for high-power laser pulses. *Rev. Sci. Instrum.* **45**, 936.

Harris, L. (1967). *The Optical Properties of Metal Blacks and Carbon Blacks.* The Eppley Foundation for Research, Newport, R.I. Monograph Series No. 1.

Harris, L., and Beasley, J. K. (1952). The infrared properties of gold smoke deposits. *J. Opt. Soc. Am.* **42**, 134.

Harris, L., and Cuff, K. F. (1956). Reflectance of goldblack deposits, and some other materials of low reflectance from 254 mμ to 1100 mμ: The scattering-unit-size in gold-black deposits. *J. Opt. Soc. Am.* **46**, 160.

Harris, L., Jeffries, D., and Siegel, B. M. (1950). The thermal stabilization and sintering of gold smoke deposits. *J. Chem. Phys.* **18**, 261.

Hawks, K. H., and Cottingham, W. B. (1970). Total normal emittance of some real surfaces at cryogenic temperatures. *Adv. Cryogen. Eng.* **16**, 467.

Heany, J. B., Steward, K. P., and Hass, G. (1983). Transmittance and reflectance of crystalline quartz and high- and low-water content fused silica from 2 μm to 1 mm. *Appl. Opt.* **22**, 4069.

Hengstberger, F. (1975). Entwurf eines Meßsystems zur absoluten Messung spektraler Verteilungen und Bau eines Absolutempfängers für elektromagnetische Strahlung im optischen Bereich. D. Sc. thesis, Technische Univ., Wien, Austria.

Hettner, G. (1946). Durchlässigkeit und Reflexionsvermögen dünner Metallschichten im langwelligen Ultrarot bei beliebigem Einfallswinkel. *Optik* **1**, 2.

Houck, J. R. (1982). New black paint for cryogenic infrared applications. *Proc. SPIE* **362**, 54.

Hsia, J. J., and Richmond, J. C. (1976). Bidirectional reflectometry. Part I. A high resolution laser bidirectional reflectometer with results of several optical coatings. *J. Res. Nat. Bur. Stand. A. Phys. Chem.* **80A**, 189.

Johnson, C. E. (1979). 'Ultra-black' coating for high absorptance of solar energy. *Dimensions NBS* **63**, 30.

Kendall, J. M., and Berdahl, C. M. (1970). Two blackbody radiometers of high accuracy. *Appl. Opt.* **9**, 1082.

Koike, C., Hasegawa, H., and Manabe, A. (1980). Extinction coefficients of amorphous carbon grains from 2100 Å to 340 μm. *Astrophys. Space Sci.* **67**, 495.

Leger, L. J. (1982). Oxygen atom reaction with shuttle materials at orbital altitudes. *Proc. NASA Shuttle Environment Workshop*, p. 151.

Liddiard, K. C. (1983). Thin-film resistance bolometer IR detectors. *Infrared Phys.* **24**, 57.

Martin, J. E., Fox, N. P., and Key, P. J. (1985). A cryogenic radiometer for absolute radiometric measurements. *Metrologia* **21**, 147.

Möstl, K. (1978). Empfängernormale zur Messung der Strahlungsleistung von Dauerstrich-Lasern. *Feinwerktechnik & Meßtechnik* **86**, 72.

Möstl, K. (1986). Ein Präzisionsradiometer für Laserstrahlung. In *Optoelectronics in Engineering*, W. Waidelich, ed. Springer-Verlag, Berlin, New York, p. 254.

Möstl, K., Brandt, F., and Xie, X. (1985). High accuracy measurements of laser pulse energies with a compensated cone calorimeter and computer-assisted data acquisition. *Proc. 11th Int. Symp. IMEKO Tech. Com. Photon-Detectors*, p. 226. IMEKO Secretariat, Budapest, Hungary.

Müller, C. (1933). Bestimmung der Konstanten σ des Stefan–Boltzmannschen Strahlungsgesetzes. *Z. Physik* **82**, 1.

Ono, A. (1979a). An absolute radiometer with uniform response: Fabrication, characteristics and analysis. *Japan. J. Appl. Phys.* **18**, 1995.

Ono, A. (1979b). Absolute radiometers with uniform sensitivity over the surface of the receiver. *Japan. J. Appl. Phys.* **18**, 697.

Pfund, A. H. (1937). The blacking of radiometers. In *Measurements of Radiant Energy*, W. E. Forsythe, ed. McGraw-Hill, New York, p. 210.

Philer, J. L., and Houck, J. R. (1971). Black paints for far infrared cryogenic use. *Appl. Opt.* **10**, 567.

Pompea, S. M., Bergener, D. W., Shepard, D. F., Russak, S., and Wolfe, W. L. (1983). Preliminary performance data on an improved optical black for infrared use. *Proc. SPIE* **400**, 128.

Pompea, S. M., Bergener, D. W., Shepard, D. F., Russak, S., and Wolfe, W. L. (1984). Reflectance measurements on an improved optical black for stray light rejection from 0.3 to 500 μm. *Opt. Eng.* **23**, 149.

Quinn, T. J., and Martin, J. E. (1985). A radiometric determination of the Stefan–Boltzmann constant and thermodynamic temperatures between -40°C and $+100$°C. *Phil. Trans. R. Soc. Lond.* **A316**, 85.

Smith, S. M. (1980). Far-infrared (FIR) optical black bidirectional reflectance distribution function (BRDF). *Proc. SPIE* **257**, 161.

Smith, S. M. (1982). Far-infrared reflectance spectra of optical black coatings. *Proc. SPIE* **362**, 57.

Smith, S. M. (1983). Analysis of 12–700 μm reflectance spectra of three optical black samples. *Proc. SPIE* **384**, 32.

Smith, S. M. (1984). Specular reflectance of optical-black coatings in the far infrared. *Appl. Opt.* **23**, 2311.

Smith, S. M., and Wolfe, W. L. (1982). Comparison of measurements by different instruments of the far-infrared reflectance of rough, optically black coatings. *Proc. SPIE* **362**, 46.

Stair, R., Schneider, W. E., Waters, W. R., and Jackson, J. K. (1965). Some factors affecting the sensitivity and spectral response of thermoelectric (radiometric) detectors. *Appl. Opt.* **4**, 703.

Stierwalt, D. L. (1966). Infrared spectral emittance measurements of optical materials. *Appl. Opt.* **5**, 1911.

Stierwalt, D. L. (1979). Infrared absorption of optical blacks. *Opt. Eng.* **18**, 147.

Stierwalt, D. L., Bernstein, J. B., and Kirk, D. D. (1963). Measurements of the infrared spectral absorptance of optical materials. *Appl. Opt.* **2**, 1169.

Stock, K. D. (1983). Effects of gusts of wind on a thermal detector of radiation. *Rev. Sci. Instrum.* **54**, 1078.

Vogel, J. (1981). An absolute radiometer for the measurement of the optical radiation scale. *Proc. 9th Int. Symp. IMEKO Tech. Com. Photon-Detectors*, p. 325. IMEKO Secretariat, Budapest, Hungary.

Wade, J. F., and Willson, W. R. (1975). Infrared properties of a highly absorbing surface. *Proc. SPIE* **67**, 59.

Wightman, T. E., and Grum, F. (1981). Low-reflectance backing materials for use in optical radiation measurements. *Color Res. Appl.* **6**, 139.

Willey, R. R., George, R. W., Ohmart, J. G., and Walvoord, J. W. (1983). Total reflectance properties of certain black coatings (from 0.2 to 20.0 micrometers). *Proc. SPIE* **384**, 19.

Willson, R. C. (1971). Active cavity radiometric scale, international pyrheliometric scale, and solar constant. *J. Geophys. Res.* **76**, 4325.

Willson, R. C. (1973). Active cavity radiometer. *Appl. Opt.* **12**, 810.

Willson, R. C. (1979). Active cavity radiometer type IV. *Appl. Opt.* **18**, 179.

Willson, R. C. (1980). Active cavity radiometer type V. *Appl. Opt.* **19**, 3256.

Woltersdorff, W. (1934). Über die optischen Konstanten dünner Metallschichten im langwelligen Ultrarot. *Z. Physik* **54**, 230.

Xu, D. (1979). *Volume Absorbing Calorimeter with Recover Mirror.* Report Nat. Inst. Metrology China. Peking, China.

Zaeschmar, G., and Nedoluha, A. (1972). Theory of the optical properties of gold blacks. *J. Opt. Soc. Am.* **62**, 348.

Zalewski, E. F., Geist, J., and Willson, R. C. (1979). Cavity radiometer reflectance. *Proc. SPIE* **196**, 152.

3

Thermal Detectors of Optical Radiation

F. HENGSTBERGER
Council for Scientific and Industrial Research
Pretoria, South Africa

3.1 INTRODUCTION

The radiant or electrical power, which is converted to heat in the detector element of an absolute radiometer, raises the temperature of the detector element above its initial operating temperature. This temperature rise can be detected with a temperature transducer and its output is then proportional (although not necessarily in a linear fashion) to the heat input. Since the detector of an absolute radiometer operating in substitution mode is merely used to compare two heat inputs at the same power level, the effect of response nonlinearities is reduced to insignificance. Only when detector signals are used instead of substituted electrical powers, as may be convenient in some correction experiments or alternative operating modes, does one have to consider response nonlinearities. (See Chapter 6.)

The thermal radiation detectors most commonly used in absolute radiometers are thermopiles, bolometers, and pyroelectric detectors. The choice between them is often governed by conflicting requirements (see Section 1.3.5) and, in many cases, there is no consensus with respect to the optimum choice for a particular application. Since they are adequately covered in the specialized literature on detectors—for example, Budde (1983), Hudson and Hudson (1975), Keyes (1977), Kingston (1978), and Smith et al. (1968)—only a brief summary of their most important properties is provided in this chapter. The temperature distribution in detector elements of thermal radiation detectors is analyzed in Chapter 4, while their frequency response is discussed in Section 1.3.7.

ISBN 0-12-340810-5

3.2 PARAMETERS USED TO CHARACTERIZE THE PERFORMANCE OF RADIATION DETECTORS

Like other optical radiation detectors, absolute radiometers can be compared in terms of their performance as detectors, according to a number of different parameters. For a long time, the term *sensitivity* was used to describe one of these parameters. Since it was not clear whether a detector with a larger output signal or with a better signal-to-noise ratio was to be regarded as more "sensitive," Jones (1953) introduced the two specific concepts of responsivity and detectivity (see Section 1.2.3) to resolve this ambiguity.

In order to compare the performance of different detectors used with different amplifiers by means of their detectivity, one has to standardize certain reference conditions of measurement. In modern practice, it is usual to refer detectivity to an amplifier bandwidth of 1 Hz and a detector area of 1 cm^2. The resulting expression for the normalized detectivity D^* (see Section 1.2.3) is based on evidence that noise varies as the square root of the amplifier bandwidth and the noise-equivalent power as the square root of the sensitive detector area. Since measurements are often made with a sufficiently narrow bandwidth for the frequency variation of noise to be insignificant, if noise is not constant with frequency in any case (as in the case of Johnson noise), the first assumption is often well justified. The same does not always hold for the second assumption although the concept is often still used in such cases for reasons of convenience and standardization.

Normalized detectivity by itself is still not a completely unbiased factor of merit for comparing the performance of different detectors. Since it refers to an amplifier bandwidth of 1 Hz, the detector time-constant τ is not taken into account at all (in contrast with Jones' (1953) original treatment). It has been suggested (Soa, 1978) that the ratio D^*/τ would be a more suitable quantity for rating purposes, because it is a positive constant over many orders of magnitude of τ. The last observation is equivalent to saying that D^* increases linearly with τ over a wide range of time-constants. All of the reasoning applied so far as well as most current and previous work on this subject deals with this linear region, which can be approximately taken to refer to time-constants smaller than one second. The region above one second is usually omitted from the treatment because detectors with small time-constants are far more important in practice and because D^* for Johnson-noise limited thermal detectors seems to remain constant with increasing τ in this region. Thus, no gain in detectivity is achieved here by increasing the time-constant, therefore, the inconvenience of a longer time-constant is not offset by any other direct advantage.

Because of the more complex structure of the detector elements in absolute radiometers and the aim to achieve a high accuracy, such instruments generally have time-constants well in excess of one second simply for technological reasons. From the preceding considerations, it is evident that Johnson-noise limited absolute radiometers will achieve optimum performance in terms of both detectivity and time-constant for time-constants near one second. It is also clear that they should be compared in their performance as detectors in terms of normalized detectivity (D^*) and not in terms of D^*/τ. Unfortunately, however, the measurement bandwidth is seldom specified for absolute radiometers. In the absence of this information, a comparison in terms of detectivity (D), noise-equivalent power (NEP), or noise-equivalent irradiance (NEI) with unspecified bandwidth (and perhaps even area) is the next best option.

Another interesting feature of thermal detectors with $\tau > 1$ s is that, at least theoretically, it may be possible to construct detectors limited in their detectivity by temperature noise rather than by Johnson noise. In this case, the increased time-constant would be offset by an increased value for D^*. No temperature-noise–limited constructions of this kind have so far been reported. However, this temperature-noise–limited performance is of interest in that it sets a fundamental limit to the minimum noise-equivalent power that can be resolved by a thermal detector. Such a performance is achieved if the detector is coupled to its thermal environment by radiation only and if other noise sources are negligible. In this case, its temperature and thus its output would fluctuate because of the statistical nature of the rate of absorption and emission of photons. For a detector element of area A, and with emissivities ε_1 and ε_2 for the front and back surfaces, respectively, this noise-equivalent power is given by (Smith et al., 1968)

$$NEP = [16(\varepsilon_1 + \varepsilon_2)A\sigma \mathrm{k} T^5 \, \Delta f]^{1/2}, \tag{3.1}$$

where σ is the Stefan–Boltzmann constant, k the Boltzmann constant, T the temperature of the receiver and of its environment, and Δf the amplifier bandwidth. Taking $A = 1\ \mathrm{cm}^2$, $T = 300$ K, $\varepsilon_1 = 1$, $\varepsilon_2 = 0$, and $\Delta f = 1$ Hz, one obtains $NEP = 6.10^{-11}$ W $= 0.06$ nW. In terms of noise-equivalent irradiance under these conditions, the preceding value is of course equivalent to $6.10^{-11}\ \mathrm{W/cm}^2$.

If a detector is to be operated in a situation where its noise is near the minimum imposed by thermodynamic constraints, its detectivity must be maximized. If, on the other hand, amplifier noise, electronic pickup, or other nonfundamental noise sources are dominant, it may be desirable to optimize the responsivity. Ease of construction, subject to a number of conflicting constraints associated with the desire for high accuracy (especially in absolute radiometers), is often more important than the ultimate detectivity or responsivity.

3.3 THERMOPILE DETECTORS

A thermopile consists of a number of thermocouples connected in series (or, less commonly, in parallel) to provide a thermoelectric voltage proportional to the temperature difference between the receiver and its thermal environment. As a temperature transducer, the thermopile should be considered a voltage source. Its optimization in detector applications has received considerable attention (Jones, 1953; Smith et al., 1968; Schley and Hoffmann, 1958, Stevens, 1970).

The responsivity s of a thermopile can be expressed as (Jones, 1953; Stevens, 1970)

$$s = \frac{n\alpha_{12}R_T\varepsilon}{(1 + \omega^2 R_T^2 C_T^2)^{1/2}}, \tag{3.2}$$

where n is the number of hot junctions, α_{12} is the Seebeck coefficient (also known as thermoelectric power) of the two materials used for each thermocouple, R_T is the thermal resistance between the receiver and the environment, ε is the emittance of the absorbing surface, ω is the angular modulation frequency of the incident beam, and C_T is the thermal capacity of the detector element. The quantity $\tau = R_T C_T$ can be interpreted as the thermal time-constant of the detector.

In order to optimize the performance of a thermopile in respect to responsivity, one will thus try to maximize α_{12}, R_T, and ε and use a modulation frequency, for which ω is as small as reasonably possible compared with $1/\tau$.

The optimization with respect to detectivity is less straightforward. Here, one has to reconcile the two contradictory requirements of maximizing R_T while minimizing the electrical circuit resistance (assuming the usual case of dominant Johnson noise). The two quantities are related via the Wiedemann–Franz law as

$$\lambda\rho/T = L, \tag{3.3}$$

λ being the thermal conductivity of the material, ρ its electrical resistivity, T its temperature, and L its Lorentz number. With this relationship, one can establish figures of merit for thermocouple materials, both for open-circuit output and for optimum matching to an amplifier (Smith et al., 1968; Stevens, 1970; Jones, 1949).

For low-frequency or DC operation, the following recommendations lead to optimum performance (Smith et al., 1968; Stevens, 1970):

(a) The materials used should have a high figure of merit.
(b) The heat losses by radiation and conduction should be equal.

(c) The ratio of the thermal to the electrical conductivity should be the same for each lead. Since the Lorentz number, L, has a very nearly constant value for most materials, this is equivalent to saying that the thermal conductance of the two arms should be very nearly equal. (The same applies for the electrical resistance.) For an equal length of both thermocouple arms, this can be achieved by varying the cross-sectional area.

Under these conditions, the ratio of the RMS voltage fluctuations due to thermal (V_T) and Johnson noise (V_J) are given by (Smith et al., 1968)

$$\frac{V_T^2}{V_J^2} = \frac{\varepsilon^2 \alpha_{12}^2}{8L}, \tag{3.4}$$

where L is an average Lorentz number of the two materials defined as (Smith et al., 1968)

$$L = \tfrac{1}{4}(L_1 + L_2) + \tfrac{1}{2}(L_1 L_2)^{1/2}. \tag{3.5}$$

For presently known thermocouple materials, Eq. (3.5) indicates that a temperature-noise-limited performance for thermopiles is not very likely to be achieved. Rather, the best that can be aspired to is a Johnson-noise limited performance.

For a thermopile designed according to the optimum conditions just stated, the noise-equivalent power is proportional to the square root of the detector area (A) and its performance can thus be assessed in terms of D^*. However, if conduction cooling predominates strongly, it is independent of A (Smith et al., 1968) and D^* cannot be used for rating purposes. In contrast with some other infrared detectors, the performance of thermopiles usually does not improve at low temperatures because of a reduction of the thermoelectric power.

The materials used for thermocouples are either metals, alloys, or semiconductors. Metallic thermocouples can be in the form of wires and electroplated or evaporated thin films. (See Section 1.3.4 for examples.) Semiconductors, while used extensively in commercial thermopiles, have so far not been reported for absolute radiometers. Alloys have only been used in the form of wires.

In some instruments (see Section 1.3.4 for examples), the thermopile sensor is mounted a small distance behind the heater–absorber component, the thermal coupling being effected by means of air conduction and convection and radiative transfer. Although a slight loss in responsivity may be incurred in this way, the additional convenience gained by being able to use commercial thermopiles as sensors can outweigh this consideration.

Since the output of all thermopile detectors in response to a DC input is inherently a DC voltage, it is subjected to drifts in the thermal voltages in the measuring circuit. The generation of an AC signal by chopping the incident beam is not considered an attractive proposition in absolute radiometers because of the relatively long time-constants and the addition of further error sources. (See Chapter 6). Consequently, good thermostatting and thermal lagging of the measuring system are essential.

3.4 BOLOMETRIC DETECTORS

A bolometer is a temperature transducer based on the change of electrical resistance with temperature. However, it is not the absolute value of temperature that is important in an absolute radiometer, but the temperature difference between the receiver and its thermal environment. Therefore, at least two resistance elements are needed: one to measure the temperature of the receiver and one to measure that of the thermal environment. Without both elements, it is difficult to distinguish temperature changes resulting from the dissipation of power in the receiver from those caused by changes in the temperature of the thermal environment. The more general subject of compensation is discussed in Section 3.7.

Bridge techniques, which are well established for all kinds of resistive transducers, can be used very effectively to indicate resistance changes, the most common versions employing Wheatstone bridge configurations. Both DC and AC bridge operations have been used (see Section 1.3.4 for examples), with the latter offering the advantage of being inherently insensitive to the changes in the thermal voltages in the measuring circuit. The optimization of bolometric detectors has also been extensively investigated (Smith et al., 1968; Chasmar et al., 1956; Barth, 1958).

The responsivity s of a bolometer used in a bridge circuit can be expressed as

$$s = \frac{F \varepsilon I R_e \alpha R_T}{(1 + \omega^2 \tau^2)^{1/2}}, \tag{3.6}$$

where F is the bridge factor ($0 < F < 1$), ε is the emissivity of the absorber, I the current flowing through the bolometer, R_e its electrical resistance, α the temperature coefficient of R_e, R_T the thermal resistance between the bolometer and its thermal environment, ω the modulation frequency of the incident beam, and τ the thermal time-constant of the bolometer. F is defined as

$$F = \frac{R_L}{R_e + R_L}, \tag{3.7}$$

R_L being the bridge resistor in series with R_e ("L" for "load" resistor). $F = 1$ can be obtained for $R_L \gg R_e$, but the configuration $R_L = R_e$ ($F = \frac{1}{2}$) is often preferred in practice, especially for absolute radiometers. Here, R_L is generally another bolometer, which is not exposed to the incident radiation. Being physically close to the measuring bolometer, it compensates for convective disturbances, pressure fluctuations, changes in the temperature of the housing, and instabilities in the bridge supply. This compensating behavior is, in most cases, considered valuable enough to forgo a factor of 2 in responsivity and $\sqrt{2}$ in detectivity. In the most general case ($R_L \neq R_e$), the expression given for the responsivity in Eq. (3.6) is based on the assumption that the heating of R_e by the bridge current is the same with or without the incident radiant power. This is not quite correct, however, and, for $\alpha > 0$, the resistance change brought about by the radiant power is actually increased by additional heating due to the bridge current (Chasmar et al., 1956). For the case $R_L = R_e$, which will be considered exclusively in the following, no such responsivity enhancement takes place and Eq. (3.6) with $F = \frac{1}{2}$ is strictly correct. Equation (3.6) indicates that s is proportional to the bolometer current I. However, an upper limit to I is set by the maximum permissible power dissipation in the bolometer. If this current is exceeded, the bolometer will disintegrate. Assuming a linear dependence of s on I is also incorrect for the reason that the part of R_T due to radiation loss will assert a major influence, tending to decrease R_T strongly at higher bolometer temperatures.

In terms of detectivity, high bolometer currents will lead to a strong increase in temperature- and possibly current-noise, which will eventually surpass the Johnson-noise contribution. Thus, there is a definite optimum operating current (and temperature) for each type of bolometer, the exact value depending on the type of predominant heat-loss mechanism and the temperature dependence of α (Smith et al., 1968; Chasmar et al., 1956; Barth, 1958). The optimum operating temperature is in the region of 10–50% above the temperature of the environment T_0 (in K). The gain in exceeding 1.1 T_0 toward the maximum is usually not too significant. Even so, such operating temperatures can only be exploited in vacuum, because of instabilities caused by convection currents in air.

It has been shown (Chasmar et al., 1956) that the NEP for bolometers is proportional to the inverse square root of their thermal impedance to the environment ($R_T^{-1/2}$). Consequently, the best performance can be achieved if R_T is a maximum (i.e., if the detector is predominantly coupled to the environment by radiation). Theory also shows that the NEP is independent of the electrical resistance of the bolometer. This fact makes it feasible to reduce conduction losses and achieve the radiation-loss dominated condition, just identified as the optimum. Since R_T is then proportional to the detector area A, such bolometers can be rated detectivity-wise in terms of D^*.

The same is not applicable for conduction-loss dominated devices. Even in this case, however, Johnson noise will exceed temperature noise for all presently known bolometer materials and an ideal thermal-detector performance is unlikely to be achieved at room temperature.

At lower temperatures, the detectivity of bolometers can benefit from a reduction in the temperature noise caused by background radiation, lower Johnson noise, and a reduction in the time-constant (because of the decrease of the specific heat with temperature). For very low temperatures, the decrease in noise may be offset by a decrease in the temperature coefficient of resistance α. At the transition to superconductivity, α assumes very large values and it is in this region that the best performances have been obtained for bolometers. In absolute radiometers, low-temperature operation has mainly resulted in the reduction of instrumental corrections while the gain in detectivity has only been slight.

Resistive tranducers consisting of metal-wire windings, metal layers, or semiconductors have been employed in absolute radiometers, with the most widely used materials being gold, platinum, nickel, and germanium. Examples for the various bolometric techniques and devices reported in the literature on absolute radiometry are given in Section 1.3.4.

3.5 PYROELECTRIC DETECTORS

Pyroelectric detectors are quite different from conventional temperature transducers in that they produce a current proportional to the rate of temperature change rather than temperature or temperature difference. The physical basis for this detection mechanism is the temperature dependence of the electrical polarization in certain materials, notably ferroelectric crystals. Below their Curie temperature, such crystals show spontaneous electric polarization, which can be aligned along the polar axis of the crystal by means of a suitable poling procedure. As the pyroelectric material is heated, the crystal lattice spacing is changed, resulting in a different amount of polarization. Consequently, the crystal surfaces, normal to the axis of polarization, are charged and if electrodes are attached to them and connected to each other, a current will flow in the circuit. This current I is given by

$$I = \mathrm{p}(T)A\,\frac{dT}{dt}, \tag{3.8}$$

$\mathrm{p}(T)$ being the pyroelectric coefficient at temperature T, A the electrode area, and dT/dt the rate of change of the crystal temperature.

Electrically, a pyroelectric detector element can be represented as a current source in parallel with a capacity and a resistance. If this element is connected across the input of an amplifier, whose input impedance can also be represented by a capacity and a parallel resistance, the equivalent circuit is a current source connected to capacity C in parallel with a resistance R. In this case, the voltage responsivity S_V is given by (Putley, 1970)

$$s_V = \frac{\varepsilon \omega p(T) A R R_T}{(1 + \omega^2 \tau_e^2)^{1/2} (1 + \omega^2 \tau_T^2)^{1/2}}, \tag{3.9}$$

where ε is the absorption coefficient, ω the modulation frequency, R_T the thermal impedance of the element to its environment, τ_e the electric time-constant equal to RC, and τ_T the thermal time-constant. s_V is the responsivity in terms of the voltage V across the amplifier input.

Depending on material parameters and constructional details, τ_e can be greater or smaller than τ_T. In the frequency region between $1/\tau_e$ and $1/\tau_T$, the responsivity is virtually constant. For lower frequencies, it is proportional to ω (being zero for $\omega = 0$), while for higher frequencies it varies as ω^{-1}. Pyroelectric detectors use either current- or voltage-mode amplification (Tiffany, 1975).

A general expression for the noise-equivalent power, although feasible (Dereniak and Brown, 1975), is complicated because of the simultaneous importance of a number of noise sources. Their relative contributions depend on the pyroelectric material, the construction, the amplifier, and the operating conditions. The most important noise sources are Johnson noise in the load feedback resistor, loss-tangent noise, and the amplifier's voltage and current noise. In case it should become possible to reduce the amplifier noise below Johnson- and temperature-noise levels, the Johnson noise would only fall below the temperature-noise level with presently known materials for thicknesses below 1 μm (Putley, 1970).

The performance of pyroelectric detectors can be expressed in terms of D^* only if temperature- or Johnson-noise are the dominant noise sources.

Since pyroelectric detectors only respond to modulated radiant flux, their use in absolute radiometers for measuring steady fluxes requires chopping of the incident beam. This procedure can provide considerable drift immunity and it also allows for the use of inherently drift-free AC amplification techniques. On the other hand, the substitution of chopped electrical power is more complicated and high levels of accuracy are not as easily achieved as with DC techniques because of the additional importance of phase errors. (See Section 6.5.6).

3.6 OTHER DETECTORS

In principle, any temperature-dependent physical quantity can be used as a temperature transducer and, thus, also as a detector in an absolute radiometer. One type of mechanism that has been used as a radiation detector, but has so far never been employed in absolute radiometers, is the Golay cell (Smith et al., 1968). It uses the deflection of a membrane caused by the expansion of an enclosed volume of gas, when heated by absorbed radiation, to produce an output signal.

3.7 COMPENSATING CONFIGURATIONS

When the temperature of the thermal environment of a temperature transducer changes, the temperature of the transducer does not follow it instantaneously. Instead, it lags by an amount that depends upon its thermal mass, its thermal impedance to the thermal environment, and the rate of change of temperature of the latter.

Suppose that the thermal environment is held at temperature T_1 for a long period of time, and then at time t_0 it is quickly raised to a temperature T_2 and held there. It will take the temperature transducer a number of time-constants after t_0 to reach a steady state with respect to the environment again. If it then gives the same output as before t_0, it is said to have first-order compensation. This form of compensation is required for bolometers but not for thermopiles. A second transducer, completely identical to the first one, except that it is not exposed to the radiation to be measured, will also provide compensation during thermal transients. This is called second-order compensation and it is highly desirable for both bolometer and thermopile radiometers.

In addition to the thermal compensation just described, compensating twin-receiver configurations also reduce the effects of pressure fluctuations and variations in the supply current (in the case of bolometers).

3.8 CONCLUSION

With the exception of some more exotic varities (e.g., Golay cells), absolute radiometers employ the full range of available thermal detectors. Taking into account the more complicated structure of their detector elements as well as the trade-offs demanded for reasons of accuracy, the performance as detectors of most of them is quite creditable. However, as of the late 1980s there remains ample scope for improvement (up to or exceeding one order of magnitude in NEI or NEP) in many cases. Considerable progress can be expected on this score from cryogenic absolute radiometers, which have so far not been optimized as thermal detectors.

REFERENCES

Barth, G. (1958). Die natürliche Nachweisgrenze beim Metallbolometer. *Optik* **15**, 694.

Budde, W. (1983). *Physical Detectors of Optical Radiation.* Academic Press, New York and London.

Chasmar, R. P., Mitchell, W. H., and Rennie, A. (1956). Theory and performance of metal bolometers. *J. Opt. Soc. Am.* **46**, 469.

Dereniak, B. D., and Brown, F. G. (1975). Pyroelectric detector evaluation. *Infrared Physics* **15**, 39.

Hudson, R. D., and Hudson, J. W. (1975). *Infrared Detectors.* Dowden, Hutchinson & Ross, Stroudsburg, Pennsylvania.

Jones, R. C. (1949). Factors of merit for radiation detectors. *J. Opt. Soc. Am.* **39**, 344.

Jones, R. C. (1953). Performance of detectors for visible and infrared radiation. *Advances in Electronics* **5**, 1.

Keyes, R. T. (1977). *Optical and Infrared Detectors.* Springer Verlag, Berlin and New York.

Kingston, R. H. (1978). *Detection of Optical and Infrared Radiation.* Springer Verlag, Berlin and New York.

Putley, E. H. (1970). In *Semiconductors and Semimetals*, Vol. 5, p. 259. Academic Press, New York.

Schley, U., and Hoffmann, F. (1958). Über das Rauschen der Strahlungsthermoelemente. *Optik* **15**, 358.

Smith, R. A., Jones, F. E., and Chasmar, R. P. (1968). *The Detection and Measurement of Infrared Radiation*, 2nd edition. Oxford University Press, London and New York.

Soa, E. A. (1978). Some remarks on parameters of thermal radiation detectors. *Exp. Technik der Physik* **26**, 203.

Stevens, N. B. (1970). In *Semiconductors and Semimetals*, Vol. 5, p. 287. Academic Press, New York.

Tiffany, W. B. (1975). Introduction and review of pyroelectric detectors. *Proc. SPIE* **62**, 153.

4

Analysis of the Temperature Distribution in a Detector Element

A. ONO
National Research Laboratory of Metrology
Tsukuba, Japan

4.1 INTRODUCTION

When analyzing the temperature distribution in a detector element, it is possible to obtain useful solutions in a relatively straightforward manner by using a number of simplifications that are very nearly fulfilled for most absolute radiometers. Figure 4.1 shows the cylindrical coordinate system used in the analysis. One assumption that is almost always valid is that the variation of the temperature in the z direction in the receiver is negligible in comparison with the temperature difference between the detector element and the environment. In most cases, it is also reasonable to assume that the heat flow through the high-absorptance coating and the substrate material is in the z direction only. This postulate applies well for receivers where the absorber layer and substrate are thin compared to the diameter of the receiver disk and where the sensor is attached to a layer with a thermal conductivity far higher than that of the substrate and absorber.

In general, the differential equation, which describes the temperature distribution $T(\rho, \phi, z)$ in the high-thermal-conductivity layer, to which the sensor is attached, is given by (Carslaw and Jaeger, 1959)

$$k_d \nabla^2 T = k_d \left[\frac{1}{\rho} \frac{\partial}{\partial \rho} \left(\rho \frac{\partial T}{\partial \rho} \right) + \frac{1}{\rho^2} \frac{\partial^2 T}{\partial \phi^2} + \frac{\partial^2 T}{\partial z^2} \right] = 0. \tag{4.1}$$

Here k_d represents the thermal conductivity of the high-thermal-conductivity layer. This equation can be solved for various appropriate thermal models of absolute radiometers, in which heat flows by conduction, convection, and radiation.

ISBN 0-12-340810-5

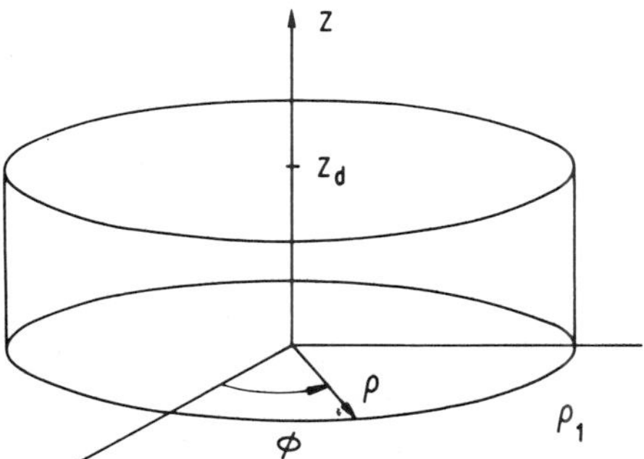

Fig. 4.1 Coordinate system used to analyze the temperature distribution in the receiver disk.

With the theoretical insight gained from solutions such as the ones presented in the following two sections, it is possible to optimize the detector design in terms of various material and thermal parameters of the receiver components. Due to difficulties in obtaining accurate values of such parameters as, for example, the thicknesses of the various layers and nonuniformities, the derived formulas are not usually employed for making quantitative corrections. For these purposes, experimental values obtained from various correction experiments are generally preferred. These experiments and the theory on which they are based will be dealt with in other chapters. However, even in this respect, the analysis of the temperature distribution provides an essential theoretical basis. It pinpoints nonequivalences in the substitution process and provides important functional relationships that are of value in fitting experimental data.

4.2 THE TEMPERATURE DISTRIBUTION IN THE DIRECTION NORMAL TO THE RECEIVER SURFACE

If one makes the assumption, $T(\rho, \phi, z) = T(z)$—i.e., that the temperature in the receiver is uniform in planes normal to z—one can obtain simple solutions, which nevertheless provide some important insights into the thermal behavior of the detector element. It is convenient to use a one-dimensional electrical network analogy (Jacob, 1949) for this analysis. By analogy with Ohm's law, the law of heat conduction can be expressed as

$$\Delta T = PR. \tag{4.2}$$

Here, ΔT is the temperature difference between the two places where the heat transfer takes place, P is the rate of heat transfer, and R is the thermal

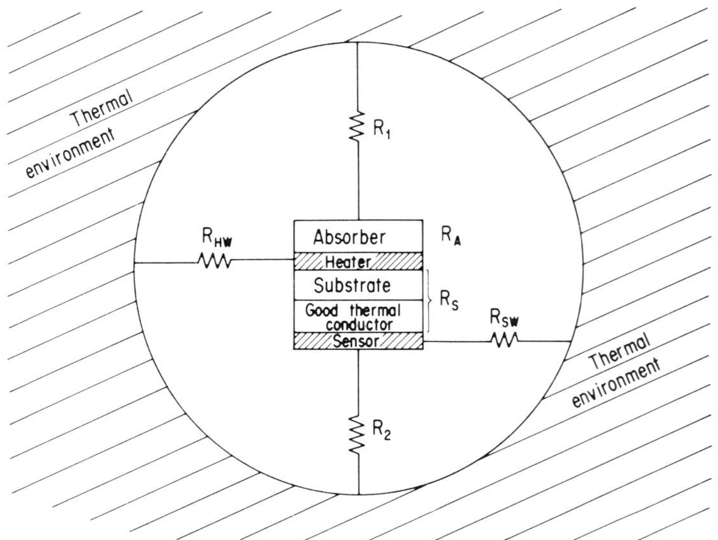

Fig. 4.2 One-dimensional thermal model of an absolute radiometer. [From Hengstberger (1977b), by permission of *Metrologia*]

resistance of the heat path. Clearly, ΔT corresponds to the potential difference, P to the current, and R to the electrical resistance in Ohm's law.

Figure 4.2 shows a schematic representation of a thermal model of an absolute radiometer, which would be applicable to most modern instruments of this kind. Networks representing the heat flows in the receiver for the cases of radiant and electrical heating are given in Figs. 4.3a and 4.3b. Details of these networks have been discussed by Hengstberger (1977b). The most important conclusions from this analysis are that when radiant and electrical heating of magnitude P_r and P_e, respectively, produce the same signal:

(a) The sensor temperatures have to be the same, i.e.,

$$T_{r1} = T_{e1}. \tag{4.3}$$

(b) In consequence, the heat flows from the sensor to the surroundings must be equal:

$$P_{r2} = P_{e2}. \tag{4.4}$$

(c) With Eq. (4.3), this means that the heater temperatures must also be identical:

$$T_{r2} = T_{e2}. \tag{4.5}$$

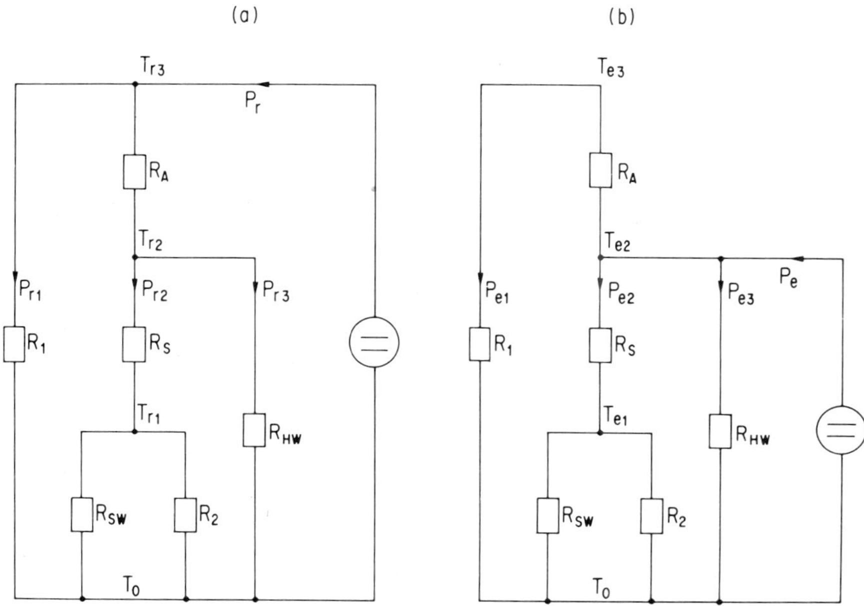

Fig. 4.3 Circuit representation of heat flow in the cases of (a) radiant heating, (b) electrical heating. [From Hengstberger (1977b), by permission of *Metrologia*]

(d) From (c), it follows that the heat flows through the heater wires are the same, i.e.,

$$P_{r3} = P_{e3}. \tag{4.6}$$

From the circuit representation in Figs. 4.3a and 4.3b, one also finds

$$T_{e3} - T_{e2} = -P_{e1}R_A, \tag{4.7a}$$

and

$$T_{r3} - T_{r2} = (P_{r2} + P_{r3})R_A. \tag{4.7b}$$

With (a)–(d), it follows that

$$T_{r3} - T_{e3} = P_e R_A > 0. \tag{4.8}$$

This means that the temperature at the top of the absorber layer is higher for radiant heating than for electrical heating. Thus, one concludes that the heat loss from that surface is nonequivalent under these forms of heating (i.e., $P_{r1} > P_{e1}$). From the network representation, it is readily established that

$$P_{r1} - P_{e1} = P_e R_A/R_1, \tag{4.9}$$

with

$$P_r - P_e = P_{r1} - P_{e1}. \tag{4.10}$$

The radiant power P_r and the electrical power P_e that produce the same signal can be related as

$$P_r = P_e(1 + R_A/R_1). \tag{4.11}$$

This formula has been widely used to assess the influence of the thermal resistance of the absorber layer on the performance of absolute radiometers (Hengstberger, 1977b; Gillham, 1962; Bischoff, 1968; Blevin and Brown, 1966; Blevin and Geist, 1974). It only applies to DC operation. For AC techniques, as used, for instance, with pyroelectric detectors, phase nonequivalencies caused by the absorber coating also play an important role (Blevin and Geist, 1974). The various thermal resistances in Fig. 4.2 can be expressed in terms of the material and thermal parameters of the receiver components as

$$R_A = z_A/k_A A \tag{4.12a}$$

$$R_S = z_S/k_S A \tag{4.12b}$$

$$R_{HW} = 1/\Sigma_i(k_{Hi} A_{Hi}/z_{Hi}) \tag{4.12c}$$

$$R_1 = 1/h_1 A \tag{4.12d}$$

$$R_2 = 1/h_2 A \tag{4.12e}$$

$$R_{SW} = 1/\Sigma_i(k_{Si} A_{Si}/z_{Si}) \tag{4.12f}$$

The definitions for the symbols just used are:

- z_A Thickness of the absorber layer
- k_A Thermal conductivity of the absorber material
- A Receiver area
- z_S Thickness of the substrate
- k_S Thermal conductivity of the substrate material
- z_{Hi} Length of heater wire i
- k_{Hi} Thermal conductivity of heater wire i
- A_{Hi} Cross-sectional area of the heater wire i
- h_1 Heat-transfer coefficient between the top surface of the absorber layer and the thermal environment
- h_2 Heat-transfer coefficient between the back surface of the receiver and the thermal environment, excluding conduction in the sensor wires
- z_{Si} Length of sensor wire i
- k_{Si} Thermal conductivity of sensor wire i
- A_{Si} Cross-sectional area of sensor wire i
- HW Heater wire
- SW Sensor wire

The heat-transfer coefficients h are measures of the rate of heat transfer per unit area and unit temperature difference between certain surfaces and their thermal environment. They incorporate all modes of heat transfer—i.e., conduction, convection, and radiation.

In the formulas for R_A, R_S, R_{HW}, and R_{SW}, conduction was assumed to be the exclusive transfer mechanism. In cases where this is not applicable (e.g., for very thin heater or sensor wires), where heat transfer from their surface by other modes can become significant (Hengstberger, 1977a), the more general formalism in terms of heat-transfer coefficients should be employed. The contribution to h by thermal radiation for small temperature changes is given by

$$h_r \approx 4\varepsilon\sigma T^3, \tag{4.13}$$

where ε is the emissivity of the surface, T the temperature, and σ the Stefan–Boltzmann constant. The contributions by air conduction and convection are usually much more difficult to calculate for the particular geometric conditions encountered. Since experimental determinations of these parameters are not too difficult, measured values are often used in calculations of this kind (Jacob, 1949; Blevin and Brown, 1966). As in the case of different input powers producing the same signal, which was previously analyzed, solutions for the same input powers ($P_r = P_e = P$), which produce different output signals, can be found easily from the networks in Fig. 4.3. If one neglects the heat loss through the heater wire ($P_{r3} = P_{e3} = 0$), the rise in the sensor temperature caused by a radiant power input $P_r = P$ follows as

$$T_{r1} - T_0 = PR_0/[1 + (R_A + R_S + R_0)/R_1] \tag{4.14}$$

where R_0 is defined as

$$R_0 = R_{SW}R_2/(R_{SW} + R_2). \tag{4.15}$$

The rise in sensor temperature due to an electrical input power $P_e = P$ is found as

$$T_{e1} - T_0 = PR_0/[1 + (R_S + R_0)/(R_1 + R_A)]. \tag{4.16}$$

Exactly the same results can be obtained by solving Eq. (4.1) under the assumption $T(\rho, \phi, z) = T(z)$. A single integration in z over the thickness of the high-thermal-conductivity layer (z_d) yields

$$k_d\left(\left.\frac{\partial T}{\partial z}\right|_{z_d} - \left.\frac{\partial T}{\partial z}\right|_0\right) = 0. \tag{4.17}$$

Substituting the boundary conditions for $\partial T/\partial z$ at z_d and 0, as derived in Geist's analysis of the temperature distribution in a detector element (Geist, 1972), results in exactly the same solutions as given previously.

4.3 THE TEMPERATURE DISTRIBUTION PARALLEL TO THE RECEIVER SURFACE

4.3.1 Method of Green's Function

In order to calculate the temperature distribution in the high-thermal-conductivity layer parallel to the receiver surface, Geist (1972) used the average temperature $U(\rho, \phi)$ defined by

$$U(\rho, \phi) = \frac{1}{z_d} \int_0^{z_d} T(\rho, \phi, z) dz. \tag{4.18}$$

Integrating (4.1) with respect to z, he obtained the equation

$$z_d k_d \left[\frac{1}{\rho} \frac{\partial}{\partial \rho} \left(\rho \frac{\partial U}{\partial \rho} \right) + \frac{1}{\rho^2} \frac{\partial^2 U}{\partial \phi^2} \right] + k_d \left(\left. \frac{\partial T}{\partial z} \right|_{z_d} - \left. \frac{\partial T}{\partial z} \right|_0 \right) = 0. \tag{4.19}$$

Under the assumption that the heat loss through the heater wires is neglible, he derived expressions for $\partial T / \partial z$ at z_d and 0 for both radiant and electrical heating. Substituting these into (4.19) resulted in a differential equation of the form

$$\frac{1}{\rho} \frac{\partial}{\partial \rho} \left(\rho \frac{\partial V}{\partial \rho} \right) + \frac{1}{\rho^2} \frac{\partial^2 V}{\partial \phi^2} - g^2 V = -Q(\rho, \phi), \tag{4.20}$$

where

$$V = U - T_0, \tag{4.21}$$

$$g^2 = (\gamma h_1 + h_2)/(z_d k_d), \tag{4.22}$$

$$\gamma = [1 + (z_S/k_S + z_A/k_A) h_1]^{-1}, \tag{4.23}$$

$$Q(\rho, \phi) = \gamma \lambda E(\rho, \phi)/z_d k_d. \tag{4.24}$$

Here, $E(\rho, \phi)$ is the power-density input to the detector, and $Q(\rho, \phi)$ represents the normalized input power. For radiant heating

$$\lambda = 1, \tag{4.25}$$

while for electrical heating

$$\lambda = 1 + h_1 z_A / k_A. \tag{4.26}$$

The Green's function $G(\rho, \phi; \rho_0, \phi_0)$ for Eq. (4.20) is given by (Geist, 1972)

$$\begin{aligned} G(\rho, \phi; \rho_0, \phi_0) = & \sum_{m=-\infty}^{\infty} \sum_{n=-\infty}^{\infty} A_{mn} I_m(g\rho) I_n(g\rho_0) e^{i(m\phi + n\phi_0)} \\ & - \frac{1}{2\pi} K_0\{g[\rho^2 + \rho_0^2 - 2\rho\rho_0 \cos(\phi - \phi_0)]^{1/2}\}, \end{aligned} \tag{4.27}$$

where the A_{mn} are determined by the boundary conditions and where I_n and K_n are the modified Bessel functions of order n. Interpreted physically, the Green's function describes the response at point (ρ, ϕ) due to a unit point source at (ρ_0, ϕ_0). With it, the temperature rise $V(\rho, \phi) = U(\rho, \phi) - T_0$ in the high-thermal-conductivity layer can be expressed as

$$V(\rho, \phi) = -\int_0^{\rho_1}\int_0^{2\pi} Q(\rho_0, \phi_0)G(\rho, \phi; \rho_0, \phi_0)\rho_0\, d\phi_0\, d\rho_0, \qquad (4.28)$$

ρ_1 being the receiver radius.

Using an appropriate boundary condition and assuming a constant heat-transfer coefficient h_e from the edge of the receiver disk to the environment, Geist (1972) also derived an expression of the Green's function for a receiver. With this function, the average temperature rise at the edge of the receiver for a point source input to the location (ρ_0, ϕ_0) is expressed as

$$\bar{V}_e(\rho_0) = \frac{1}{2\pi}\int_0^{2\pi} V(\rho_1, \phi)d\phi, \qquad (4.29)$$

$$= \frac{P\gamma\lambda I_0(g\rho_0)}{2\pi g\rho_1 z_d k_d[\xi I_0(g\rho_1) + I_1(g\rho_1)]}, \qquad (4.30)$$

where P is a quantity of input power and

$$\xi = h_e/gk_d. \qquad (4.31)$$

The voltage across the radial thermocouple, where thermocouple junctions are attached uniformly on the edge of the receiver, is given by

$$v_e = \int_0^{2\pi} W(\phi)V(\rho_1, \phi)d\phi. \qquad (4.32)$$

Here, $W(\phi)$ is the sensitivity (per unit angle) of the thermopile as a function of position around the edge of the receiver. Assuming that $W(\phi)$ is independent of ϕ, the voltage is expressed as

$$v_e = 2\pi W V_e. \qquad (4.33)$$

The responsivity distribution is defined by

$$F(\rho_0) = v/P. \qquad (4.34)$$

Using the approximations, $I_0(x) = 1 + x^2/4$ and $I_1(x) = x/2$ for $x \ll 1$, the responsivity distribution of well-constructed radiometers with the condition of $g\rho \ll 1$ (i.e., small responsivity variations) is expressed as

$$F_e(\rho_0) = \frac{2\pi W\gamma\lambda(1 + g^2\rho_0^2/4)}{G_e(1 + g^2\rho_1^2/4) + \gamma G_1 + G_0}, \qquad (4.35)$$

where G_e is the thermal conductance between the edge of the receiver and the environment defined by

$$G_e = 2\pi\rho_1 z_d h_e, \tag{4.36}$$

G_1 is that between the top surface of the receiver and the environment defined by

$$G_1 = \pi\rho_1^2 h_1, \tag{4.37}$$

and G_2 is that between the back surface of the receiver and the environment defined by

$$G_2 = \pi\rho_1^2 h_2. \tag{4.38}$$

One can demonstrate with (4.30) and (4.35) that the absorption of radiant and electrical power in different parts of the receiver represents an error source that can be factored into two separate effects (Geist, 1972). The first one, the "excess-emittance effect," is due to the thermal resistance of the absorber layer as was discussed in the previous section. The effect is represented by the different values of λ given by (4.25) and (4.26) for radiant and electrical heating, respectively. The other one, also called the "relative responsivity effect," is due to different input-power distributions over the receiver surface between the radiant heating and the electrical heating. It represents an interaction between the variation of the responsivity across the detector surface, $F_e(\rho_0)$, and the spatial distribution of the input-power density, $E(\rho_0, \phi_0)$.

4.3.2 Method of Annular Power Source

Most absolute radiometers are constructed circularly, and the temperature sensors are distributed axially symmetrically as in the cases of a radial thermopile and of a thermopile uniformly attached to the back of the receiver. For such axially symmetric absolute radiometers and for the purpose of investigating the responsivity distribution over the receiver, it is general enough to use annular power sources concentric with the receiver perimeter.

It is also possible to seek solutions to Eq. (4.1) for axially symmetrical temperature distributions, disregarding temperature differences in the z direction. Such solutions would be applicable to thin receivers with constant heat-transfer coefficients h_1 and h_2 over their surface. This case is characterized by the condition $T(\rho, \phi, z) = T(\rho)$. Boundary conditions at the edge of the receiver disk and at the location of the annular power source follow from the law of energy conservation and the continuity of temperature. This approach has been used by Ono (1979a). In this case, the heat-transfer equation in the steady state is reduced to

$$\frac{1}{\rho}\frac{\partial}{\partial\rho}\left(\rho\frac{\partial V}{\partial\rho}\right) - \frac{1}{\beta^2}V = 0, \tag{4.39}$$

where $V(\rho) = T(\rho) - T_0$ and β is a characteristic length defined by

$$\beta = [kz/(h_1 + h_2)]^{1/2}. \tag{4.40}$$

Here k represents the average thermal conductivity of the receiver and z its thickness. Under the same conditions, the parameter g used in the previous section would be the reciprocal of β.

The temperature distributions over the receiver are illustrated in Fig. 4.4 for annular sources of a given power having different radii. For an infinite value of β, i.e., $h_1 = h_2 = 0$ (see Fig. 4.4a), the temperature distribution is constant inside the annulus and is described by the logarithm outside that. For finite values of β (see Fig. 4.4b), the temperature distribution is described by the modified Bessel functions I_0 and K_0: inside the annulus (i.e., for $\rho \leq \rho_0$), as

$$V(\rho) = C_1 I_0(\rho/\beta), \tag{4.41}$$

and outside the annulus (i.e., for $\rho \geq \rho_0$), as

$$V(\rho) = C_2 K_0(\rho/\beta) + C_3 I_0(\rho/\beta) \tag{4.42}$$

where C_1, C_2, and C_3 are constants determined by the boundary conditions. Equation (4.42) leads to exactly the same result as the average temperature rise at the edge of the receiver given by Eq. (4.30) (Ono, 1979a).

Figure 4.4a shows that, for an infinite value of β, the temperature rise of the edge of the receiver is independent of the radius of the annular source. Then a perfect uniformity of response is expected for a radial thermopile. This situation is understood from the fact that, wherever the energy is given on the receiver, all energy flows through the thermopile legs without a loss. For a finite value of β (see Fig. 4.4b), the temperature of the edge of the receiver decreases as the radius of annular source decreases, where a part of the energy absorbed by the receiver is lost to the environment before reaching the thermopile legs.

In the case of a temperature sensor attached uniformly to the back of the receiver, the average temperature rise $\bar{V}_b$ on this surface is the important quantity. $\bar{V}_b$ is given as

$$\bar{V}_b(\rho_0) = \frac{1}{\pi\rho_1^2}\int_0^{\rho_1}\int_0^{2\pi} V(\rho)\rho \, d\phi \, d\rho, \tag{4.43}$$

$$= \frac{P}{G}\left[1 - \frac{\xi I_0(\rho_0/\beta)}{\xi I_0(\rho_1/\beta) + I_1(\rho_1/\beta)}\right], \tag{4.44}$$

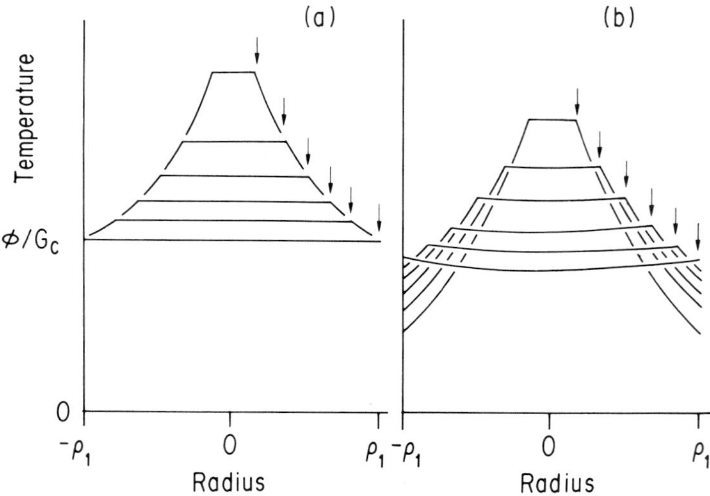

Fig. 4.4 Temperature distributions over the receiver surface (a) without, (b) with heat losses from the surface. The arrows indicate the radii of the annular power sources. [From Ono (1979a), by permission of *Metrologia*]

where G is the total thermal conductance between the receiver and the environment defined by

$$G = G_e + G_1 + G_2. \tag{4.45}$$

The voltage across the uniformly attached thermopile to the back of the receiver is expressed as

$$v_b = \int_0^{\rho_1} \int_0^{2\pi} w(\rho, \phi) V(\rho) \rho \, d\phi \, d\rho, \tag{4.46}$$

where $w(\rho, \phi)$ is the sensitivity per unit area of the thermopile as a function of position on the back of the receiver. Assuming that $w(\rho, \phi)$ is independent of ρ and ϕ, the voltage is expressed as

$$v_b = \pi \rho_1^2 w \bar{V}_b. \tag{4.47}$$

When $\rho_1/\beta \ll 1$, Eq. (4.44) is expressed as a series expansion, and the responsivity distribution is given as approximately

$$F_b(\rho_0) = \frac{\pi \rho_1^2 w}{G} \left[1 - \frac{\xi(1 + \rho_0^2/4\beta^2)}{\xi\left(1 + \dfrac{\rho_1^2}{4\beta^2}\right) + \dfrac{1}{2}\dfrac{\rho_1}{\beta}} \right]. \tag{4.48}$$

Equation (4.48) indicates that the responsivity distribution has a maximum at the center of the receiver and a minimum at the edge of the receiver. A perfect uniformity of the responsivity is expected for $\xi = 0$ (i.e., no heat loss from the edge of the receiver to the environment).

4.3.3 Numerical Method

Ono (1979a) has reported a numerical method of analysis based on the division of a receiver into small parts that have a thermal resistance R between each other. A simple model of this kind is shown in Fig. 4.5. Here, the thermal resistance from the central plate to the environment is R_1 and that from each peripheral plate to the environment is R_0. For a radial thermopile, the responsivity variation can be calculated as

$$\delta = R/4R_1, \tag{4.49}$$

δ being defined as the difference between the maximum response and the minimum response divided by the maximum response. The relationship

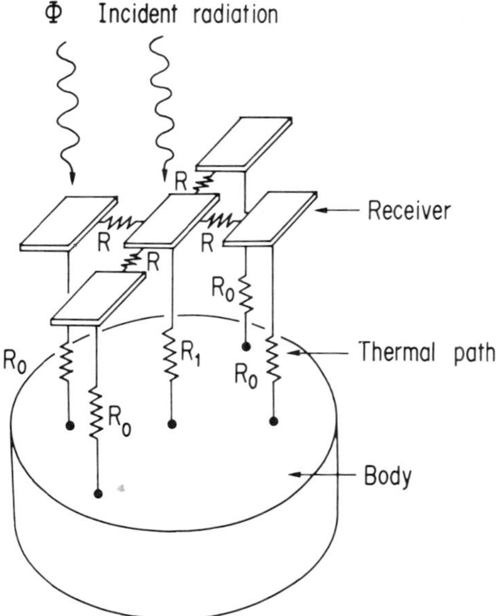

Fig. 4.5 A thermal model of absolute radiometers where the receiver disk is divided into five plates. It is used to investigate the responsivity variations on the receiver. [From Ono (1979a), by permission of *Metrologia*]

shows that a good uniformity of response can be obtained with small R and large R_1.

In the case of a uniformly attached temperature sensor to the back of the plates, the responsivity variation follows as

$$\delta = \frac{R}{5R_1 + R} \frac{R_1 - R_0}{R_1}. \tag{4.50}$$

A perfect uniformity can be expected here if the thermal resistances to the environment from all plates are equal; i.e., if $R_1 = R_0$.

Extending this technique, Ono (1979a) has analyzed receivers with radial thermopiles using a model with 37 plates. In particular, probable factors causing additional nonuniformity were investigated. These were:

(a) Variations in the thermal resistances,
(b) Variations in sensor responsivities,
(c) Nonuniform receiver thickness.

4.4 GUIDELINES FOR MINIMIZING RESPONSIVITY VARIATIONS

From Eqs. (4.49) and (4.50), it is evident that a reduction of the heat losses from the receiver to the environment (i.e., an increase in R_1) always reduces the responsivity variations. Ono (1979a) has proposed several possible ways to achieve this:

(a) By operating the detector in vacuum to eliminate heat losses through the air. At room temperature, heat losses through the air decrease rapidly below 10 Pa (10^{-1} Torr) and become almost negligible at 10^{-2} Pa (10^{-4} Torr).

(b) By using low-emissivity materials for the back of the receiver to avoid heat losses by thermal radiation.

(c) By placing a hemispherical mirror with an aperture in front of the receiver to reduce heat losses by thermal radiation.

(d) By lowering the operating temperature of the detector to reduce heat losses by thermal radiation. At liquid-nitrogen temperature (77K), there is a reduction by a factor of 50 as compared to room temperature.

Experimental results by the same author (Ono, 1979b, 1979c) have confirmed the validity of this approach. Responsivity variations still existing are estimated by Eq. (4.35), for a temperature sensor attached uniformly on the edge of the receiver, and by Eq. (4.48), for a temperature sensor attached uniformly to the back of the receiver.

REFERENCES

Bischoff, K. (1968). Ein einfacher Absolutempfänger hoher Genauigkeit. *Optik* **28**, 183.

Blevin, W. R., and Brown, W. J. (1966). Black coatings for absolute radiometers. *Metrologia* **2**, 139.

Blevin, W. R., and Geist, J. (1974). Influence of black coatings on pyroelectric detectors. *Appl. Opt.* **13**, 1171.

Carslaw, H. S., and Jaeger, J. C. (1959). *Conduction of Heat in Solids*, 2nd edition. Oxford University Press.

Geist, J. (1972). *Fundamental Principles of Absolute Radiometry and the Philosophy of This NBS Program.* National Bureau of Standards Tech. Note 594-1. U.S. Government Printing Office, Washington, D.C.

Gillham, E. J. (1962). Recent investigations in absolute radiometry. *Proc. Roy. Soc. (London) Ser. A.* **269**, 249.

Hengstberger, F. (1977a). *A Fresh Look at the Lead-heating Effect in Absolute Radiometry.* CSIR Research Report 333. Council for Scientific and Industrial Research, Pretoria, South Africa.

Hengstberger, F. (1977b). An improved theory of the instrumental corrections for absolute radiometers. *Metrologia* **13**, 69.

Jacob, M. (1949). *Heat Transfer*, Vol. 1. J. Wiley & Sons, New York.

Ono, A. (1979a). A design of thermal detectors of radiation with uniform sensitivity over the surface of the receiver. *Metrologia* **15**, 127.

Ono, A. (1979b). Absolute radiometer with uniform responsivity over the surface of the receiver. *Jpn. J. Appl. Phys.* **18**, 697.

Ono, A. (1979c). An absolute radiometer with uniform response: fabrication, characteristics and analysis. *Jpn. J. Appl. Phys.* **18**, 1995.

5

Environmental Corrections in Absolute Radiometry

L. P. BOIVIN
National Research Council
Ottawa, Canada

5.1 INTRODUCTION

Absolute radiometers can be used for the accurate spectral calibration of other radiometers and, in conjunction with filters, for the calibration of lamps for photometric or radiometric quantities. When used to calibrate lamps, absolute radiometers may require corrections for effects external to the radiometer head itself. These effects are: (a) diffraction at apertures located between the source and the radiometer head, (b) absorption of radiation by the water-vapor and carbon-dioxide components of air, and (c) the use of filters of nonnegligible thickness between the source and the radiometer head. In this chapter, we shall be concerned with the calculation of these corrections (or the estimation of the magnitudes of the errors that can be caused by these effects if corrections are not applied). Since the same theoretical concepts are involved when discussing the diffraction at the radiometer aperture, that correction will also be analyzed in this chapter. However, it should be clearly understood that this particular correction and the correction for diffraction at other apertures inside the radiometer head are actually part of the group of instrumental corrections covered in Chapter 6. (See Section 6.5.5.)

5.2 DIFFRACTION

There are two cases to consider: (1) diffraction by the radiometer aperture itself and (2) diffraction by baffles placed between the source and the radiometer aperture.

ISBN 0-12-340810-5

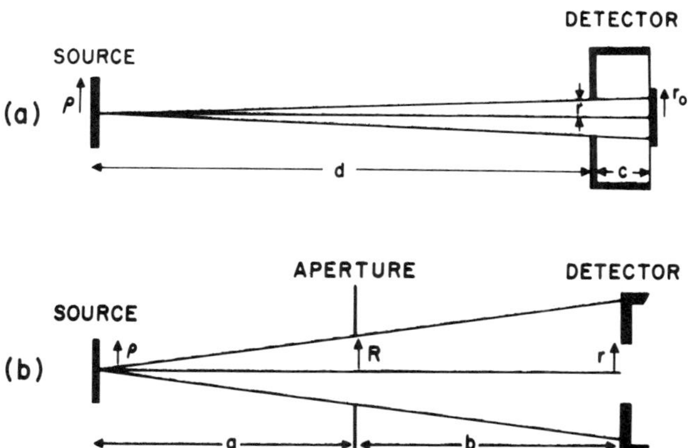

Fig. 5.1 Diffraction effects in absolute radiometry: definition of symbols used. (a) Diffraction by the radiometer aperture. (b) Diffraction by circular baffles.

An absolute radiometer having a uniformly sensitive element of radius r_0 and an aperture of radius r is used to measure the irradiance in the plane of the aperture due to a source at a distance d from the aperture. In the first case (Fig. 5.1a), diffraction at the aperture r results in a fraction of the flux transmitted by the aperture to fall outside the sensitive element of radius r_0, although by design, and geometrically speaking, all of the flux transmitted by the aperture r should be collected by the sensitive element. In the second case (Fig. 5.1b), diffraction at the baffle aperture R results in some flux being diffracted into the radiometer aperture r. As the effects in both cases are small, and because the radiation considered in radiometry is usually incoherent, the two cases can be treated independently. In both cases, we shall calculate the corrections for a point source or an extended source of monochromatic or complex radiation. For these calculations, we shall assume that the baffles and apertures have edges that are circular, smooth, and infinitely thin. We shall also discuss the reduction of diffraction errors by means of specially shaped apertures. Finally, we shall consider briefly the case where there are a number of apertures between the source and the detector. We shall not consider the case of thick apertures. Such apertures are to be avoided in practice because they can cause errors due to reflections from the edges (Gillham, 1962).

5.2.1 Diffraction at the Radiometer Aperture

In Fig. 5.1a, the source of radius ρ emits incoherent, monochromatic radiation of wavelength λ. The ratio of the flux incident on the sensitive element of radius r_0, taking into account diffraction at the aperture of radius

r, to that which would be received on a purely geometric basis, is given by (Steel et al., 1972)

$$F_1(u, w, w') = \tfrac{1}{2} \int_0^{w+w'} I(v, w, w')J(u, v)v \, dv, \tag{5.1}$$

where

$$u = \frac{2\pi}{\lambda} \frac{r^2(c+d)}{cd}, \qquad w = \frac{2\pi}{\lambda} \frac{r\rho}{d}, \qquad w' = \frac{2\pi r r_0}{\lambda c}.$$

$J(u, v)$ is the intensity distribution of diffracted radiation in the receiver plane for a point source; $I(v, w, w')$ is an autocorrelation function (Steel et al., 1972). The derivation of Eq. 5.1 assumes that $u > w' - w$, which implies that, geometrically, any ray from the source passing through the aperture r will be collected by the receiver of radius r_0.

In order to evaluate $F_1(u, w, w')$, we need to know $J(u, v)$ only in the shadow region. For the cases usually encountered in radiometry, $u \gg 1$, $v \gg 1$, and $v \gg u$, so that the following approximation can be used (Steel et al., 1972),

$$J(u, v) \simeq \frac{2}{\pi v}\left[\frac{1}{(v+u)^2} + \frac{1}{(v-u)^2}\right]. \tag{5.2}$$

We shall define the diffraction error by the expression $\varepsilon(u, w, w') = F_1(u, w, w') - 1$. The correction factor is given by $(1/F_1) = 1/(1 + \varepsilon)$.

5.2.1.1 Point Source of Monochromatic Radiation

For a point source, it can be shown that the diffraction error is given approximately by (Blevin, 1970)

$$\varepsilon = \frac{-2w'}{\pi(w'^2 - u^2)} = \frac{-\lambda r_0 c}{\pi^2 r\left[r_0^2 - \left(\dfrac{c+d}{d}\right)^2 r^2\right]}. \tag{5.3}$$

If $w' \geq 1.3u$ and $u \geq 15\pi$, this expression will give a value of ε that differs from the true value by less than 3% (Boivin, 1975a). These conditions are usually met in radiometry. In other cases, the value of ε may be obtained from graphs or tables, such as those given by Blevin (1970), Boivin (1975a), and Fussell (1974). Figure 5.2 shows a graph from which ε can be obtained for a large range of u and w'.

Numerical Example

A radiometer has a sensitive element of radius $r_0 = 5$ mm, with a limiting aperture of radius $r = 3$ mm located 5 cm from the sensitive element. The point source, of wavelength $\lambda = 1\ \mu$m, is 100 cm from the aperture. We have

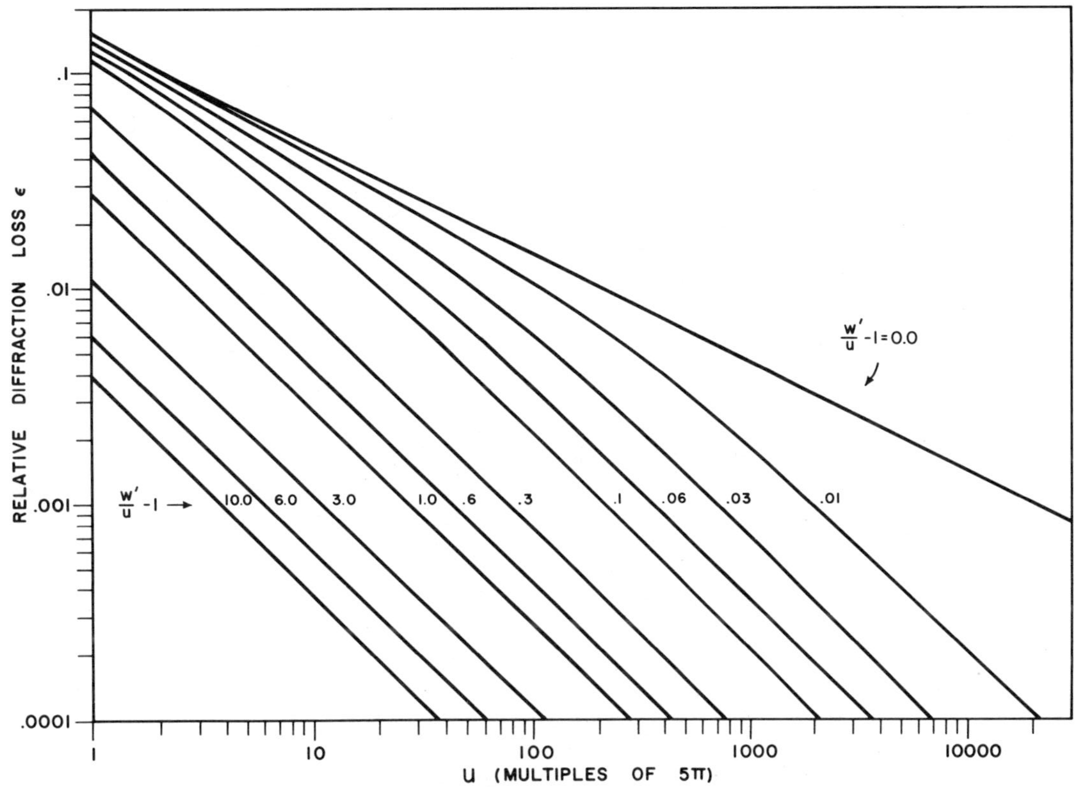

Fig. 5.2 Diffraction by the radiometer aperture. The diffraction error ε as a function of u (for constant w'/u) in the case of a point source of monochromatic radiation. [From Boivin (1975a), by permission of *Applied Optics*]

$w' = 1885$ and $u = 1188$, so that Eq. 5.3 is applicable. The diffraction error is thus calculated to be -0.056%.

5.2.1.2 Point Source of Complex Radiation

For complex radiation, diffraction errors can be calculated by replacing the monochromatic wavelength λ in Eq. 5.3 (and in the expressions for u and w') by an effective wavelength λ_e defined by (Blevin, 1970)

$$\lambda_e = \int_{\lambda_1}^{\lambda_2} \lambda^q \phi(\lambda) d\lambda \bigg/ \int_{\lambda_1}^{\lambda_2} \phi(\lambda) d\lambda, \tag{5.4}$$

where $\phi(\lambda)$ is the product of the spectral distribution of the source, the spectral responsivity of the radiometer, and the spectral transmittance of any filter used; and (λ_1, λ_2) is the effective bandwidth of the source-filter-radiometer combination. The parameter q has a value between 0.5 and 1, depending on the relative magnitudes of u and w'. For the case where Eq. 5.3 is applicable, $q = 1$, since ε is proportional to λ in that case. In other cases, Blevin (1970) has shown that the average of the λ_e's calculated for $q = 1$ and $q = 0.5$ can be used with negligible error for calculating diffraction errors.

Effective wavelengths have been calculated for several cases. Blevin (1970) has calculated the effective wavelengths for a blackbody at an absolute temperature T, used with a nonselective detector, and found values of $\lambda_e = 0.479/T$ and $0.532/T$ for $q = 0.5$ and 1, respectively. Boivin has calculated the effective wavelengths for tungsten radiation, limited to the approximate spectral range (0.3–2.7 μm) by means of a glass filter, for the case of a nonselective detector (Boivin, 1975a) and for a typical silicon photodiode (Boivin, 1976). Blevin (1970) has also calculated the effective wavelengths corresponding to several $V(\lambda)$ filtered sources, such as would be encountered when making luminous intensity measurements with absolute radiometers. In those cases, the difference in λ_e for $q = 0.5$ and $q = 1$ is negligible. The values of λ_e corresponding to $V(\lambda)$ filtered blackbody radiation at 2045 K (platinum point) and 2856 K (CIE illuminant A) are about 582 and 572 nm, respectively.

5.2.1.3 Extended Source of Monochromatic or Complex Radiation

The general case given by Eq. 5.1 can be quite complicated to calculate. However, a simple approximation can be derived that is sufficiently accurate in most cases encountered in radiometry. We have (Steel et al., 1972)

$$\varepsilon(u, w, w') \simeq -(1/(2\pi w)) \ln\left[\frac{(w' + w)^2 - u^2}{(w' - w)^2 - u^2}\right), \tag{5.5a}$$

for $(w + w')^2 - u^2 > 6 \times 10^3$.

This expression can be further simplified to more clearly define the variation of ε with w (which is proportional to the diameter of the source). To a good degree of approximation, we can write

$$\varepsilon(u, w, w') \simeq -\frac{2w'(w'^2 - u^2)}{\pi[(w'^2 - u^2)^2 - w^2 w'^2]}. \tag{5.5b}$$

Thus, ε increases slowly as the source size increases. Taking the same numerical example as before, but considering a circular source of 1.5-cm diameter instead of a point source, the diffraction error is calculated to be $-0.059\,\%$, which differs very little from the value obtained for a point source.

In Eq. 5.5, ε is proportional to λ, so that the effective wavelengths for $q = 1$ can be used to calculate diffraction errors with complex radiation.

5.2.2 Diffraction at Circular Baffles

A radiometer aperture having a radius r is located in the fully irradiated region projected by a circular baffle of radius R. See Fig. 5.1b. We shall calculate the ratio of the flux received by the aperture r, taking into account diffraction at the baffle R, to the flux that would be received geometrically. This ratio is given by (Steel et al., 1972)

$$\mathrm{F}_2(u, w, w') = \frac{1}{2}\left(\frac{u}{w'}\right)^2 \int_0^{w+w'} \mathrm{I}(v, w, w')\mathrm{J}(u, v)v\, dv, \tag{5.6}$$

where

$$u = \frac{2\pi}{\lambda} R^2 \frac{(a+b)}{ab}, \quad w' = \frac{2\pi R r}{\lambda b}, \qquad \text{and} \quad w = \frac{2\pi R \rho}{\lambda a}.$$

$\mathrm{I}(v, w, w')$ is defined as before. $\mathrm{J}(u, v)$ is the intensity distribution in the fully irradiated region due to a point source. The exact expression for $\mathrm{J}(u, v)$ is somewhat complex. However, as was the case for the shadow region, a simpler expression for $\mathrm{J}(u, v)$ can be used for the type of calculations described here, where typically $u \gg 1$ and $v \ll u$. The following expression is quite valid (Boivin, 1975b) for the central portion (approximately $w'/u < \frac{1}{3}$) of the irradiated region:

$$\mathrm{J}(u, v) = \frac{4}{u^2}\left[1 + J_0^2(v) + \frac{v^2}{u^2} J_1^2(v) - 2J_0(v)\cos\left(\frac{u}{2} + \frac{v^2}{2u}\right) - 2\frac{v}{u} J_1(v)\sin\left(\frac{u}{2} + \frac{v^2}{2u}\right)\right], \tag{5.7}$$

in which J_0 and J_1 are Bessel functions of order zero and one, respectively.

The expression for F_2 (Eq. (5.6)) was derived assuming $u > w' + w$, which means that the detector aperture lies wholly in the fully irradiated region. Equation (5.6) is also restricted to the case $w' > w$; however, this is not a limitation, since the diffraction errors for $w > w'$ can be derived by reciprocity (Steel et al., 1972).

As before, we shall define the diffraction error ε as being equal to $F_2 - 1$ and the diffraction correction is then given by $(1/F_2) = 1/(1 + \varepsilon)$.

5.2.2.1 Point Source of Monochromatic or Complex Radiation

For a point source, the diffraction error is given approximately by (Boivin, 1975b)

$$\varepsilon(u, 0, w') = \frac{2}{\pi w'} - \frac{4}{w'}\sqrt{\frac{2}{\pi w'}}\sin\left(w' - \frac{\pi}{4}\right)\cos\left(\frac{u}{2} + \frac{w'^2}{2u}\right). \tag{5.8}$$

The variation of ε with w' is shown in Fig. 5.3 for $u = 1500\pi$. Note the rapid oscillation of ε about the mean value $2/\pi w'$. It can be shown (Boivin, 1975b)

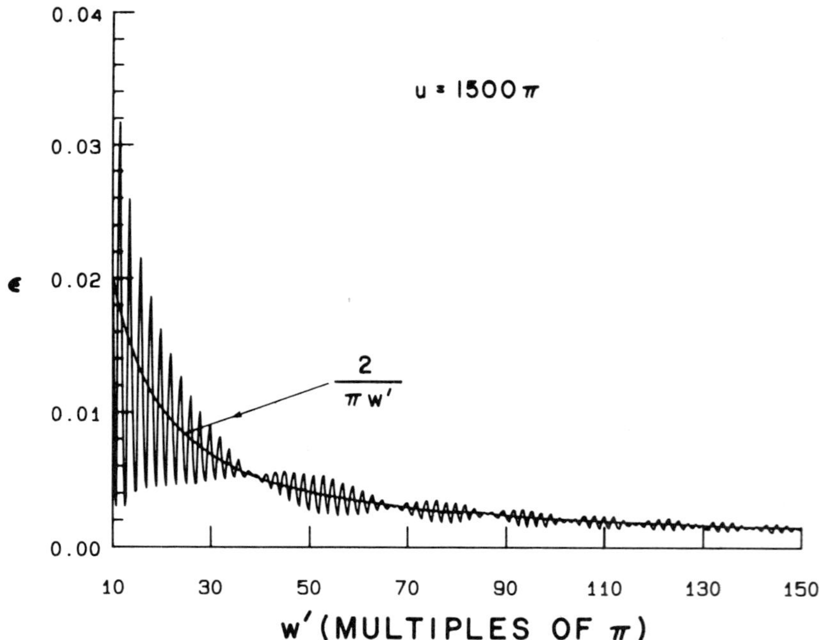

Fig. 5.3 Diffraction by circular baffles. The diffraction error ε as a function of w', in the case of a point source of monochromatic radiation. [From Boivin (1975b), by permission of *Applied Optics*]

that for radiation having a bandwidth of at least 2 nm, the oscillations cancel out, so that the resulting diffraction error is given by

$$\varepsilon(u, 0, w') = \frac{2}{\pi w'} = \frac{b\lambda}{\pi^2 R r}, \tag{5.9}$$

evaluated for the central wavelength of the radiation bandwidth.

Since ε given by Eq. 5.9 is proportional to the wavelength λ, diffraction errors in the case of complex radiation can be evaluated by using the effective wavelength given by Eq. 5.4, for $q = 1$, in this expression.

Equation 5.8 (or 5.9) is applicable in the central portion of the irradiated region—i.e., for $w'/u < \frac{1}{3}$ approximately. (Geometrically, w'/u is the ratio of the diameter of the radiometer aperture to the diameter of the irradiated region in the plane of the radiometer aperture.) For $w'/u > \frac{1}{3}$, the mathematical treatment used is not applicable and the calculations become very complicated. Close to the edge of the shadow (i.e., for $0.75 \leq w'/u < 1$ approximately), other approximations can be used. However, all results must be obtained by numerical integration (Boivin, 1977) and no simple formula is available for ε, as is the case in the central region.

Figure 5.4 shows the variation of ε throughout the irradiated region for a typical case (3000 K tungsten radiation and a silicon diode detector): the

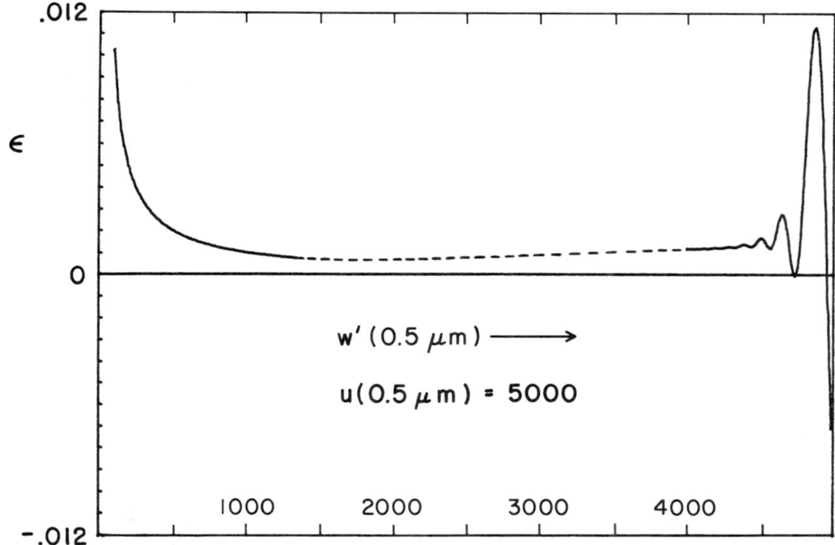

Fig. 5.4 The diffraction error ε for a point source of complex radiation (tungsten 3000 K) and a typical silicon diode detector. The graph shows the variation of ε throughout the irradiated region. [From Boivin (1977), by permission of *Applied Optics*]

solid portion of the curve, for $w' > 4000$, has been obtained by numerical integration; the solid portion of the curve for $w' < 1500$ has been obtained from Eq. 5.9; and the dotted portion has been obtained by interpolation. Thus, the diffraction error is positive throughout the irradiated region, except very near the shadow edge. The diffraction error is large near the center and close to the edge, but achieves a broad minimum approximately midway ($w'/u \approx \frac{1}{2}$ between the center and the edge. The edge region is particularly bad, because diffraction errors are large and oscillate rapidly with w', even for large-bandwidth radiation. Furthermore, in the case of an extended source, vignetting effects would be observed near the edge. For these reasons, it is best to ensure that $w'/u \leq \frac{1}{2}$ in practice, and we shall consider only the central region in the following discussions.

5.2.2.2 Extended Sources

For an extended source, the diffraction error can be calculated from Eq. 5.6 by numerical methods, using the approximation for $J(u, v)$ given by Eq. 5.7 and the exact expression for $I(v, w, w')$ given by Steel et al. (1972). Some typical results (Boivin, 1976) are shown in Fig. 5.5; it is apparent that for

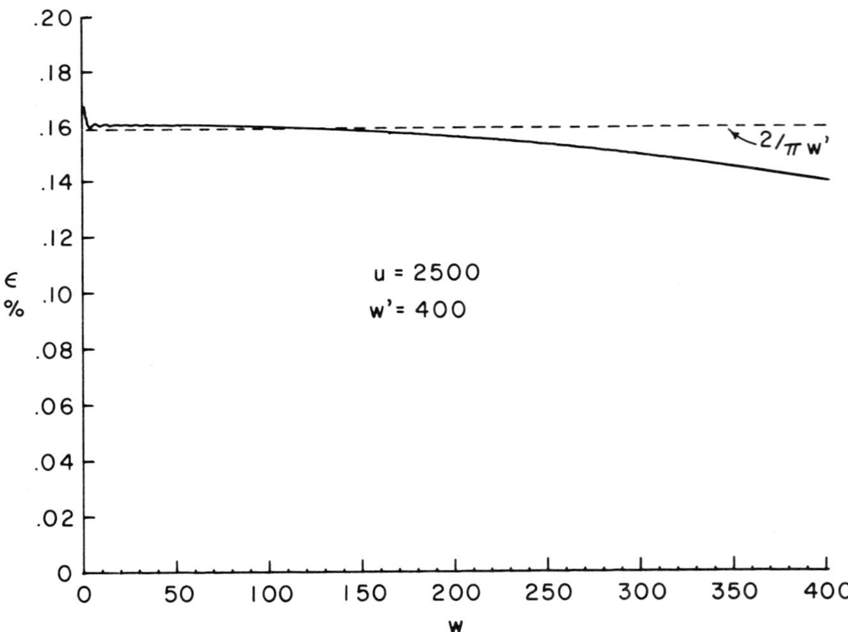

Fig. 5.5 The diffraction error ε as a function of w (proportional to source radius) for an extended source of monochromatic radiation, calculated by numerical integration. [From Boivin (1976), by permission of *Applied Optics*]

$w < w'$, ε is practically independent of the source size; even for $w \approx w'$, ε differs from the point source value by no more than 10 to 15%. Furthermore, the small oscillations in ε for small w would be removed (Boivin, 1976) in practice, due to the finite bandwidth of most sources (except lasers). The same behavior is observed for other values of u and w'. Since diffraction errors are usually small, it is sufficiently accurate to assume that for $w < w'$, ε is independent of w and given by

$$\varepsilon = \frac{2}{\pi w'} = \frac{\lambda b}{\pi^2 R r} \qquad \text{for} \quad w < w'. \tag{5.10}$$

For $w \geq w'$, the required diffraction errors must be derived by reciprocity (Steel et al., 1972). This means that the roles of the source and the detector can be interchanged, and the diffraction errors computed for that case will be the same as those we wish to calculate. As a result, we have

$$\varepsilon = \frac{2}{\pi w} = \frac{\lambda a}{\pi^2 R \rho} \qquad \text{for} \quad w \geq w'. \tag{5.11}$$

For $w \geq w'$, then, the diffraction error is not independent of the source size, but varies linearly with $1/\rho$.

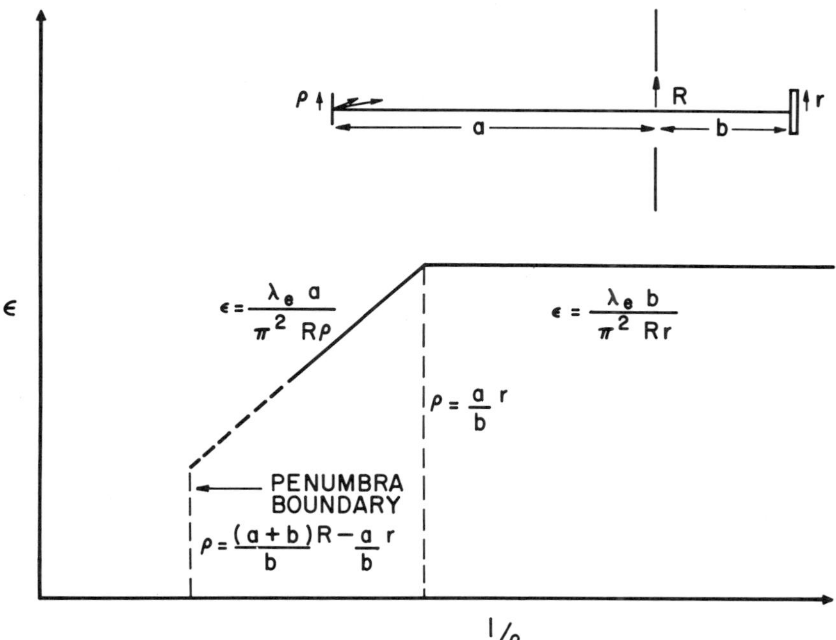

Fig. 5.6 Graphic summary of the diffraction error ε in the case of an extended source of complex radiation. [From Boivin (1976), by permission of *Applied Optics*]

Equations 5.10 and 5.11 will be valid for $u \gg (w + w')$, that is, for the central region. These expressions can be used in the case of complex radiation by substituting the effective wavelength $\lambda_e(q = 1)$ for λ.

The preceding results are summarized in graphic form in Fig. 5.6. The lower portion of the sloping line is shown dashed to reflect the fact that as one approaches the boundary of the penumbra, the condition $u \ll w + w'$ will no longer hold and Eq. 5.11 will no longer be applicable.

5.2.3 Reduction of Diffraction Errors by Means of Toothed Apertures

In some cases, we may wish to eliminate diffraction errors rather than calculate corrections for them. This may be particularly useful when a series of apertures is involved. It has been shown (Boivin, 1978) that toothed apertures can be used to reduce diffraction errors by at least an order of magnitude compared with smooth circular apertures of the same diameter. Figure 5.7 shows a schematic diagram of a toothed aperture, where R is the nominal radius of the aperture, Δ the tooth depth, 2θ the apex angle of the teeth, and 2ϕ the angular separation of the teeth. Radiation diffracted by the edge of the aperture will propagate in a direction perpendicular to the edges

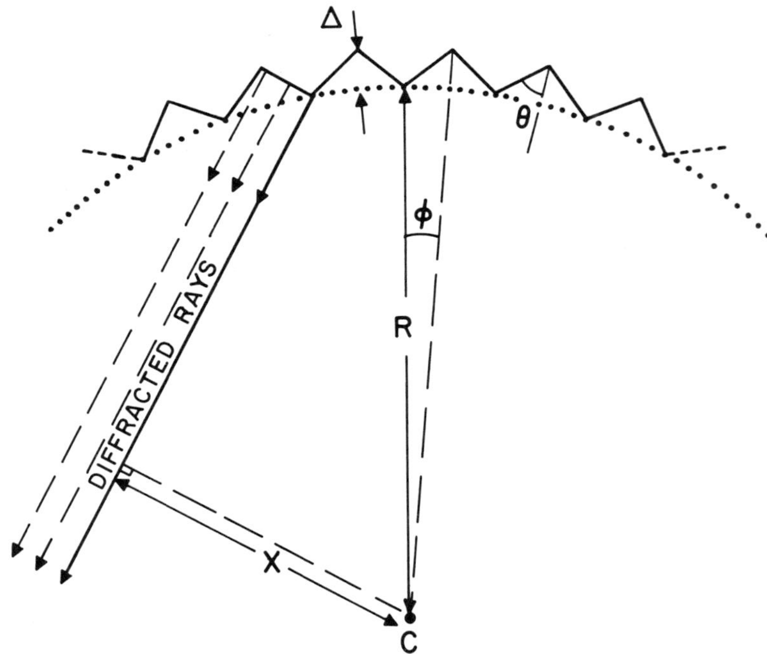

Fig. 5.7 Configuration of a toothed aperture, with definition of symbols. [From Boivin (1978), by permission of *Applied Optics*]

of the teeth. Consequently, in an observation plane at a distance b from the aperture, there will be a circular region in which diffracted flux is excluded: the diameter of this diffraction-free region is defined by the inner envelope of the diffracted rays, and is given by

$$D = 2\left(\frac{a+b}{a}\right)X,$$
$$X = R\cos(\theta + \phi), \tag{5.12}$$
$$\tan\theta = R\sin\phi/[R(1-\cos\phi) + \Delta],$$

where a is the aperture-source distance.

In practice, it is found that if the toothed aperture is designed such that the diameter of the diffraction-free region (as just calculated) is somewhat bigger than the diameter of the radiometer aperture used, the diffraction error is reduced by a factor of 10 or better, compared with a smooth aperture of the same diameter. In order to ensure good results (Boivin, 1978), the tooth depth should not be less than the value given in Eq. (5.13):

$$\Delta_{\min} = \frac{2\lambda ab}{R(a+b)}. \tag{5.13}$$

For irregular apertures (i.e., circular apertures with rough edges), Eq. 5.13 gives an estimate of the degree of edge "roughness" required to result in reduced diffraction errors; that is, if the edge roughness is equal to or greater than this value, the resulting diffraction error is expected to be much smaller than that calculated in the preceding sections for a perfectly smooth aperture.

As an example, Fig. 5.8 shows photographs of the diffraction patterns corresponding to two apertures having a diameter of 7 mm and having either a smooth circular edge (left) or a toothed edge (right) for a point source of monochromatic light. The central region of the diffraction pattern for the toothed aperture has only a small amount of diffracted flux compared with that of the smooth aperture, as can be judged from the contrast in the fringes. The residual diffracted flux in the central region of the diffracted pattern of the toothed aperture is caused mainly by edge imperfections in the aperture, which had been made by a photoetching process (Boivin, 1978).

The use of toothed apertures is specially relevant if a series of apertures is involved. Such situations are frequently encountered in practice. Consider the case of a series of apertures of equal diameters. If the apertures are not too close to one another, and if the detector aperture is not greater than the diffracting aperture diameter, the net diffraction error will be the sum of the diffraction errors that each aperture, if it were alone, would cause. This is demonstrated in Fig. 5.9, where the results of some experimental measurements are shown. Up to four apertures were used, and the diffraction errors

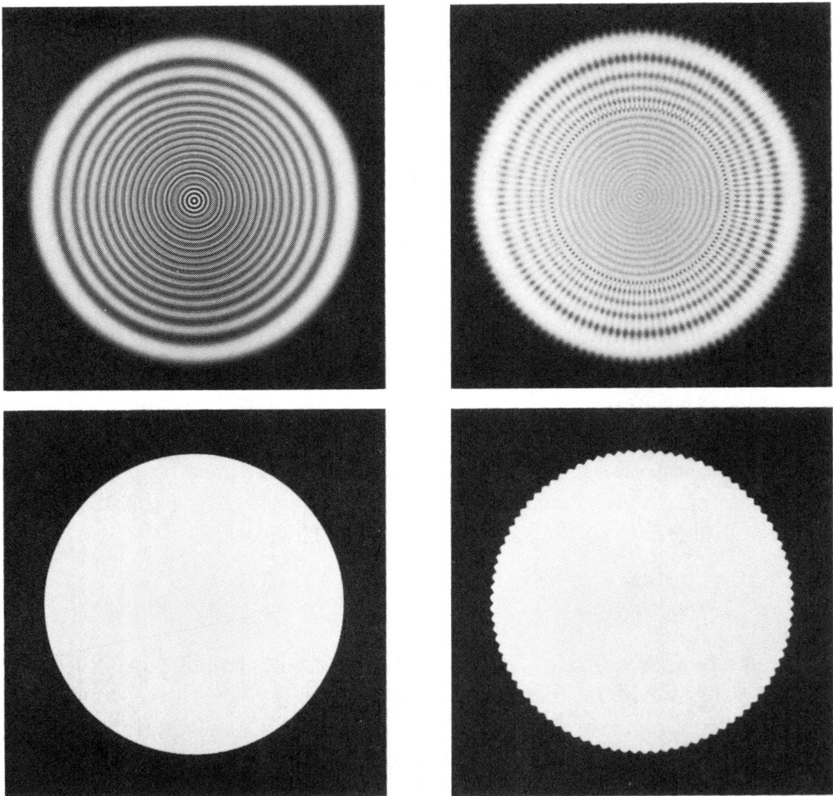

Fig. 5.8 Photographs of the diffraction patterns corresponding to 7-mm diameter circular apertures having either a smooth edge (left) or a toothed edge (right) for a point source of monochromatic radiation.

were measured in the sequence a, $a + b$, $a + b + c$, and $a + b + c + d$. In the first experiment, smooth circular 7-mm diameter apertures were used. In the second experiment, toothed 7-mm diameter apertures were used; Fig. 5.9b shows the configuration of these apertures, together with ray tracings indicating the theoretical diffraction-free region. The dotted circle shows the relative size of the detector aperture, corresponding to a toothed aperture located at d. In Fig. 5.9a, the solid curve gives the theoretical prediction for smooth apertures; the squares give the measured diffraction errors for the smooth apertures, whereas the circles give the measured diffraction errors for the toothed apertures. It is evident that the use of toothed apertures can result in greatly reduced overall diffraction errors when a series of apertures is used.

Square apertures are special cases of toothed apertures of the type considered here, for which $\theta = \phi = 45°$ and $\Delta = (\sqrt{2} - 1)R$ (where $L = 2R$

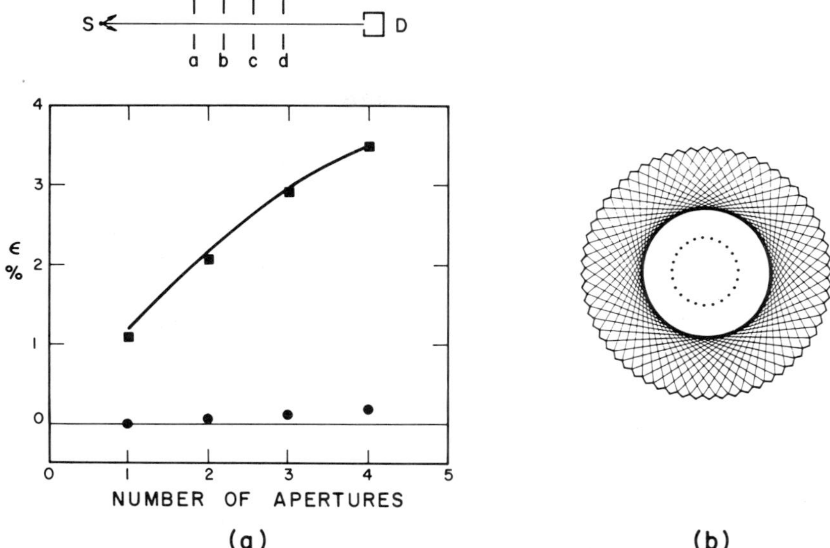

Fig. 5.9 (a) Comparison of diffraction errors by a series of apertures having either smooth edges (squares) or toothed edges (circles). The solid line is the theoretical prediction for smooth apertures. [From Boivin (1978), by permission of Applied Optics] (b) Configuration of the toothed apertures used in (a), also showing the diffracted rays and the relative diameter of the radiometer aperture (dotted circle) when the toothed aperture is at position *d* in (a).

is the side of the square, and R is the radius of the inscribed circle). A priori, one would expect a square aperture to cause smaller diffraction errors than a circular aperture. This is indeed what one observes in practice. However, the diffraction error caused by a square aperture is still much greater than that obtained with a properly designed toothed aperture.

5.3 ATMOSPHERIC ABSORPTION

Radiometric measurements that are carried out in air and that include a certain portion of the infrared spectrum must take into account the absorption of radiation by the water vapor and CO_2 components of air. Water vapor has seven absorption bands between 0.94 and 6.3 μm, whereas carbon dioxide has eight absorptions bands between 1.4 and 15 μm. If we denote by $A(\lambda)$ the spectral absorption of either H_2O or CO_2 in any absorption band, then the relative decrease in radiometer response due to that band will be given by

$$\alpha = \frac{\int_{\lambda_1}^{\lambda_2} A(\lambda)E(\lambda)S(\lambda)d\lambda}{\int_{\lambda_0}^{\lambda_F} E(\lambda)S(\lambda)d\lambda}, \tag{5.14}$$

where (λ_1, λ_2) denotes the wavelength limits of the absorption band, (λ_0, λ_F) denotes the effective bandwidth of the source–detector combination, and $E(\lambda)$ and $S(\lambda)$ are the spectral radiance of the source and spectral responsivity of the detector, respectively; the transmittance of any filter used can be included with $E(\lambda)$. Calculating an absorption-band correction from Eq. 5.14 would be a very complicated task, due to the complexity of the structures of the absorption bands; furthermore, for the small absorber concentrations usually encountered in radiometry, the available data (measured or calculated) is incomplete and may not be very accurate. Thus, if the atmospheric absorption has to be accurately known (say, better than 10%), it should be measured experimentally. Or else the need for a correction can be obviated altogether by enclosing the radiometric system in a dry nonabsorbing gas, such as dry nitrogen. The latter approach has usually been adopted when atmospheric absorption effects were expected to be large. However, in other cases, atmospheric absorption effects need to be known to only modest accuracy; in such cases, it is possible to calculate an absorption correction as follows. In Eq. 5.14, we assume that $S(\lambda)$ is approximately constant over the spectral range $(\lambda_0 \lambda_F)$; this is usually true for absolute radiometers. We also make the approximation $\int_{\lambda_1}^{\lambda_2} A(\lambda)E(\lambda)d\lambda \simeq \bar{E}\int_{\lambda_1}^{\lambda_2} A(\lambda)d\lambda$, where $\bar{E} = \int_{\lambda_1}^{\lambda_2} E(\lambda)d\lambda/(\lambda_2 - \lambda_1)$. This approximation will be valid if $E(\lambda)$ does not vary too rapidly inside the interval (λ_1, λ_2). Thus, we have

$$\alpha \simeq \frac{\bar{A}\int_{\lambda_1}^{\lambda_2} E(\lambda)d\lambda}{\int_{\lambda_0}^{\lambda_F} E(\lambda)d\lambda}, \tag{5.15}$$

where

$$\bar{A} = \frac{\int_{\lambda_1}^{\lambda_2} A(\lambda)d\lambda}{(\lambda_2 - \lambda_1)}.$$

$\bar{A}$ is the average fractional absorption; i.e., it gives the fraction of all radiation within the wavelength range (λ_1, λ_2) that will be absorbed by the particular CO_2 or H_2O band. Now, the total band absorptions ($\int A\,d\lambda$) for the various CO_2 and H_2O bands have been measured experimentally by several workers. Thus, from this data, and knowing the spectral distribution of the source used, one can determine approximately the total absorption by CO_2 and H_2O vapor for a particular case. The accuracy of these calculated absorptances should be between 10 and 20%.

5.3.1 Water Vapor

The absorption bands for water vapor are at 0.94, 1.1, 1.38, 1.87, 2.7, 3.2, and 6.3 μm. The total band absorptions $\int A\,d\lambda$ as a function of absorber concentration have been measured experimentally by several workers. We shall use here the results of measurements by Howard et al. (1956b) and Burch et al. (1963).

We wish to estimate the total absorption by the water-vapor component of air (at normal atmospheric pressure—760 mm Hg—and 20°C), using the total band absorptions and Eq. 5.15. Since such an approach is only approximate, we shall retain only the strongest absorption bands of water vapor, namely, the bands at 1.38, 1.87, 2.7, and 6.3 μm.

Howard et al. (1956b) and Burch et al. (1963) have measured the total band absorptions as a function of the absorber concentration w, expressed in precipitable centimeters of water. The relationship between w and the partial pressure p of water vapor is given by

$$w = pl \frac{M}{\mathrm{R}T\rho}, \tag{5.16}$$

where p is the partial pressure of water vapor in mm Hg, l is the absorbing path length in cm, M is the molecular weight of water in grams, R is the gas constant, T is the absolute temperature in Kelvin, and ρ is the density of water. Thus, at 20°C, $w = 9.85 \times 10^{-7}\, pl$.

Howard et al. (1956b) found that for total band absorptions of less than about 20%, the following empirical relationship was valid:

$$\bar{A} \propto (w)^{1/2} \text{ (weak band)} \tag{5.17}$$

at constant total pressure (atmospheric here). For stronger absorptions, they found the relationship

$$\bar{A} = \mathrm{K}_1 \log w + \mathrm{K}_2 \text{ (strong band)}, \tag{5.18}$$

where K_1 and K_2 are constants (the total pressure being constant). However, Howard et al. (1956b) made their investigations for values of w not less than 0.01, except for the 2.7-μm band, where w was as small as 0.002. Burch et al. (1963) have used values of w as small as 0.002 for all the bands they studied, namely, the 1.87, 2.7, and 6.3-μm bands. However, they have not obtained empirical formulas fitting their experimental data.

The smallest value of w considered in the previously mentioned studies ($w \approx 0.002$) corresponds to a pl product of about 2000. Values of pl smaller than this are frequently, if not usually, encountered in practice. Blevin and Brown (1969) have made measurements of total absorption by water vapor in the range $pl = 0$ to 5000. They found that for pl less than approximately 1500 torr-cm, the total absorption varies linearly with pl. This behavior is to be expected from Beer's law, for very small absorber concentrations. Thus, Eqs. (5.17) and (5.18) cannot be used to extrapolate Howard et al.'s (1956b) or Burch et al.'s (1963) data to small absorber concentrations. Nor can the measurements of Blevin and Brown (1969) be used in the general case because these were *total*-absorption (i.e., all-bands) measurements made for one type of radiation (2854 K tungsten) and one type of filter (ordinary plate glass).

The total-band absorption $\bar{A}$ for small absorber concentrations ($w < 0.002$) can be obtained by graphic extrapolation as follows. From the work of Blevin and Brown (1969), it is assumed that for $w < 0.0015$, $\bar{A}$ will vary linearly to zero for all four bands considered here. For a given band, the values of $\bar{A}$ from Howard et al. (1956b) or Burch et al. (1963) for $0.002 < w < 0.007$ were plotted against w, and a curve fitted through the points, such that the curve conforms to a linear decrease to zero for $w < 0.0015$.

Figure 5.10 shows the curves thus obtained for the four bands and $0 < w < 0.007$. The data for the 6.3 and 1.87 μm bands were obtained from Burch et al. (1963), whereas the data for the 2.7 and 1.38 μm lines were

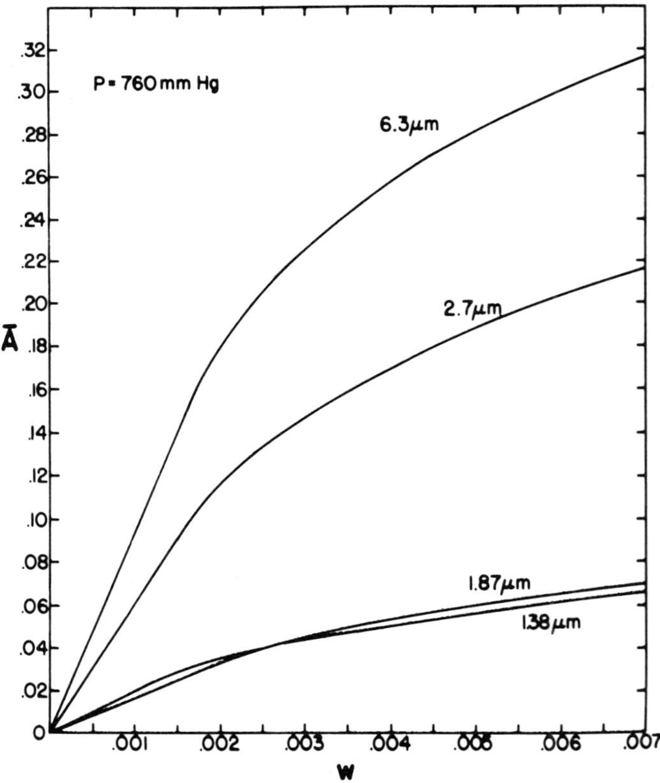

Fig. 5.10 Absorption by water vapor. Average total band absorption $\bar{A}$ as a function of absorber concentration (in precipitable cm of water) for the strongest bands.

obtained from Howard et al. (1956b). The band limits used in Fig. 5.10 (and to be used to evaluate band absorptions) are from Howard et al. (1956b):

$$\begin{aligned} 6.3\ \mu\text{m}&: 4.88\text{–}8.70\ \mu\text{m} \\ 2.7\ \mu\text{m}&: 2.27\text{–}2.99\ \mu\text{m} \\ 1.87\ \mu\text{m}&: 1.70\text{–}2.08\ \mu\text{m} \\ 1.38\ \mu\text{m}&: 1.25\text{–}1.54\ \mu\text{m} \end{aligned} \tag{5.19}$$

Numerical Example

Consider the case studied by Blevin and Brown (1969), i.e., a tungsten lamp operating at 2854 K filtered by a 3-mm-thick plate glass. We shall assume (1) blackbody radiation at 2800 K, and (2) that the plate gives uniform transmission from 0.35 to 3 μm and zero transmission outside these limits. Setting $p = 8$ mm Hg and $l = 100$ cm, we have $w = 0.008$, and (from Fig. 5.10) the values of $\bar{A}$ are 0.049, 0.013, and 0.016 for the 2.7-, 1.87-, and 1.38-μm bands, respectively. These values are then weighted for the relative amounts of blackbody radiation inside the wavelength limits of the bands (given by Eq. 5.19) and then summed to give a total absorptance of 1.05%. The absorptance measured by Blevin and Brown (1969) was about 0.84%. This is a reasonable agreement, considering that the assumptions just made would tend to overestimate the relative amounts of radiation included within the wavelength limits of the absorption bands and, hence, result in a calculated absorptance that would be higher than the measured one.

5.3.2 Carbon Dioxide

The absorption bands of carbon dioxide in the near infrared are at 1.4, 1.6, 2.0, 2.7, 4.3, 4.8, 5.2, and 15 μm. As before, we shall retain only the strongest bands in order to calculate total absorptances. These are the bands at 2.7, 4.3, and 15 μm. All the other bands are at least 20 to 50 times weaker than these, except the 2.0-μm band, which is about 10 times weaker. The total absorption $\int A\, d\lambda$ of these bands as a function of absorber concentration w has been determined by several workers. The work of Burch et al. (1962) is particularly relevant here because it included the range of very small absorber concentrations usually encountered in radiometry. The absorber concentration is defined here as

$$w = \frac{pl}{760}\left(\frac{273}{273 + T}\right), \tag{5.20}$$

where p is the partial pressure of CO_2 in mm Hg, l is the absorbing path length in cm, and T is the temperature in °C. The unit for w is the atmos-cm.

Thus, for a one-meter path length in air at 20°C and normal atmospheric pressure, $w \approx 0.03$ atmos-cm. We shall now consider each absorption band separately.

5.3.2.1 The 2.7 μm Band

It is apparent from Burch et al.'s (1962) experimental results that in the range of w of interest, the variation of the total absorption $\int A\, d\lambda$ with w obeys neither the square-root law given by Howard et al. (1956a) nor the limiting linear law expected for very small w. However, the following empirical equation fits the data quite well in the range of w indicated:

$$\bar{A} = 0.156(w)^{0.8}, \qquad 0.01 \le w \le 1, \qquad \lambda = 2.63\text{–}2.87\ \mu\text{m}. \tag{5.21}$$

The preceding band limits are from Howard et al. (1956a). Equation 5.21 assumes an atmospheric pressure of 760 mm Hg.

5.3.2.2 The 4.3 μm Band

For this band, Burch et al. (1962) have derived the following empirical equation for the range of w indicated:

$$\bar{A} = 0.65(w)^{0.54}, \qquad 0.01 < w < 0.1, \qquad \lambda = 4.00\text{–}4.63\ \mu\text{m}, \tag{5.22}$$

where an atmospheric pressure of 760 mm Hg is assumed; the band limits are from Howard et al. (1956a). The empirical equation

$$\bar{A} = 0.33(w)^{0.25}, \qquad 0.1 < w < 1, \qquad \lambda = 4.00\text{–}4.63\ \mu\text{m}, \tag{5.23}$$

can be used to extend the range of w upwards to $w = 1$.

5.3.2.3 The 15 μm Band

This band is fairly complex. It was treated by Howard et al. (1956a) as only one band, whereas Burch et al. (1962) have investigated it as three separate contiguous bands. Only two of these (13.89–14.99 μm and 14.99–16.21 μm) contribute significantly to the total band absorption and, hence, only these two shall be considered here. Burch et al. (1962) have not given empirical formulas for these two bands. However, we find that the following equation gives, with satisfactory accuracy, the total absorption by these two bands:

$$\bar{A} = 0.935(w)^{0.6}, \qquad 0.01 < w < 1, \qquad \lambda = 13.89\text{–}16.21\ \mu\text{m}. \tag{5.24}$$

Equations 5.21 through 5.24 can be used to calculate the total absorption correction for atmospheric CO_2 for path lengths of about 0.3 to 30 meters. $\bar{A}$ gives the fraction of all radiation within the indicated band limits that will be absorbed. The absorption coefficients $\bar{A}$ for each band must be weighted according to the spectral distribution of the radiation measured (including the spectral transmission of any filter used). The overall accuracy of the absorption corrections so calculated should be in the range of 10 to 20%. This should be sufficient in most cases, as the CO_2 absorption corrections are usually quite small.

Numerical Example

Consider the same example as before, i.e., a tungsten source at 2854 K, which we shall approximate by a blackbody at 2800 K, used with a plate-glass filter cutting off all radiation outside the 0.35–3.0-μm range; thus, only the 2.7-μm absorption band need be considered. At atmospheric pressure and for a path length of 100 cm, $w \approx 0.031$, from which we calculate that the total absorption is 0.03%. This is quite small and shows that, in many cases, absorption by atmospheric CO_2 is negligible.

5.4 REFRACTION

It is well known that introducing a plane parallel plate of glass between a source and a detector shortens the effective source–detector distance. This effect is not important in relative radiometry, where two sources are compared, because the filter glass is usually included with both sources. However, in absolute radiometry, the effective distance between the radiometer aperture and the reference surface of the source must be accurately known. It is thus important to correct for the distance shortening effect of filters located between the source and the detector.

5.4.1 Normal Incidence

A parallel plate of thickness t and refractive index n placed perpendicularly to the axis joining the center of the source to the center of the radiometer aperture will effectively shorten the true separation d between the source and the radiometer as shown in Fig. 5.11. The amount by which the source–radiometer distance has been shortened is given by Smith (1966) as

$$\Delta = \left(\frac{n-1}{n}\right)t. \tag{5.25}$$

If E_{true} denotes the irradiance that would be measured with a filter of negligible thickness, then

$$E_{\text{true}} = \left(\frac{d-\Delta}{d}\right)^2 E, \tag{5.26}$$

where E is the irradiance that will be measured with a filter of thickness t. Equation 5.26 assumes the validity of the inverse square law relating the

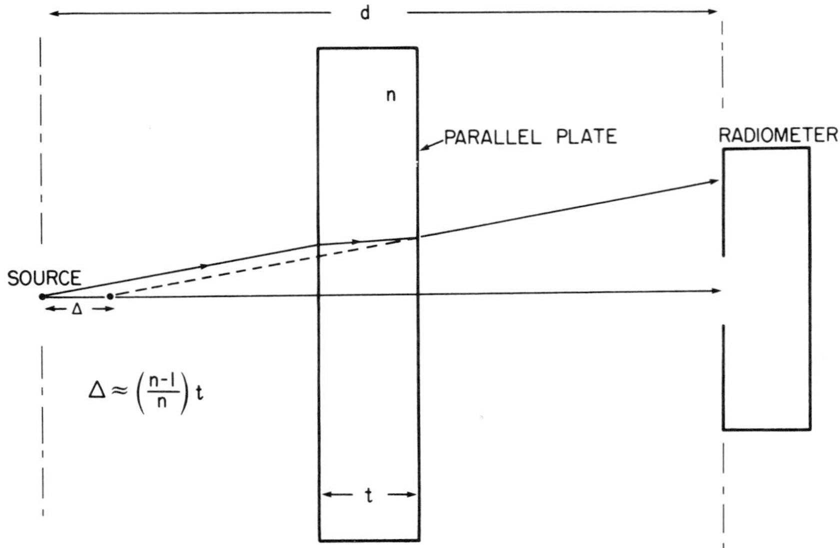

Fig. 5.11 A parallel plate of refractive index n and thickness t will effectively shorten by an amount Δ the true separation d between a source and a radiometer.

variation of irradiance with distance from the source. This will not always be true, especially when dealing with relatively large sources and relatively short distances d. In such cases, it is desirable to minimize the thickness of filters used. Furthermore, d represents the true separation between the radiometer aperture and the radiation *source*. However, in many radiometric measurements, distances are set relative to a reference surface on the source that may be separated from the actual source of radiation (e.g., filament) by several centimeters. This should be taken into account in applying Eq. 5.26.

Numerical Example

An irradiance standard is set up with a distance of exactly 50 cm between the measurement plane and a reference surface located 3 cm in front (on the detector side) of the lamp filament. A filter having a thickness of 12 mm and a refractive index of 1.5 is interposed between the source and the radiometer. The value of Δ given by Eq. 5.25 is 0.4 cm so that

$$\frac{E_{\text{true}}}{E_{\text{measured}}} = \left(\frac{53 - 0.4}{53}\right)^2 = 0.985, \qquad \text{i.e., a } 1.5\% \text{ correction.}$$

E_{true} gives the irradiance that would be measured using a filter having the same transmittance but a negligible thickness.

5.4.2. Tilted Filters

Consider now the case of plane parallel filters of thickness t, of refractive index n, and tilted with respect to the axis by an angle i. The tilting of filters is frequently used in practice to avoid interreflections between the filter and other elements. If the tilt angle is not greater than about 20°, no polarization problems should be encountered.

The effect of a tilted plate on the effective distance between a source and a detector is somewhat more complicated than in the previous case, due to the astigmatism introduced by the plate. For sagittal rays (i.e., rays in a plane parallel to the tilt axis), the apparent position of the source (as seen from the detector side of the plate) does not change as the plate is tilted. However, for meridional rays (i.e., rays in a plane perpendicular to the tilt axis), the apparent position of the source moves toward the detector as the plate is tilted.

It can be shown (Smith, 1966) that

$$\Delta_s = \left(\frac{n-1}{n}\right)t, \qquad \text{for sagittal rays and}$$

$$\Delta_m = \left(\frac{n-1}{n}\right)t\left(1 + \frac{(n+1)\tan^2 i}{n^2}\right), \qquad \text{for meridional rays.}$$

The apparent position of the source will also be shifted transverse to the axis when the plate is tilted. However, this shift will have a negligible effect on the measured irradiance. The axial displacement to be used in evaluating the correction given by Eq. 5.26 can be taken to be the average of the sagittal and meridional displacements. That is,

$$\Delta = \frac{\Delta_s + \Delta_m}{2} = \left(\frac{n-1}{n}\right)t\left(1 + \frac{(n+1)\tan^2 i}{2n^2}\right). \tag{5.27}$$

Thus, even for tilt angles of up to 20°, Δ will vary by less than 10%.

REFERENCES

Blevin, W. R. (1970). Diffraction losses in radiometry and photometry. *Metrologia* **6**, 39.

Blevin, W. R., and Brown, W. J. (1969). Corrections in radiometry for absorption by atmospheric water vapour, *Metrologia* **5**, 28.

Boivin, L. P. (1975a). Diffraction losses associated with tungsten lamps in absolute radiometry. *Appl. Opt.* **14**, 197.

Boivin, L. P. (1975b). Diffraction corrections in radiometry: comparison of two different methods. *Appl. Opt.* **14**, 2002.

Boivin, L. P. (1976). Diffraction corrections in the radiometry of extended sources. *Appl. Opt.* **15**, 1204.

Boivin, L. P. (1977). Radiometric errors caused by diffraction from circular apertures: edge effects. *Appl. Opt.* **16**, 377.

Boivin, L. P. (1978). Reduction of diffraction errors in radiometry by means of toothed apertures. *Appl. Opt.* **17**, 3323.

Boivin, L. P. (1980). Calibration of incandescent lamps for spectral irradiance by means of absolute radiometers. *Appl. Opt.* **19**, 2771.

Burch, D. E., Gryvnak, D. A., and Williams, D. (1962). Total absorptance of carbon dioxide in the infrared. *Appl. Opt.* **1**, 759.

Burch, D. E., France, W. L., and Williams, D. (1963). Total absorptance of water vapour in the near infrared. *Appl. Opt.* **2**, 585.

Fussell, W. B. (1974). *Tables of diffraction losses.* NBS Technical Note 594-8. U.S. Government Printing Office, Washington, D.C.

Gillham, E. J. (1962). Recent investigations in absolute radiometry. *Proc. R. Soc. London, Ser. A* **269**, 249.

Howard, J. N., Burch, D. E., and Williams, D. (1956a). Infrared transmission of synthetic atmospheres. II. Absorption by carbon dioxide. *J. Opt. Soc. Am.* **46**, 237.

Howard, J. N., Burch, D. E., and Williams, D. (1956b). Infrared transmission of synthetic atmospheres. III. Absorption by water vapor. *J. Opt. Soc. Am.* **46**, 242.

Smith, W. J. (1966). *Modern Optical Engineering*, Section 4.8. McGraw-Hill, New York.

Steel, W. H., De, M., and Bell, J. A. (1972). Diffraction corrections in radiometry. *J. Opt. Soc. Am.* **62**, 1099.

6

Instrumental Corrections in Absolute Radiometry

F. HENGSTBERGER

Council for Scientific and Industrial Research
Pretoria, South Africa

6.1 INTRODUCTION

In order to recognize the difference between the environmental and instrumental corrections for absolute radiometers, it is useful to consider the main components involved in a typical measurement with such instruments. These are depicted schematically in Fig. 6.1.

The environment in which a measurement is performed is a very difficult concept to define. It is generally accepted as including such parameters as the temperature, pressure, composition, thermal conductivity, and humidity level of the medium (e.g., air) in which the measuring system is placed. The physical dimensions of the measuring room or measuring station, the spectral reflectance of its walls, the strength and distribution of electric and magnetic fields in it, ambient light and radiation levels, vibrations, and draughts are further important aspects associated with the environment. These parameters can influence every part of the actual measuring system, i.e., the output of the radiation source, the properties of the radiation path (e.g., the transmittance of an optical filter), and the output of the absolute radiometer.

In absolute radiometry, the group of so-called environmental corrections (see Chapter 5) includes corrections for all the factors that change the radiance in a beam of radiant energy between the radiation source and the radiometer head, without any change taking place in the radiance of the source itself. Corrections for the effect of the medium as well as for optical components in the radiation path (e.g., baffles or optical filters), which are thus part of the measurement environment in a wider sense, also fall into this

ISBN 0-12-340810-5

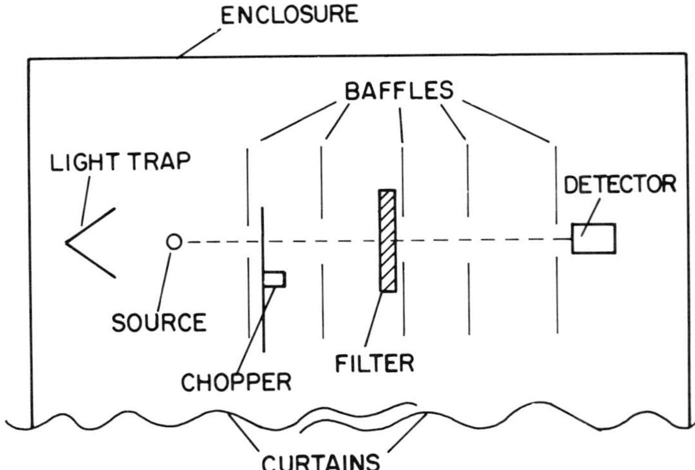

Fig. 6.1 Components of a typical measurement system.

category. The group of "instrumental corrections," on the other hand, represents corrections for parasitic power sources and sinks inside the radiometer head and for all factors that change the beam radiance inside the radiometer head or that cause equal amounts of radiant power passing through the radiometer aperture and of electrical power dissipated in the heating element of an absolute radiometer to produce different detector outputs.

6.2 THE RESPONSIVITY OF AN ABSOLUTE RADIOMETER

According to its definition (see Section 1.2.3), the responsivity of an optical radiation detector is the ratio of the detector output (quantity Y) to the detector input (quantity X). Both the detector input and the detector output can take various forms and the concept of responsivity therefore needs a number of further specifications before it can be employed unambiguously. In terms of the detector input, it has to be stated whether the responsivity is to be expressed for incident radiant power (power or flux responsivity), irradiance (irradiance responsivity), incident photons (quantum efficiency), or some other radiometric quantity. The detector output, on the other hand, can be a current, a voltage, a resistance change, the number of photoelectrons, and so on. Therefore, responsivity can be expressed in a wide variety of units such as $[V/W]$ and $[A/(W/m^2)]$.

Since the radiance in a beam of optical radiation intercepted by a detector is a function of wavelength, direction, polarization, spatial position, and time (see Section 1.2), it is clear that the reflectance of a thermal-radiation detector in the most general case may also depend on all these parameters. In addition, the detector's responsivity to "power converted to heat" has been found to depend on the spatial position at which the conversion takes place, on the power level to which the whole detector is exposed, on environmental parameters such as temperature, on electric and magnetic fields, as well as on detector-specific operating parameters (e.g., bias voltage, bridge operating voltage, chopping frequency). The general subject of detector responsivity, the specific terminology in use, the related concept of detector linearity, and methods for measuring responsivity have been treated extensively by Budde (1983) and will not be addressed in detail in this context.

In the following discussion, it will be assumed that the environmental parameters and the detector operating parameters are being held constant and that the responsivity referred to is applicable to a particular set of these parameters, which is clearly defined and documented. It will further be assumed that the source radiance, the detector responsivity, and the detector reflectance do not vary with time and that the detector reflectance is independent of the polarization and direction of the incident radiation. Within certain limitations, these approximations hold quite well for the detector elements of most absolute radiometers in most measurement situations. A polar coordinate system with its origin at the center of the front surface of the detector element (Fig. 4.1) will be used in further analysis.

In these coordinates and under the stated conditions, the power responsivity of an absolute radiometer for an area element located at the position (r, ϕ) and of area $r\,dr\,d\phi$ being supplied with a power per unit area of $\mathrm{E}(r, \phi)$ is given by

$$S(r, \phi, P) = \frac{1}{E(r, \phi)} \frac{\partial^2 Y(r, \phi, P)}{r\,dr\,d\phi} \tag{6.1}$$

The denominator of Eq. (6.1) represents the total power dP supplied to the area element in the form of heat, while the heat responsivity of the absolute radiometer is assumed to depend only on the coordinates (r, ϕ) and the applied power level. Assuming additivity of the contributions of the various area elements, the detector output Y for the total power P supplied to it over its whole front surface (radius R) follows as

$$Y(E) = \int_0^{2\pi} \int_0^R S(r, \phi, P) E(r, \phi) r\,dr\,d\phi. \tag{6.2}$$

Introducing a mean responsivity $\bar{S}(E)$ defined by

$$\bar{S}(E) = \frac{\int_0^{2\pi} \int_0^R S(r, \phi, P) E(r, \phi) r\,dr\,d\phi}{\int_0^{2\pi} \int_0^R E(r, \phi) r\,dr\,d\phi} \tag{6.3}$$

the radiometer output $Y(E)$ can also be expressed as

$$Y(E) = \bar{S}(E)P, \tag{6.4}$$

with

$$P = \int_0^{2\pi} \int_0^R E(r, \phi) r \, dr \, d\phi. \tag{6.5}$$

6.3 A COMPREHENSIVE THEORY OF THE INSTRUMENTAL CORRECTIONS FOR ABSOLUTE RADIOMETERS

A list of the symbols used in this section is given together with their definitions in Table 6.1.

As Eq. (6.4) shows, $\bar{S}(E)$ is the mean responsivity of an absolute radiometer in terms of the total input power P applied to the detector element in the form

TABLE 6.1 Definition of Symbols

Symbol	Definition
$\mathrm{E}(r, \phi, \lambda)$	Power distribution per unit area across the detector element in the absence of parasitic power sources
$\mathrm{E}'(r, \phi, \lambda)$	Power distribution per unit area across the detector element in the presence of parasitic power sources
P	$P = \int_r \int_\phi \int_\lambda \mathrm{E}(r, \phi, \lambda) r \, dr \, d\phi \, d\lambda$
P'	$P' = \int_r \int_\phi \int_\lambda \mathrm{E}'(r, \phi, \lambda) r \, dr \, d\phi \, d\lambda$
k_e	Lead-heating correction
$\bar{\rho}_r$	Mean reflectance of a detector element for an irradiance distribution $\mathrm{E}(r, \phi, \lambda)$
$\rho(r, \phi, \lambda)$	Spectral reflectance of a detector element as a function of coordinates (r, ϕ)
k_{1r}	Case-heating correction
k_{2r}	Scattering correction
k_{3r}	Diffraction correction
$\mathrm{Y}(E')$	Detector signal due to applied power per unit area $\mathrm{E}'(r, \phi, \lambda)$
$\bar{\mathrm{S}}(E')$	$\bar{\mathrm{S}}(E') = \mathrm{Y}(E')/P'$
$P' \sum_i \varepsilon_i(P')$	Sum of "equivalent" heat losses from a detector element
$P' \sum_i N_i(P')$	Sum of "nonequivalent" heat losses from a detector element
ξ	Fraction of total applied power lost from a detector element by emission from the absorber
η	Fraction of total applied power lost from a detector element by conduction and convection from the absorber
C_{FB}	Feedback correction

of heat. If one distinguishes the quantities referring to electrical heating of the absolute radiometer by the subscript "e" and the ones referring to radiant heating by "r," and if one applies Eq. (6.4) separately to each form of heating, one finds

$$Y(P'_e) = P'_e \bar{S}_e(E'_e), \tag{6.6a}$$

$$Y(P'_r) = P'_r \bar{S}_r(E'_r). \tag{6.6b}$$

For electrical heating, the total power P'_e supplied to the detector element as heat consists of the electrical power P_e dissipated in the heating element and the part of the electrical power dissipated in the current leads ($k_e P_e$), which reaches the detector element by conduction, i.e.,

$$P'_e = (1 + k_e)P_e. \tag{6.7}$$

In a similar way, the total power P'_r converted to heat in the detector element during radiant heating is made up of the direct radiant power P_r incident on the detector element minus the reflected fraction $\rho_r P_r$, plus the power $k_{1r} P_r$ supplied to the detector element from the parts of the radiometer head exposed to the incident beam due to emission, convection, or conduction, plus the power $k_{2r} P_r$ scattered from parts of the radiometer head onto the detector element, plus the extra power $k_{3r} P_r$ reaching or not reaching the detector element due to diffraction at apertures in the radiometer head, i.e.,

$$P'_r = (1 - \bar{\rho}_r + k_{1r} + k_{2r} + k_{3r})P_r, \tag{6.8}$$

with

$$\bar{\rho}_r = \frac{\int_0^\infty \int_0^{2\pi} \int_0^R \rho_r(r, \phi, \lambda) E_r(r, \phi, \lambda) r\, dr\, d\phi\, d\lambda}{\int_0^\infty \int_0^{2\pi} \int_0^R E_r(r, \phi, \lambda) r\, dr\, d\phi\, d\lambda} \tag{6.9}$$

and

$$P_r = \int_0^\infty \int_0^{2\pi} \int_0^R E_r(r, \phi, \lambda) r\, dr\, d\phi\, d\lambda. \tag{6.10}$$

Here $\bar{\rho}_r$ represents the effective reflection factor of the detector element for the incident spectral irradiance distribution $E_r(r, \phi, \lambda)$. It should be noted that only the relative spectral irradiance distribution is required for the calculation of $\bar{\rho}_r$.

If a measurement is conducted with an absolute radiometer and the signal $Y(E'_r)$ observed in response to radiant heating is matched by an identical signal $Y(E'_e)$ produced by electrical heating, i.e., if $Y(E'_r) = Y(E'_e)$, then it follows from Eqs. (6.6), (6.7), and (6.8) that the unknown radiant power P_r can be expressed in terms of the measured electrical power P_e as

$$P_r = \frac{1 + k_e}{1 - \bar{\rho}_r + k_{1r} + k_{2r} + k_{3r}} \frac{\bar{S}_e(E'_e)}{\bar{S}_r(E'_r)} P_e. \tag{6.11}$$

Similarly, in the case of an absolute radiometer operated with a feedback-controlled electrical heating power, the signal $Y(E'_{e1})$ due to electrical heating alone is matched to the signal due to additional heating by a radiant power P'_r and an appropriately reduced amount of electrical power (P'_{e2}), yielding

$$(1 + k_e)P_{e1}\bar{S}_e(E'_{e1}) = (1 + k_e)P_{e2}\bar{S}_e(E'_{e2} + E'_r) + (1 - \bar{\rho}_r + k_{1r} + k_{2r} + k_{3r})P_r\bar{S}_r(E'_{e2} + E'_r). \tag{6.12}$$

Since, for absolute radiometers with small instrumental corrections, P'_{e1} is approximately equal to $P'_{e2} + P'_r$ and $S_e(E'_{e1})$ is therefore practically equal to $S_e(E'_{e2} + E'_r)$, the difference between the two quantities being a second-order correction, (6.12) can be rewritten with negligible error as

$$P_r = \frac{1 + k_e}{1 - \bar{\rho}_r + k_{1r} + k_{2r} + k_{3r}} \frac{\bar{S}_e(E'_{e1})}{\bar{S}_r(E'_{e1})}(P_{e1} - P_{e2}). \tag{6.13}$$

In both cases, therefore, the measured electrical power or power difference must be corrected by means of the correction terms appearing on the right-hand side of Eqs. (6.11) and (6.13) to yield the value of the unknown radiant power P_r. While the correction terms k_e (lead-heating correction), $\bar{\rho}_r$ (reflection correction), k_{1r} (case-heating correction), k_{2r} (scattering correction), and k_{3r} (diffraction correction) can either be determined by appropriate correction experiments (see Section 6.5) or by calculation, some further theoretical analysis is required before the ratio of the mean electrical and radiant responsivities $\bar{S}_e(E'_e)$ and $\bar{S}_r(E'_r)$ can be expressed in terms of experimentally measurable quantities. As has been shown by Hengstberger (1977c), this can be achieved by dividing the heat losses from the detector element of an absolute radiometer in thermal equilibrium into two groups. These are distinguished by either being the same (equivalent) for the same detector output under radiant and electrical heating, or being different (nonequivalent). If the sums of the equivalent heat losses are denoted by $P'_e \sum_i \varepsilon_{ei}$ and $P'_r \sum_i \varepsilon_{ri}$ for the two forms of heating and the sums of the corresponding nonequivalent heat losses as $P'_e \sum_i N_{ei}$ and $P'_r \sum_i N_{ri}$, the energy conservation principle yields

$$P'_e = P'_e\left(\sum_i \varepsilon_{ei} + \sum_i N_{ei}\right), \tag{6.14a}$$

$$P'_r = P'_r\left(\sum_i \varepsilon_{ri} + \sum_i N_{ri}\right). \tag{6.14b}$$

Equations (6.14) express the fact that under equilibrium conditions, the amount of power supplied to the detector element (left-hand side) equals the amount of power lost from it (right-hand side). They also imply that

$$\sum_i \varepsilon_{ei} + \sum_i N_{ei} = 1, \tag{6.15a}$$

and

$$\sum_i \varepsilon_{ri} + \sum_i N_{ri} = 1, \tag{6.15b}$$

which merely expresses the fact that the ε and N terms are fractional amounts adding up to unity.

If it is accepted that there are heat losses, which are the same for equal detector outputs under radiant and electrical heating, then their equality can be expressed mathematically as

$$P'_e \sum_i \varepsilon_{ei} = P'_r \sum_i \varepsilon_{ri} \tag{6.16a}$$

or, alternatively, with the aid of Eqs. (6.15) as

$$P'_e\left(1 - \sum_i N_{ei}\right) = P'_r\left(1 - \sum_i N_{ri}\right). \tag{6.16b}$$

If one rewrites Eq. (6.11) in the form

$$P'_e \bar{S}_e(E'_e) = P'_r \bar{S}_r(E'_r) \tag{6.17}$$

and compares this equation with (6.16a) and (6.16b), one finds the identities

$$P'_r/P'_e = \bar{S}_e(E'_e)/\bar{S}_r(E'_r) = \sum_i \varepsilon_{ei} \Big/ \sum_i \varepsilon_{ri}, \tag{6.18a}$$

$$P'_r/P'_e = \bar{S}_e(E'_e)/\bar{S}_r(E'_r) = \left(1 - \sum_i N_{ei}\right)\Big/\left(1 - \sum_i N_{ri}\right). \tag{6.18b}$$

These equations express the ratio of the mean responsivities of the absolute radiometer for electrical and radiant heating in terms of ratios of equivalent or nonequivalent fractional heat losses from the detector element, which are, at least in principle, measurable.

As shown in Chapter 4, the heat losses from the surface (usually the rear surface) of the detector element to which the sensor is attached can be identified as equivalent under both forms of heating. These losses are generally by conduction, convection, and radiation and no practical method for measuring them has so far been devised or suggested. The heat losses through the wires attached to the electrical heating element are likewise shown to be equivalent (see Chapter 4) and their measurement would seem to present a further unresolved problem. Therefore, the determination of the mean responsivity ratio for radiant and electrical heating would seem to be a very difficult task to perform via measurements of the equivalent heat-loss ratio (6.18a). However, the identification of equivalent heat losses is an indispensable precondition for the validity and applicability of the derived relationship between the responsivity and the heat-loss ratios.

Since Eq. (6.18a) does not seem to be a very attractive proposition for determining the needed responsivity ratio, it is obvious that Eq. (6.18b) is the only available alternative. With the equivalent heat losses already identified, the only remaining heat losses from the detector element are by conduction, convection, and emission from the front surface covered by the absorber layer. If one denotes the fractions of the applied power lost from that surface by emission by ξ_r and ξ_e for both forms of heating, respectively, and those lost by conduction and convection in the same way as η_r and η_e, one can write

$$\frac{\bar{S}_e(E'_e)}{\bar{S}_r(E'_r)} = \frac{1 - \sum_i N_{ei}}{1 - \sum_i N_{ri}} = \frac{1 - \xi_e - \eta_e}{1 - \xi_r - \eta_r}. \tag{6.19}$$

As will be shown in Section 6.5, the nonequivalent losses are indeed somewhat easier to assess experimentally and it is therefore possible to finally express the relationship between the unknown radiant power and the measured electrical power as

$$P_r = \frac{1 + k_e}{1 - \bar{\rho}_r + k_{1r} + k_{2r} + k_{3r}} \frac{1 - \xi_e - \eta_e}{1 - \xi_r - \eta_r} P_e. \tag{6.20}$$

There is still one proviso, however, in that the corrections in (6.20) are only valid for the particular irradiance distribution $E(r, \phi, \lambda)$ for which they were determined. In the formal definitions of these corrections (see Section 1.3.3), this standard irradiance distribution is equivalent to CIE Illuminant A (CIE, 1986) and produces a spatially uniform irradiation over the area of the radiometer aperture. This particular source was chosen because of its ready availability for performing the required correction experiments and in order to achieve some standardization for comparing the nonequivalence corrections of different absolute radiometers. Apart from these considerations, it could theoretically be any source that happens to be used for the measurement of the individual terms in the nonequivalence correction. In order to find the nonequivalence correction for another type of source (e.g., a laser source), one obvious solution would be to remeasure the individual correction terms for that source. However, since correction experiments of this type are difficult and time-consuming to perform, this approach is not very attractive. Hengstberger (1979) has shown that this redetermination can be avoided and that the nonequivalence correction for another type of irradiance distribution E'_{r1} can be calculated from the value for the reference source E'_{r0} and other measurable source and detector parameters. For this purpose, Eq. (6.20) is rewritten as follows

$$\begin{aligned} P_{r1} &= \frac{1 + k_e}{1 - \bar{\rho}_{r1} + k_{1r1} + k_{2r1} + k_{3r1}} \frac{\bar{S}_e(E'_e)}{\bar{S}_r(E'_{r1})} P_e \\ &= \left[\frac{1 + k_e}{1 - \bar{\rho}_{r0} + k_{1r0} + k_{2r0} + k_{3r0}} \frac{\bar{S}_e(E'_e)}{\bar{S}_r(E'_{r0})} \right] \\ &\quad \times \left[\frac{1 - \bar{\rho}_{r0} + k_{1r0} + k_{2r0} + k_{3r0}}{1 - \bar{\rho}_{r1} + k_{1r1} + k_{2r1} + k_{3r1}} \frac{\bar{S}_r(E'_{r0})}{\bar{S}_r(E'_{r1})} \right] P_e \\ &= \left[\frac{1 + k_e}{1 - \bar{\rho}_{r0} + k_{1r0} + k_{2r0} + k_{3r0}} \frac{1 - \xi_e - \eta_e}{1 - \xi_r - \eta_r} \right] \\ &\quad \times \left[\frac{1 - \bar{\rho}_{r0} + k_{1r0} + k_{2r0} + k_{3r0}}{1 - \bar{\rho}_{r1} + k_{1r1} + k_{2r1} + k_{3r1}} \frac{\bar{S}_r(E'_{r0})}{\bar{S}_r(E'_{r1})} \right] P_e . \end{aligned} \tag{6.21}$$

This differs from Eq. (6.20) by the addition of the term in the second set of square brackets. The new term reduces to unity in the case of an irradiance distribution equivalent to the reference distribution. In order to investigate the responsivity ratio in the new correction term, it is useful to express it according to the original definition of the mean responsivities, Eq. (6.3), as

$$\begin{aligned} \frac{\bar{S}_r(E'_{r0})}{\bar{S}_r(E'_{r1})} &= \frac{\int_0^\infty \int_0^{2\pi} \int_0^R S_r(r, \phi, P) E'_{r0}(r, \phi, \lambda) r \, dr \, d\phi \, d\lambda}{\int_0^\infty \int_0^{2\pi} \int_0^R S_r(r, \phi, P) E'_{r1}(r, \phi, \lambda) r \, dr \, d\phi \, d\lambda} \\ &\quad \times \frac{\int_0^\infty \int_0^{2\pi} \int_0^R E'_{r1}(r, \phi, \lambda) r \, dr \, d\phi \, d\lambda}{\int_0^\infty \int_0^{2\pi} \int_0^R E'_{r0}(r, \phi, \lambda) r \, dr \, d\phi \, d\lambda} . \end{aligned} \tag{6.22}$$

It is immediately obvious from the form of this equation that the functions $S_r(r, \phi, P)$, $E'_{r0}(r, \phi, \lambda)$, and $E'_{r1}(r, \phi, \lambda)$ in it need only be known on a relative scale. Further, since this merely amounts to a second-order effect, one can approximate $E'_{r0}(r, \phi, \lambda)$ and $E'_{r1}(r, \phi, \lambda)$ by $E_{r0}(r, \phi, \lambda)$ and $E_{r1}(r, \phi, \lambda)$, respectively. If the functions $S_r(r, \phi, P)$ and $E_r(r, \phi, \lambda)$ are such that they can be written as products of independent functions of the spatial, spectral, and power variables, i.e., if

$$S_r(r, \phi, P) = S_{1r}(r, \phi) S_{2r}(P) \tag{6.23a}$$

and

$$E_r(r, \phi, \lambda) = E_{1r}(r, \phi) E_{2r}(\lambda), \tag{6.23b}$$

further simplifications become possible in the calculation of the responsivity ratio according to (6.22), which then reduces to

$$\frac{\bar{S}_r(E'_{r0})}{\bar{S}_r(E_{r1})} = \frac{\int_0^{2\pi}\int_0^R S_{1r}(r,\phi)E_{1r0}(r,\phi)r\,dr\,d\phi}{\int_0^{2\pi}\int_0^R S_{1r}(r,\phi)E_{1r1}(r,\phi)r\,dr\,d\phi} \times \frac{\int_0^{2\pi}\int_0^R E_{1r1}(r,\phi)r\,dr\,d\phi}{\int_0^{2\pi}\int_0^R E_{1r0}(r,\phi)r\,dr\,d\phi}. \tag{6.24}$$

In that case both the power-dependent part $S_{2r}(P)$ of the responsivity function and the wavelength-dependent part $E_{2r}(\lambda)$ of the spectral-irradiance function disappear from the responsivity ratio, which then depends solely on the spatial-response function $S_{1r}(r, \phi)$ and the spatial-irradiance distribution functions $E_{1r0}(r, \phi)$ of the reference source (for which the corrections were originally determined) and $E_{1r1}(r, \phi)$ of the new source. All of these are reasonably easy to determine on a relative scale and the correction term can therefore be readily calculated.

According to Crommelynck (1982), the measured electrical power in a feedback-controlled system should be multiplied by a further correction term c_{FR}, which depends on the exact operating mode and tends toward unity for an accurate servo system. For further details regarding this correction term, Sections 1.3.3, 1.3.6, and 6.5 should be consulted.

If one thus relates the unknown radiant power P_r to the measured electrical power P_e via the instrumental correction function $C(c_i)$, where c_i symbolizes the list of individual instrumental corrections, in the form

$$P_r = C(c_i)P_e \tag{6.25}$$

then

$$C(c_i) = \left(\frac{1 + k_e}{1 - \bar{\rho}_{r0} + k_{1r0} + k_{2r0} + k_{3r0}} \frac{1 - \xi_e - \eta_e}{1 - \xi_r - \eta_r}\right) \times \left[\frac{1 - \bar{\rho}_{r0} + k_{1r0} + k_{2r0} + k_{3r0}}{1 - \bar{\rho}_{r1} + k_{1r1} + k_{2r1} + k_{3r1}} \frac{\bar{S}_r(E'_{r0})}{\bar{S}_r(E'_{r1})}\right] C_{FB} \tag{6.26}$$

with $\bar{S}_r(E'_{r0})/\bar{S}_r(E'_{r1})$ given by Eq. (6.22) or, if applicable, Eq. (6.24).

If it is not the incident radiant power P_r but the irradiance E_r that is of interest, it can be calculated from the value derived for the radiant power as

$$E_r = \frac{P_r}{A_{TO}} \frac{1}{[1 + \alpha(T - T_0)]^2}, \tag{6.27}$$

where A_{T0} is the area of the radiometer aperture measured at a temperature T_0, T is the temperature of the radiometer aperture during the operation of the absolute radiometer, and α is the linear thermal expansion coefficient of the aperture material.

The instrumental correction function $C(c_i)$ derived in this way contains only parameters and functions, which can be determined experimentally in all absolute radiometers designed to permit a full instrumental characterization. It incorporates corrections for all the instrumental imperfections considered important at present, subject to the limitations (e.g., no dependence of the reflectance on time, polarization, and direction) spelled out at the outset. However, although the results were derived by neglecting some parameters, (parameters generally considered less important), their incorporation in the theory would be relatively straightforward and could be considered in cases where they are likely to be significant.

6.4 DISCUSSION OF OTHER CORRECTION FORMULAS

While individual instrumental corrections for absolute radiometers were discussed in the literature right from the invention of these instruments, the exact form of the measurement equation or correction formula was hardly ever stated. It seems reasonably clear that the measured electrical power P_e was usually divided by the absorptance $(1 - \bar{\rho})$ of the detector and later also multiplied by $(1 + k_e)$ to apply a lead-heating correction. In what analytical form other corrections such as the case-heating and nonequivalence corrections were applied is a lot less certain in most cases. One can speculate that the instrumental correction function used was of either of two forms:

$$C(c_i) = 1 + \sum_i c_i \tag{6.28}$$

or

$$C(c_i) = \prod_i (1 + c_i). \tag{6.29}$$

Although the two versions agree in first order, there is a residual uncertainty due to the occurrence of second-order terms (products of two instrumental correction factors) in (6.29). Therefore, the question arises, which one, if either, of the two intuitive versions is correct.

This question can only be answered by deriving the instrumental correction function theoretically with the aid of the physical principles involved in measurements with an absolute radiometer. The first attempt to formulate such a theoretical framework is due to Geist (1971). He made use of the responsivity concept to relate the power supplied to the detector element to the observed signal and derived the instrumental correction function by equating the signals for electrical and radiant heating as

$$P_r = \frac{1 - \xi_e - \delta_e}{1 - \xi_r - \delta_r} \frac{S_e}{S_r} P_e + \frac{S'_e P'_e - S'_r P'_r}{S_r(1 - \xi_r - \delta_r)}. \tag{6.30}$$

If the lead-heating and case-heating contributions to the detector output are not denoted as $S'_e P'_e$ and $S'_r P'_r$, respectively, as in Geist's theory, but as $k_e S_e P_e$ and $k_r S_r P_r$, respectively (to use the same formalism as in Section 6.3, while accepting the validity and equivalence of both formulations), Geist's formula can be rewritten in a more compact form as (Hengstberger, 1977c)

$$P_r = \frac{1 + k_e + \xi_e - \delta_e}{1 + k_r - \xi_r - \delta_r} \frac{S_e}{S_r} P_e. \tag{6.31}$$

Equation (6.31) is formally very close to Eq. (6.19) (the correction formula derived in Section 6.3) and seems to agree with it in the first order. However, as pointed out by Hengstberger (1977c), there are inconsistencies in the derivation by Geist, which needed to be addressed. Nevertheless, his contribution provided all the concepts necessary for deriving an instrumental correction function for absolute radiometers and served as an invaluable basis for further work on this subject.

Willson (1973) formulated an instrumental correction function through a quasi-equilibrium–power-balance analysis. It was based on the principle that the sums of the power sources and the power sinks have to be the same in the cases of both radiant and electrical heating. A term was included in this analysis for the power input to the heat capacity of the detector element due to temperature drifts. Depending on the direction of the drift, this factor can either act as a source or a sink in the power balance, and the analysis was therefore labelled a "quasi-equilibrium" analysis. Using the same symbols for quantities as in Section 6.3, Willson's measurement equation had the form

$$P_r = \frac{(P_{e1} - P_{e2}) + P_{NE} - k_{1r} P_r - k_{2r} P_r (P_c - P'_c) - (P_{c1} - P'_{c1}) + C\left(\frac{dT}{dt} - \frac{dT'}{dt}\right)}{(1 - \bar{\rho}_c + \rho\bar{\rho}_c)}, \tag{6.32}$$

where P_{el} and P_{e2} are the electrical powers during electrical and radiant heating; P_{NE} is the difference between the radiant powers emitted from the front surface during electrical and radiant heating; P_{1r} is the case-heating power; P_{2r} is the radiant power supplied by scattered radiation; P_c and P'_c are the power conducted to the heat sink via the thermal resistance during electrical and radiant heating, respectively; $C(dT/dt)$ and $C(dT'/dt)$ are the powers expended or supplied during a temperature drift of the detector element during electrical and radiant heating, respectively; $\bar{\rho}_c$ is the detector reflectance; and ρ is the reflectance of the detector's field of view.

If one converts Eq. (6.32) to an equilibrium form for comparison with Eq. (6.20), the term $C(dT/dt - dT'/dt)$ vanishes by definition (T = constant, T′ = constant) and the terms $(P_c - P'_c)$ and $(P_{c1} - P'_{c1})$ vanish because in that case the powers represented by these terms are equivalent (i.e., equal) under both forms of heating. If one further rewrites the equation's remaining terms in the form

$$P_r = \frac{P_{\text{e1}} - P_{\text{e2}}}{(1 - \bar{\rho}_{\text{c}} + \rho\bar{\rho}_{\text{c}} + k_{1\text{r}} + k_{2\text{r}})} + \frac{P_{\text{NE}}}{(1 - \bar{\rho}_{\text{c}} + \rho\bar{\rho}_{\text{c}} + k_{1\text{r}} + k_{2\text{r}})}$$
$$= \frac{1}{(1 - \bar{\rho}_{\text{c}} - \rho\bar{\rho}_{\text{c}} + k_{1\text{r}} + k_{2\text{r}})}\left(1 + \frac{P_{\text{NE}}}{P_{\text{e1}} - P_{\text{e2}}}\right)(P_{\text{e1}} - P_{\text{e2}}), \tag{6.33}$$

the correspondence with the terms in Eq. (6.20) is obvious. The only differences are the absence of the lead-heating correction and of the diffraction correction and the extra term $\rho\bar{\rho}_{\text{c}}$ in the denominator to correct for the interreflections between the detector and its field of view.

Brusa and Frölich (1975) stated that their instrumental correction function had the form (converted to the terminology of Section 6.3)

$$\text{C}(c_i) = \frac{(1 + k_{\text{e}})(1 - k_{2\text{r}})(1 - k_{3\text{r}})}{(1 - \bar{\rho}_{\text{r}})} c_{\text{NE}}, \tag{6.34}$$

with c_{NE} being the nonequivalence correction. Apart from the missing case-heating correction and the missing diffraction correction, which was added according to later publications (Brusa, 1983; Brusa and Fröhlich, 1986), Eq. (6.34) is a first-order equivalent of (6.18).

Hengstberger (1977c) removed the inconsistencies in Geist's (1971) theory and derived a formula equivalent to (6.20), with the exception of the scattering and diffraction terms. He later also derived a correction term for the nonuniform responsivity of the detector element (Hengstberger, 1979) along the lines of Eq. (6.21) and (6.22).

Crommelynck (1982) used the equilibrium power balance to derive a very comprehensive expression for the instrumental correction function. In the terminology used in Section 6.3, it can be expressed as

$$\text{C}(c_i) = \frac{1 + k_{\text{e}}}{1 - \bar{\rho}_{\text{r}} + k_{1\text{r}} + k_{2\text{r}} + k_{3\text{r}}} \frac{\bar{S}_{\text{e}}}{\bar{S}_{\text{r}}} C_{\text{FB}}. \tag{6.35}$$

It has basically all the same ingredients as Eq. (6.26) except that, (1) the ratio of the mean responsivities is not expressed in terms of the heat losses and (2) a correction term for the nonuniform responsivity is not explicitly contained in it.

Since the different approaches, symbols, and terminology used by various authors to derive instrumental correction functions were extremely confusing even to experts in the field, it is very satisfying indeed that the various versions, when brought to a common denominator, all seem to be consistent. Even some versions that do not have the correct analytical form are now seen to be correct in the first order. However, it would certainly improve the flow of communication in the field if a common terminology, common symbols for the correction factors, and a common analytical form of the correction function were used by everybody active in this area of metrology.

6.5 CORRECTION EXPERIMENTS AND CALCULATIONS

6.5.1 The Lead-Heating Correction

The lead-heating correction is a correction for the extra power $k_e P_e$ that reaches the detector element of an absolute radiometer when an electrical power P_e is dissipated in its heating element due to the power dissipated in the heater leads. Not all the power dissipated in the heater leads adds to the heater power P_e; $k_e P_e$ is the fraction of this power that does. According to its definition (see Section 1.3.3) the lead-heating correction is the ratio of $k_e P_e$ and P_e, namely k_e.

Since the lead-heating correction can be made quite small and since it is also easily and accurately measurable in a properly constructed absolute radiometer, it could be argued that it deserves little discussion. Due to the high accuracy with which it can be measured, it is, however, also just about the only correction, for which second-order effects can be readily detected. For this reason, it is very instructive both theoretically and experimentally and its detailed discussion is justified.

When electrical power was supplied to the early absolute radiometers by a single pair of heater wires and the dissipated power was calculated from the measured current flowing through the heater and from its known resistance, it would have been impossible to measure this correction experimentally in the finished instrument. It would only have been feasible to estimate an upper limit for k_e by measuring the lead resistances and calculating the ratio of the powers dissipated in the leads and the heating element, respectively.

Although there was some awareness of this potential error source, the experimenters had more pressing problems to deal with (e.g., the reflection correction and the absorption correction) and it was not until 1937 that the lead-heating correction was discussed in detail (Guild, 1937). Although Coblentz and Emerson had used separate potential and current leads attached to each side of the heating element of their absolute radiometers as far back as 1916, Guild was the first to employ these two pairs of leads in his absolute radiometers for the measurement of the lead-heating correction. His experimental arrangements for the two parts of the experiment are depicted in Figs. 6.2a and 6.2b, respectively.

The two current leads are marked C_1 and C_2; the two potential leads are P_1 and P_2. The circuit in Fig. 6.2a is the one normally used in the operation of

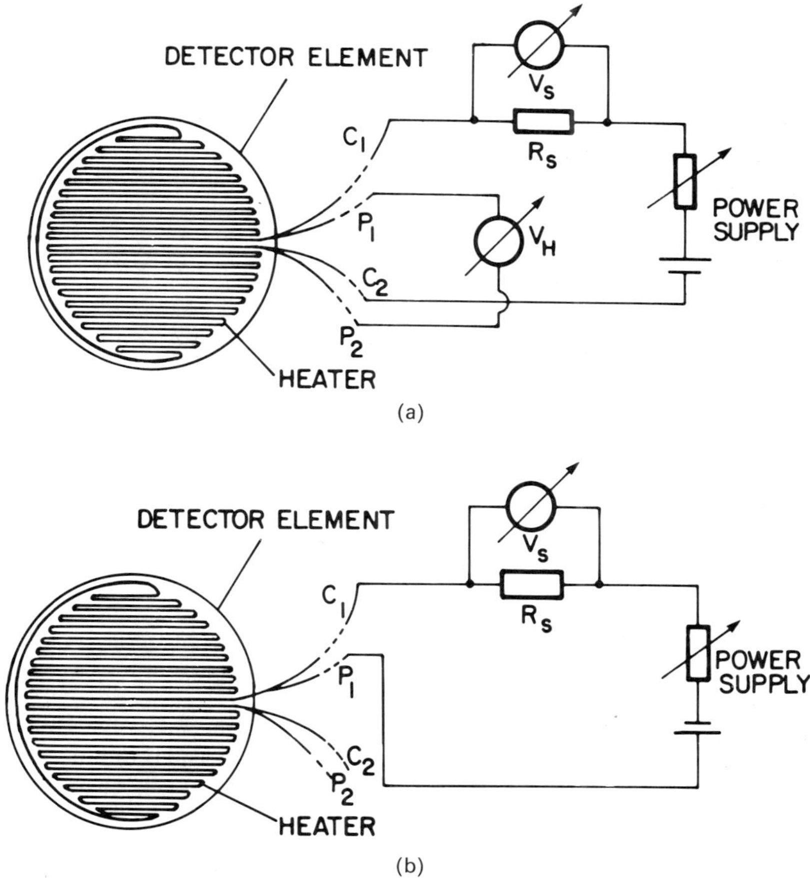

Fig. 6.2 First (a) and second (b) parts of Guild-type lead-heating experiment.

the absolute radiometer, with the heater power supplied by a variable power supply. The heater current flowed from the power supply via a standard resistor (R_s) to current lead C_1, then through the heater and eventually back to the power supply via current lead C_2. The heater current was determined by measuring the voltage (V_s) across R_s and dividing this value by R_s. The heater voltage (V_h) was measured across the potential leads P_1 and P_2 and the power dissipated in the heater followed as $P_e = P_h V_s / R_s$. The detector output $Y(P_e + k_e P_e)$ due to this power P_e plus the lead-heating power $k_e P_e$ was noted.

In order to determine the lead-heating power contributed to the detector element in the previous experiment, the circuit was then reconfigured as depicted in Fig. 6.2b. No current flowed through the heater itself in this case.

Instead, the same current V_s/R_s as used in the previous experiment was passed through one current lead (C_1 or C_2) to the junction with the associated potential lead, through which it flowed back to the power supply. Assuming that the same thickness of wire used for both the potential and current leads means that the same power is dissipated in them for the same current flowing through them, that it is conducted in the same way to the detector element and sensed with the same spatial responsivity at both ends of the heater, the observed signal $Y(k_e P_e)$ will be solely due to the unknown lead-heating power. If one further assumes for convenience and since it is a second-order effect that the responsivity of the absolute radiometer is linear with power between 0 and P_e, the two observed signals can be expressed as

$$Y(P_e + k_e P_e) = \bar{S}_e(P_e + k_e P_e), \tag{6.36a}$$

and

$$Y(k_e P_e) = \bar{S}_e k_e P_e. \tag{6.36b}$$

The lead-heating correction k_e follows from this equation as

$$k_e = Y(k_e P_e)/[Y(P_e + k_e P_e) - Y(k_e P_e)]. \tag{6.37}$$

In order to remove some of the uncertainties associated with the just mentioned assumptions, Gillham (1962) devised a new heater lead configuration, which is used with few exceptions in all modern absolute radiometers constructed after Gillham's 1962 publication. It consists of two pairs of current leads and one pair of potential leads for each heating element. The arrangement and the two experiments required for the measurement of the lead-heating correction with the new technique are depicted in Figs. 6.3a and 6.3b.

In order to derive the full benefit of this heater lead configuration, it should be ensured by means of trim resistors (R_T) in series with the current leads that the resistance measured between the potential lead on each side of the detector element and the two current leads associated with it are the same. The resistance measurements for matching each pair of current leads should be performed close to the point where the heater current is split between the two current leads. In this way, it will be assured that there are equal currents flowing through each of the two current leads on the same side of the heating element during the normal operating mode, which is depicted in Fig. 6.3a.

In the first experiment conducted in the lead configuration of Fig. 6.3a, power is supplied to the heater via both current leads on each side. The detector output $Y(P_e + k_e P_e)$, the heater power P_e, and the heater current are recorded. There are a number of different possibilities for conducting the

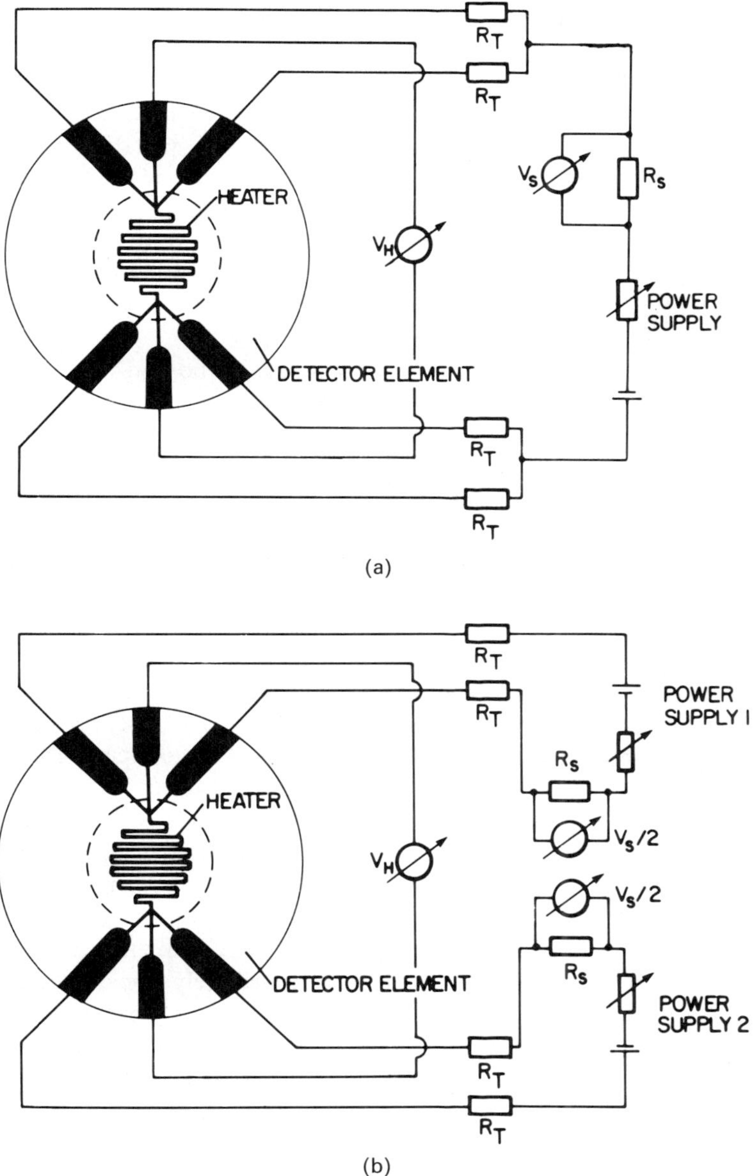

Fig. 6.3 First (a) and second (b) (version 1) parts of Gillham-type lead-heating experiment.

second part of the experiment. In the version shown in Fig. 6.3b, the same current as used in the first part is passed through each of the four current leads, without flowing through the heater. The resulting detector output $Y(k_e P_e)$ is recorded and, provided that the output is linear with power, the calculation is as discussed for Guild's experiment.

This way of conducting the second part of the lead-heating experiment requires the use of two independent standard resistors and power supplies. The latter also have to be floating and well isolated to avoid the flow of leakage currents through the heater. This can be checked by connecting a voltmeter across the potential leads and ascertaining that the observed voltage remains zero.

In another possible version of the second part of the lead-heating experiment, the contribution from the current leads on each side of the detector element could be determined separately and added to obtain the total correction. In this way, only one power supply and one standard resistor are required and the leakage problem is also avoided. However, the number of measurements needed is increased.

Yet another possibility is depicted in Fig. 6.4. In this method, all four current leads are heated simultaneously, involving only a single power supply and standard resistor, but the price to be paid for the shortcut is a small, unavoidable current leakage through the heater. By measuring the heater voltage, it is possible to correct both for the power dissipated in the heater and for the reduction in the current through two of the current leads, provided the heater resistance and the resistances of the four current leads are known.

Geist (1971) suggested a further variation in which the detector is irradiated during the second part of the lead-heating experiment to avoid a second-order error due to the nonlinearity of the detector. However, as Hengstberger (1977c) has pointed out, this is not completely true since in that case one still has to know the ratio $\bar{S}_e(P'_e)/S_e(P'_e)$ and thus, in effect, the detector linearity. This conclusion is not immediately apparent and only follows from an exact analysis of the experiment using the theoretical concepts developed in Section 6.3. However, the lead-heating correction measured in this way will have less of a nonlinearity error than without irradiating the detector in the second part of the experiment.

Geist's (1972) suggestion highlighted his correct perception that the use of the detector signal instead of the values of the substituted electrical powers makes the results dependent on the detector linearity, albeit in second order. The magnitude of the effect was demonstrated by Hengstberger (1977b), who analyzed several approaches in detail and illustrated his theoretical conclusion with the results of an extensive series of measurements. Similar measurements by other authors (Gentile and Rastello, 1980; Etchechoury and Cogno,

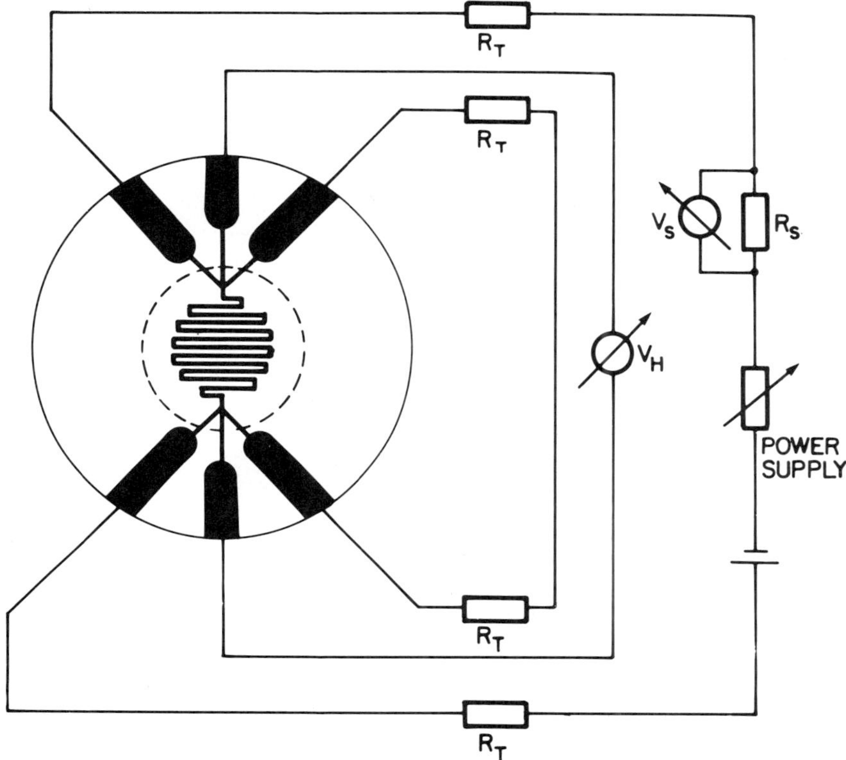

Fig. 6.4 Second part of Gillham-type lead-heating experiment (version 2).

1982) confirmed these results. Figure 6.5 shows the lead-heating correction for an absolute radiometer (Hengstberger, 1977b) as a function of heater power with and without taking into account the detector nonlinearity.

As Fig. 6.5 shows, the error in the value of the measured lead-heating correction is directly proportional to the power level used in the first part of the experiment (the x-axis of Fig. 6.5). If the response nonlinearity is not taken into account, the error can already amount to 10% of the measured correction at a power level of 20 mW and to even more at higher power levels. For lead-heating corrections in the order of 0.1–0.01% this may indeed be negligible, but for high k_e values and high power levels in the correction experiment, it could be significant.

The approach used for correcting the experimental data for the response nonlinearity of the detector for the version of the lead-heating experiment depicted in Fig. 6.3b is as follows. Equations (6.36) are rewritten with

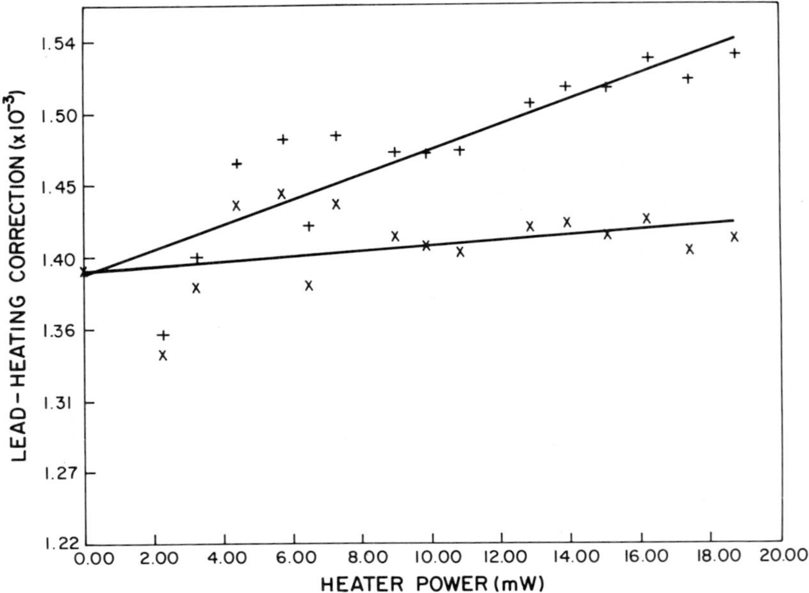

Fig. 6.5 Lead-heating correction as a function of heater power (+ = uncorrected; × = corrected).

responsivities, which are functions of the power level. This changes Eq. (6.37) to

$$k_e = Y(k_e P_e) \Big/ \left[Y(P_e + k_e P_e) \frac{\bar{S}_e(k_e P_e)}{\bar{S}_e(P_e + k_e P_e)} - Y(k_e P_e) \right]. \tag{6.38}$$

If the nonlinearity of the responsivity by electrical heating is determined as described in Section 6.5.11, the correction of the lead-heating result according to Eq. (6.38) is a straightforward matter.

As already mentioned, a nonlinearity correction can be avoided if power ratios are used instead of signal ratios to calculate the correction term. As the power P_e is already known from the first part of the lead-heating experiment, it is merely necessary to substitute an electrical power (in the circuit configuration of Fig. 6.3a), which produces the same signal as observed in the second part of the experiment. With that lead-heating power P_{LH} determined, k_e follows simply as P_{LH}/P_e.

Once the lead-heating correction has been determined and if the resistances of the four current leads are known, it is also possible to calculate the percentage of the total power dissipated in the current leads that reaches the detector element. Hengstberger (1977b) found a value of 10% for his instruments, which employed 20-μm-thick gold wires as current leads.

One remaining imperfection in all the versions of the lead-heating experiments discussed so far is the fact that the current through two of the current leads in the second part of the experiment is in the opposite direction to the one in the first part. This problem has received no further attention so far either theoretically or experimentally, although Gillham had already mentioned it in 1962. Gillham suggested that each junction could be treated as a four-terminal network since a connection could be made to it through the two current leads and the associated potential lead as well as through one of the leads on the other side of the heater. He used the transfer impedance Z between two pairs of terminals (current lead no. 1 and potential lead against current lead no. 2 and one lead from the other side of the heater, which is independent of the lead resistance and depends only on the junction) and assumed that the junction was symmetrical about the potential lead. Assuming further that the transfer impedance was the same for the junctions on both sides of the heater, he stated that the electrical power supplied to the detector element, already corrected for the lead-heating effect, should be corrected again by the fractional amount $2Z/R_H$, R_H being the resistance of the heater. According to Gillham's evidence, the effect amounted to less than 0.1% for his radiometers. However, it would seem that the phenomenon would warrant a more detailed study with today's more advanced measurement techniques.

6.5.2 The Reflection Correction

The correction for the fraction of the incident radiant power lost from the absorber by reflection is the oldest instrumental correction, having been applied and discussed already by both Ångstrom (1893) and Kurlbaum (1894). As described in detail in Chapter 2, the different types of absorber and the absorber materials used in absolute radiometers ensure that as large a portion of the incident radiation as possible is absorbed (converted to heat). Thus, the reflected fraction of the incident beam will typically not be higher than a few percentage points, and in many cases even much less than that figure. There are few examples of detector elements reflecting specularly, even though the walls of cavity or cone absorbers may be coated with specularly reflecting absorber materials. When measuring the reflection correction, it will therefore normally be necessary to collect and measure the diffuse radiation reflected into the full hemispherical solid angle of $2\pi sr$.

A number of different techniques are available to perform meaurements of diffuse reflectance. The first of these is depicted in Fig. 6.6.

Here, the absorber of the detector element is placed as the sample against a port of an integrating sphere and a detector (usually photoelectric) is mounted on a second port not far from the sample port, with a screen placed

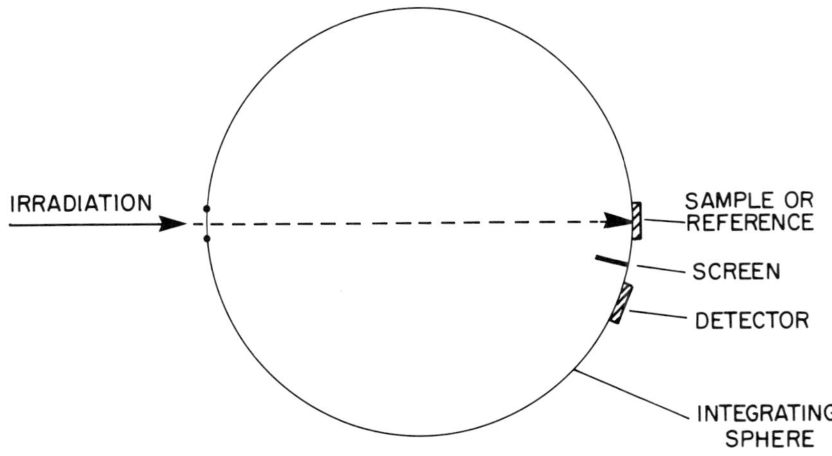

Fig. 6.6 Possible measuring system for diffuse reflectance using an integrating sphere.

between them to prevent any direct radiation from the sample reaching the detector. The ratio of the detector signals when the sample is irradiated with a collimated, narrow beam normal to its surface to the detector signal when the sample is replaced by a diffuse, white reflectance standard such as barium sulphate is then equal to the diffuse sample reflectance. A zero reading with the sample port open and facing a completely darkened room, which should ideally be equal to the dark reading of the detector (i.e., with both the sample and source ports closed and, therefore, no irradiation of the sample), should be subtracted from both the sample and the reference (white standard) signal. In this way, it can be assured that there is no significant amount of stray light, which can have a very detrimental effect on the measurement accuracy at low values of reflectance. The detector linearity should likewise be checked by an independent method—e.g., a double-aperture technique (a discussion of suitable methods can be found in Budde, 1983) and the change in the sphere efficiency on changing from the sample to the reference should be taken into account. The source for irradiating the sample could be of the same type as the source to be measured by the absolute radiometer. In that case, the measured reflectance will equal the effective reflectance $\bar{\rho}_r$ for that source as referred to in Section 6.3. It could also be the monochromatic output beam from a monochromator, in which case the reflectance could be measured as a function of wavelength. An additional advantage of this approach is that sample heating, which can affect the measurement accuracy at infrared wavelengths, is minimized.

A measurement of the same type can also be performed on many commercial spectrophotometers if they have a diffuse reflectance attachment.

The sphere used in that case usually has both a sample port and a reference port, so that the sample and the reference can be mounted simultaneously. Similar precautions should be observed as in the previous case.

Sphere coatings based on materials such as barium sulphate decrease in reflectance in the near-infrared region and have to be replaced by either sulphur-based coatings (Tkachuk and Kuzina, 1978) or diffusely reflecting gold coatings (Labsphere, 1987). Specular mirrors for the collection of the reflected radiation are also widely used at longer wavelengths (e.g. Gillham, 1953; Brandenberg, 1964; Blevin and Brown, 1965; Wood et al., 1976).

Techniques with cone-shaped detectors, which collect the reflected radiant power directly over a whole hemisphere (Blevin and Geist, 1974) have also become popular, especially for the measurement of very low reflectances, such as those of cavities and cones. This method, for which uncertainties in the order of 0.002 to 0.01% are claimed (Zalewski et al., 1979; Willson, 1980; Brusa and Fröhlich, 1986) is depicted schematically in Fig. 6.7.

At such low values of reflectance and for the relatively high power levels needed to resolve them, the effect of sample heating and the separation of the reflected and emitted portions become important factors affecting the accuracy of the measurement. Techniques involving a chopped laser beam and phase-sensitive detection of the detector signal both in phase and in quadrature with the chopped signal have been used for this purpose (Brusa and Fröhlich, 1986).

In Section 6.3, the reflectance function of the absorber has been treated as a function of both wavelength and coordinates (r, ϕ). While the wavelength dependence has received considerable attention, the spatial dependence

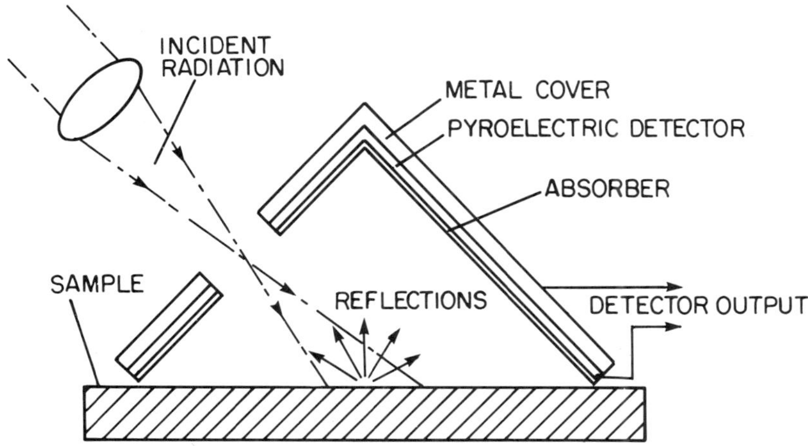

Fig. 6.7 Possible measuring system for diffuse reflectance using a cone-shaped detector.

seems to have been somewhat neglected so far. The reflectance variation for a narrow beam as a function of the coordinates of the spot where it strikes the absorber could conceivably contribute as large an uncertainty as the variation of the reflectance with wavelength, especially for the complicated absorber structures used to achieve very low reflectances.

6.5.3 The Case-Heating Correction

A case-heating correction (see also Section 1.3.3) has to be applied only if the radiant power supplied to the detector element is not confined to an area smaller than the radiometer aperture. In that case, parts of the radiometer head, other than the detector element, may be heated by the absorption of parts of the incident beam. Power from these heated parts (which reaches the detector element by emission, convection, or conduction) can necessitate a case-heating correction if the compensating detector element is either not receiving an identical amount of this power or does not provide complete compensation.

Techniques for measuring the correction factor k_r usually have the objective of heating the parts of the radiometer head responsible for the effect with a beam of known radiant power, while at the same time ensuring that no direct or scattered radiation from the beam passes through the radiometer aperture.

Gillham (1962) selected another approach in which an electrical heater was fixed to the irradiated area around the radiometer aperture. The case-heating correction was then determined by supplying this heater with the same amount of electrical power as was estimated to be absorbed over that area during radiant-power measurement. Gillham found that there was no measurable effect for two of his absolute radiometers and that the effect for his cavity-type radiometer was $+0.0025$ for one of the twin units and -0.0025 for the other. He concluded that this seemed to indicate that the effect was mainly due to imperfect compensation between the two detector elements and only to a very small degree to a temperature nonuniformity in the radiometer head.

In another of the reported techniques (Hengstberger et al., 1977; Gao zhizhong et al., 1983), the radiometer aperture is closed with a plastic or rubber plug not much larger than the aperture, which is carefully shielded from direct irradiation by the source by additional shields. The rest of the beam for which the correction is sought is left to impinge on all parts of the radiometer head that are exposed to it during normal operation. By observing the radiometer output as a function of time from the start of the exposure, and converting this information into the corresponding case-heating power function by electrical substitution, it is possible to derive a correction factor. The electrical substitution has to be performed under identical conditions to avoid the effect of a changed responsivity due to the closed aperature. From the time-dependence of the observed case-heating, the correction can be calculated for different cycle lengths between electrical and radiant heating during the normal operation of the absolute radiometer. Alternatively, the same cycle length as during normal operation should be employed during the measurement of the case-heating correction to compensate for the effect of different time-constants of the heated environment and of the detector element.

Crommelynck (1982) measured the case-heating effect by replacing the radiometer aperture with an identical part without a hole, scanning over the surface of the aperture with a small, powerful laser beam (200 mW), and measuring the detector output as a function of the position where the laser beam struck the surface. His response function depended only on the radius from the center (ρ) and had the form

$$\mathrm{f}(\rho) = a + b\rho + c\rho^2. \tag{6.39}$$

He calculated the correction as

$$k_{2r} = \frac{\int_r^R 2\pi \mathrm{f}(\rho)\rho\, d\rho}{\int_0^r 2\pi \mathrm{f}'(\rho)\rho\, d\rho}, \tag{6.40}$$

with r being the radius of the radiometer aperture, R the beam diameter, and $\mathrm{f}'(\rho)$ the beam power density.

Crommelynck also remarked on the significant difference in the function $\mathrm{f}(\rho)$ as determined in air and in vacuum. In vacuum, the case-heating correction as determined by this method decreased from the value in air (0.3 %) to approximately 0.01 %, indicating the dominance of heat transfer by air conduction and convection. It is an imperfection of this method that the part without the aperture hole will have a different conduction pattern for the absorbed heat than the part used in the actual operation and that heat transfer from the filled-in portion will modify the effect to an unknown degree. A comparison of the results outside the area of the radiometer aperture obtained by this and the previous method (plugged aperture) would indicate whether these imperfections have a significant influence on the results or not.

All of the described methods are also imperfect in the sense that they allow only contributions due to the heating of the radiometer aperture to be observed. Contributions from other parts of the radiometer head cannot be measured in this way. However, since the contribution from the radiometer aperture is usually by far the largest contribution anyway, the error introduced, because the other contributions have been neglected, will very likely not be very significant. Nevertheless, some new experimental alternatives would certainly be a welcome addition to the available arsenal.

6.5.4 The Scattering Correction

A correction for scattering (diffuse reflection) at the edges of baffles, apertures, and other parts of the detector head exposed to the measured beam was mentioned as a potential error source by Kurlbaum as early as 1898. Since the radiometer aperture is usually reflective to reduce the case-heating effect, a fraction of the radiant power reflected from it can find its way through the radiometer aperture by scattering at other parts of the radiometer head. Reflection from the edge of the radiometer aperture is a further error source that can be included with this correction (Quinn and Martin, 1985).

Gillham devised a method to measure the reflection from the edge of the radiometer aperture, which was based on placing the detector element at different distances from the radiometer aperture. When repeating measurements of the same parallel beam at these different distances, he reasoned that the ratio of the measured powers would give an indication for the error contribution from the aperture edge, since it would decrease with increasing distance. However, because of the simultaneous change in the collected diffracted portion and possible other distance effects, it would be difficult to separate the various contributions and the effectiveness of the technique is thus debatable.

A valuable technique for measuring the contribution from the radiation reflected from the radiometer aperture was reported by Crommelynck (1982) and Brusa and Fröhlich (1986). Here, the detector element is replaced by, for example, a silicon photodiode, which cannot detect the case-heating power, and the radiometer aperture is irradiated with a collimated laser beam at different locations and with slightly different angles of incidence. The scattering contribution from the reflective radiometer aperture is then simply the average ratio of the detector output when the aperture is irradiated to the output when the radiation falls directly on the detector. Crommelynck's (1982) technique was a refinement of this basic version in that he did a differential measurement consisting of a scan over the whole accessible aperture area using the standard aperture and an identical scan with a blackened aperture. From the differential data he calculated the correction using a formula identical to Eq. (6.40). This technique would seem to provide a sound basis for comprehensive characterizations of absolute radiometers in terms of the scattering correction.

6.5.5 The Diffraction Correction

The diffraction effects at apertures in the beam path between the source and the radiometer head are classified in the context of this work as environmental corrections, which are therefore covered in Chapter 5. On the other hand, the diffraction correction k_{3_r} for further apertures inside the radiometer head is included in the instrumental corrections in line with the definitions given for these two groups of corrections in Section 6.1. Like scattering at the edges of apertures (which is, however, much less significant for apertures located outside the radiometer head and is therefore not discussed separately under environmental corrections), a diffraction correction is included in both groups of corrections.

For this reason, it is not necessary to repeat the conclusions for apertures in the beam path reached in Chapter 5 (Section 5.2.2) and they apply with equal validity in the context of this section. However, as pointed out already in Chapter 5, the correction for the diffraction at the radiometer aperture is clearly part of the instrumental corrections. Since it was discussed fully in Chapter 5, for reasons of economy, reference for this part of the diffraction correction k_{3_r} should be made to Section 5.2.1.

6.5.6 The Nonequivalence Correction

As spelled out in the definition of this correction term (see Section 1.3.3), it is caused by the difference in the temperature distributions in the detector element under radiant and electrical heating. As a consequence of this, equal detector outputs for radiant and electrical heating do not imply the conversion to heat of equal amounts of electrical and radiant power.

The difference in the temperature distribution in the detector element is brought about because the radiant power is converted to heat on top or inside of the absorber layer, while the electrical power is supplied to the detector element under the absorber layer. This fact was already mentioned by Kurlbaum in 1900; Guild (1937) also discussed it in considerable detail. Guild also constructed two different versions of his absolute radiometer with the express intention of maximizing the nonequivalence effect in one and minimizing it in the other to get an indication of the magnitude of the effect. (See Section 1.3.4.4.) He failed to detect a difference and concluded that the correction was of a negligible magnitude for his radiometers.

It has been shown in Chapter 4 that a simplified analysis of the temperature distribution normal to the surface of the detector element can be made under the assumption that the temperature is uniform in planes parallel to that surface. For this simplified thermal model of the detector element, the nonequivalence correction can be expressed in terms of the thermal parameters of the detector element as

$$\frac{1 - \xi_e - \eta_e}{1 - \xi_r - \eta_r} = 1 + R_A/R_1, \tag{6.41a}$$

where R_A is the thermal resistance of the absorber layer and R_1 the thermal resistance between the absorber layer and the thermal environment of the detector element. By expanding the denominator of (6.41a) in a series, one can get the approximate identity

$$(\xi_r + \eta_r) - (\xi_e + \eta_e) = R_A/R_1. \tag{6.41b}$$

One way of calculating the nonequivalence correction is therefore via a known value of the ratio R_A/R_1. In order to measure R_A, Gillham employed another conclusion from the electrical-circuit representation of an absolute radiometer (Fig. 4.3) according to which the difference in the surface temperature of the absorber layer for equal detector outputs is given by

$$\Delta T = R_A R_1/(R_A + R_1)P'_r \sim R_A P'_r. \tag{6.42}$$

Since ΔT can be relatively easily measured with suitable techniques that not only have to be based on the ratio of the emitted powers but also have to take into account the radiant power reflected during radiant heating, and since P'_r is equal in the first order to P_e, R_A is not too difficult to determine. Gillham (1962) used a thermopile detector combined with a potassium bromide filter, which only transmitted the emitted portion and not the reflected one, to measure the excess radiant flux emitted from the absorber surface during radiant heating. For small temperature differences, the emitted flux difference and, thus, the net thermopile signal S_1 is proportional to

$$4\sigma\varepsilon T^3 \, \Delta T \tag{6.43}$$

where ΔT can be substituted from Eq. (6.42). Replacing the potassium chloride filter by a glass plate, which only transmitted the reflected part and not the emitted one, he determined the signal S_2 for the reflected part, which was proportional to $\bar{\rho}_r P_r$. Therefore, the signal ratio from the two measurements was sufficient for the calculation of R_A in terms of known values for $\bar{\rho}_r$, ε, T, and σ. A similar technique was reported by Blevin and Brown (1966).

In a modern version of this experiment, one would presumably irradiate the detector uniformly with a laser beam and measure the temperature change between radiant and electrical heating with a calibrated pyrometer or thermal imaging system operating at a wavelength different from the laser wavelength.

While a determination of the thermal resistance of the absorber layer is not too difficult in principle, the measurement of the thermal resistance from the absorber layer to the environment is a much more complex problem. For this reason, Gillham estimated R_1 from the geometrical and thermal properties of his radiometers and of the surrounding air. (See also Chapter 4.) With the excess temperature at the top of the absorber layer under radiant heating already known, a knowledge of R_1 also implies a knowledge of the difference between the nonequivalent heat losses $(\xi_r + \eta_r) - (\xi_e + \eta_e)$.

It follows that the nonequivalence correction could be calculated in this case according to either the left-hand side or the right-hand side of Eq. (6.41b). This is to be expected of course since the two sides of Eq. (6.41) are basically only different formulations of the same basic relationship. The formulation in terms of the nonequivalent leat losses is, however, the more general of the two, since the expression in terms of the ratio R_A/R_1 is only valid for the simplified, one-dimensional thermal-heat-flow model.

Since the beginning of the 1970s, there has been a healthy trend away from calculating corrections and toward a full experimental characterization wherever possible. For the nonequivalence correction, this has meant that calculations based on a measured value of R_A and an estimated value of R_1 have been replaced by other techniques, with the obvious intention of basing the determination purely on experimental data.

The most popular technique for this purpose seems to be the measurement of a stable source with the radiometer head either in air or in vacuum and the use of the air-to-vacuum signal or power ratios for electrical and radiant heating for the calculation of the nonequivalence correction. Brusa and Fröhlich (1986) include the correction for the heating of the radiometer aperture (main part of the case-heating correction) in their definition of the nonequivalence correction and state: "Thus the air-to-vacuum ratio of the sensitivity of an instrument measured with the sun as a source minus the air-to-vacuum ratio measured with electrical heating is equivalent to the correction for nonequivalence." In his summary of the conclusions of a conference titled "Advances in Absolute Radiometry," Geist (1985) stated: "A good example of this trend is the vacuum-to-air correction for the electrical substitution radiometers without windows. Ten years ago, the correction for this effect was not measured, and it was a subject of much speculation and argument. Now it is just the subject of careful measurement." The problem with both of the quoted statements is that they imply that the measurement of the air-to-vacuum signal or power ratios per se is a determination of the nonequivalence correction. Since a proof for this assumption does not seem to have been published anywhere, its validity needs to be examined in some detail in this context. If the measurements in air and in vacuum are conducted in such a way that the total power supplied to the detector element in the radiant-heating mode in air (P'_{r1}) produces the same detector output as a substituted electrical power P_{e1} and that the corresponding amounts in vacuum (using an unchanged source power P_r) are P'_{r2} and P'_{e1}, respectively, then the results of the two measurements can be written in the terminology of Section 6.3 as

$$\bar{S}_{eA} P'_{e1} = \bar{S}_{rA} P'_{r1}, \tag{6.44a}$$

$$\bar{S}_{eV} P'_{e2} = \bar{S}_{rV} P'_{r2}. \tag{6.44b}$$

The ratios $\bar{S}_{eA}/\bar{S}_{rA}$ and $\bar{S}_{eV}/\bar{S}_{rV}$ are of course the desired nonequivalence corrections for air and vacuum, respectively. Combining the two equations yields

$$\left(\frac{\bar{S}_{eA}}{\bar{S}_{rA}}\right)\Bigg/\left(\frac{\bar{S}_{eV}}{\bar{S}_{rV}}\right) = \frac{P'_{e2}}{P'_{e1}}\frac{P'_{r1}}{P'_{r2}} = \frac{P'_{e2}}{P'_{e1}}\frac{1 - \bar{\rho}_r + k_{1r} + k_{2r} + k_{3r}}{1 - \bar{\rho}_r + k'_{1r} + k_{2r} + k_{3r}}. \quad (6.45)$$

The ratio P'_{r1}/P'_{r2} in this equation differs from unity because the case-heating contribution in vacuum (k'_{1r}) and air (k_{1r}) is different as demonstrated by Crommelynck (1982). (See Section 6.5.3.) This was most probably the reason why Brusa and Fröhlich (1986) included the case-heating correction in their nonequivalence correction, an approach that is consistent with their empirical correction formula (see Section 6.4) but that is not desirable in the context of the strict expression for the instrumental corrections as derived in Section 6.3. If the values of ρ_r, k_{1r}, k'_{1r}, k_{2r}, and k_{3r} are already known at this stage, Eq. (6.45) shows that the nonequivalence correction can be quite easily separated from the case-heating correction and that a mixed correction factor for the two effects is not necessary. Equation (6.45) also shows that the measurements in air and in vacuum do not directly yield a value for the nonequivalence correction; they merely supply a value for the ratio of the nonequivalence corrections in air and in vacuum. Further assumptions or measurements are needed to derive the values of the nonequivalence corrections themselves in either air or vacuum.

If one considers the formulation of the nonequivalence correction in the form $(1 + R_A/R_1)$ in air and $(1 + R'_A/R'_1)$ in vacuum, then it is quite clear, because of the absence of air conduction and convection in vacuum, that $R'_1 > R_1$. If it can be shown that $R'_1 \gg R_1$, which will be true in most cases, and that $R'_A \sim R_A$, which is definitely not true in many cases (see Chapter 2), then one will be justified in setting the nonequivalence correction in vacuum approximately equal to unity and in deriving the nonequivalence correction in air directly from Eq. (6.45). Since it is known that a specular black glazing paint was used for the absolute radiometers employed by Fröhlich and Brusa (Brusa, 1983), the condition $R'_A \sim R_A$ was very likely met in this case and the authors were therefore justified in assuming $R'_A/R'_1 \ll R_A/R_1$ and calculating the nonequivalence correction in air under this assumption from (6.45). It is, however, not completely clear from their publication (Brusa and Fröhlich, 1986) whether they used a formula equivalent to (6.45) in their calculations as they refer to the nonequivalence correction as "the air-to-vacuum ratio of the sensitivity with the sun as a source minus the air-to-vacuum ratio measured with electrical heating." The validity of this formulation is not immediately apparent in the light of the preceding conclusions.

While the use of the electrical air-to-vacuum power ratio according to (6.45) may thus yield a value for the nonequivalence correction in air under the assumptions of $R'_A \sim R_A$ and $R'_1 \gg R_1$, a sweeping generalization to the effect that this procedure always yields a correct value for the nonequivalence correction is certainly wrong. It is known that the thermal resistance of a number of absorber materials with a porous structure (e.g., metal blacks and 3M Nextel Black Paint; see Chapter 2) increases considerably in vacuum and, therefore, the assumption $R'_A/R'_1 \ll R_A/R_1$ cannot always be made, making the calculation of the nonequivalence correction in air from Eq. (6.45) impossible without further data.

However, even in a case where the nonequivalence correction cannot be derived from air-to-vacuum ratios, a solution can be found by combining the technique for measuring the thermal resistance of the absorber layer with the results of the measurements in air and in vacuum. If the excess temperature at the top of the absorber layer for radiant heating is determined in vacuum ($\Delta T'$) and the thermal resistance of the absorber layer in vacuum (R'_A) is derived from it in the same way as described earlier for air, the term R'_A/R'_1 can be quite easily determined, in contrast to the situation in air. In vacuum, the only contribution to R'_1 is by emission from the absorber layer and R'_1 follows therefore immediately as

$$R'_1 = 1/(4\sigma\varepsilon T^3 \mathrm{A}), \tag{6.46}$$

where A is the effective radiating area of the detector element, ε its emissivity, and T its temperature. With R'_A and R'_1 determined, the nonequivalence correction for vacuum is given according to (6.41a) as $(1 + R'_A/R'_1)$ and the nonequivalence correction for air follows via Eq. (6.45).

It should also be noted that the nonequivalence correction in vacuum could be further reduced by cooling the detector element to a lower temperature T during the vacuum measurements. According to (6.46), the thermal resistance from the front surface of the absorber to the environment (R'_1) increases with the third power of $1/T$ and an increase by a factor of about 55 is achieved by lowering the detector temperature from room temperature to the temperature of liquid nitrogen. In that case, it would be much easier to meet the condition $R'_A/R'_1 \ll R_1/R_1$ and the nonequivalence correction could be calculated from the air-to-vacuum power ratios for most absolute radiometers without having to determine the thermal resistances R'_A and R'_1. This reduction in the value of the nonequivalence correction with the third power of the temperature of the detector element, combined with the fact that they operate in vacuum, is also responsible for its vanishingly small value for cryogenic absolute radiometers working at the temperature of liquid helium.

When radiant and electrical heating are alternated relatively rapidly (as in pyroelectric absolute radiometers, which are usually operated at chopping rates well in excess of 1 Hz), the nonequivalence can no longer be expressed simply in terms of equilibrium heat losses or thermal resistances as in (6.41). It has been shown (Blevin and Geist, 1974a) that the correct expression for this case is

$$(1 + R_A/R_1)\sec(\phi_r - \phi_e), \tag{6.47}$$

where $(\phi_r - \phi_e)$ is the difference between the phase lags during radiant and electrical heating.

Depending on the chopping frequency, thermal capacity and thermal resistance of the coating, and the thermal properties of the surrounding medium, $(\phi_r - \phi_e)$ can amount to a few degrees even for gold-black absorbers (Blevin and Geist, 1974a). Thus, the phase term could add significantly to the nonequivalence correction in such cases and the magnitude of the phase difference should therefore be verified experimentally for pyroelectric absolute radiometers.

6.5.7 The Correction for Nonuniform Responsivity

In order to correct for a change in the spatial distribution of irradiance across the detector element according to Eq. (6.24), a knowledge of the function $S_{1r}(r, \phi)$ on a relative scale is required. In order to measure this function experimentally, an experimental arrangement similar to the one depicted in Fig. 6.8 could be used.

Here a beam of radiation focused onto the plane of the detector element is scanned across the area of the radiometer aperture. In order to compensate for changes in the radiant power of the beam, a monitor detector should be employed and readings of both the radiometer and the monitor detector should be taken at regular radial and angular intervals. Although it is not

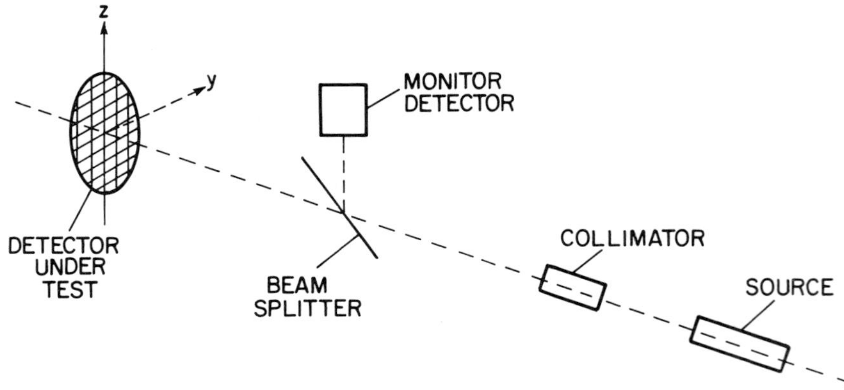

Fig. 6.8 Possible experimental arrangement for measuring response nonuniformity across the surface of a detector element.

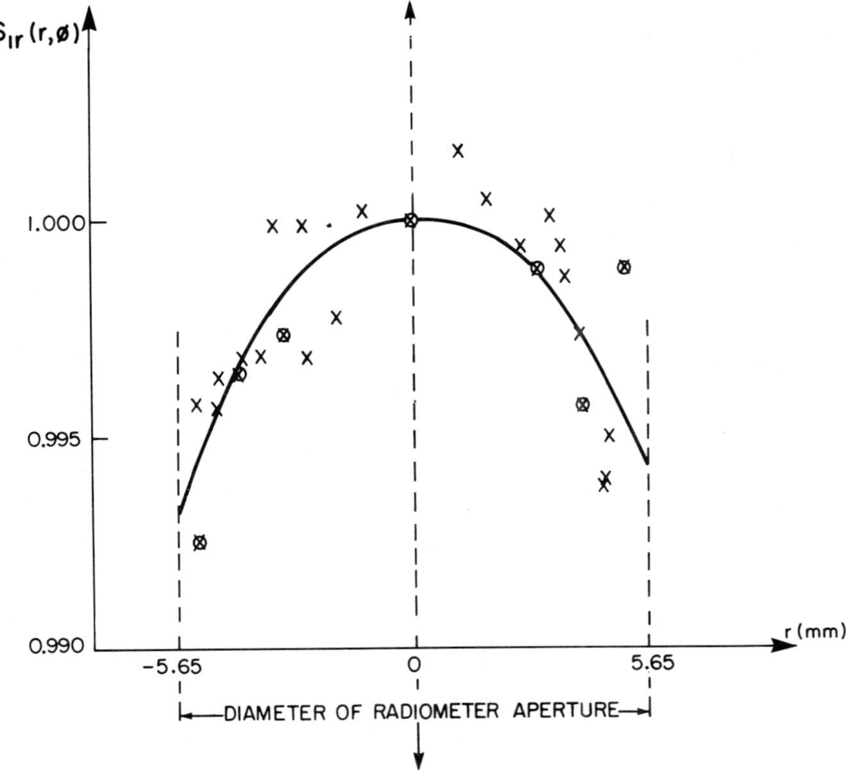

Fig. 6.9 Measured response nonuniformity of an absolute radiometer at the NPRL.

necessary for the calculation, the spatial responsivity function calculated from the measured data could be normalized to unity at the maximum for ease of presentation and comparison. Figure 6.9 shows the results of such measurements for the absolute radiometer at the National Physical Research Laboratory (NPRL), South Africa (Hengstberger, et al. 1985). The responsivity curve fitted to the experimental data had the form

$$S_{1r}(r, \phi) = 1 - br^2, \tag{6.48}$$

with $b = 0.00022$.

Using this equation to calculate a correction according to (6.24) for a hypothetical uniform beam of 3-mm diameter ($E_{1r1}(r, \phi) = E$ for $0 < r < 1.5$ mm; $E_{1r1}(r, \phi) = 0$ elsewhere) when the basic set of instrumental corrections was determined for a spatially uniform beam overfilling the radiometer aperture of 11.3-mm diameter ($E_{1r0}(r, \phi) = E_0$ for $0 < r < 5.65$ mm; $E_{1r0}(r, \phi) = 0$ elsewhere) yields a value of 0.9967. Naturally, a correction for vanishing case-heating and scattering corrections as well as for a possibly different spectral distribution ($\bar{\rho}_r$) and diffraction correction would also have to be applied according to the term in the second square bracket of (6.21).

6.5.8 The Dual-Detector Correction

This correction only has to be applied if one detector element in a compensated configuration is exposed to the radiant power being measured, while the compensating element is used for the substitution of the electrical power. It is necessitated by the fact that the two detector elements have different responsivities and different thermal resistances to the thermal environment. The correction is easily determined by supplying a known amount of electrical power P_{e1} to the detector element used for the radiant-power measurement and measuring the required amount P_{e2} having to be substituted in the compensating detector element to produce the same detector output. The power ratio P_{e1}/P_{e2} is then the dual-detector correction, with which the electrical power substituted in the compensating detector element has to be multiplied to find the equivalent electrical power one would have had to substitute in the detector element used for the radiant-power measurement. The instrumental corrections for the detector employed for the radiant power measurement should then be used to find the incident radiant power.

It is recommended that the value of the dual-detector correction should be included in the basic list of instrumental correction factors specified for an absolute radiometer, even if it is not used in the mode requiring this correction. The value can be of interest when judging the quality of the thermal compensation achieved for a particular instrument.

6.5.9 The Feedback Correction

This correction (c_{FR}) only needs to be applied if an absolute radiometer is used in an active mode, where a feedback-controlled servo system performs the substitution of the electrical reference power. (See also Section 1.3.6.) A practical servo system always requires a small error signal at its input, either because of its operating principle (e.g., in a proportional regulator) or because of device or component imperfections such as offset voltages, bias currents, and so on. For the servo system depicted in Fig. 1.51, the error signal E(s) is given by

$$\mathrm{E}(s) = \mathrm{R}(s)/[1 + \mathrm{KG}(s)\mathrm{H}(s)]. \tag{6.49}$$

If the error signal merely introduced a constant power offset for each power level, then a differential measurement (as, for example, with and without the instrument exposed to the radiant power) would eliminate the effect and no

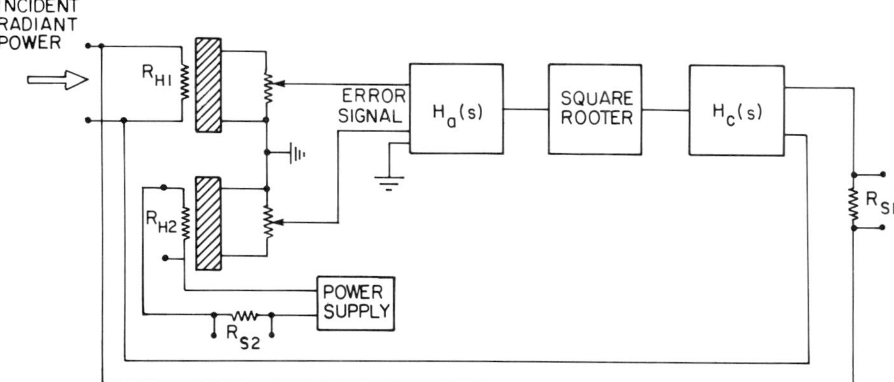

Fig. 6.10 Servo system (including square-root element) for a compensated absolute radiometer with both detector elements operating at the same power level. [After Crommelynck (1982)]

correction would be required. Unfortunately, however, the error signal in a servo system of the type considered here introduces a power error proportional to the measured power level, which affects the power difference by exactly the same factor as a single power. The required correction for this effect can be made negligible in most cases by a proper choice of components and system parameters. Crommelynck (1982) has derived correction factors for different operating modes of absolute radiometers subject to the requirement that the feedback system is linearized by the inclusion of a square-root element in the loop. An example for an absolute radiometer where the irradiated and the compensating detector element work at the same power level and where the irradiated detector is in the feedback loop is depicted in Fig. 6.10.

The feedback correction for this case (using successive measurements with the shutter open and closed) is given by Crommelynck (1982) as

$$C_{\mathrm{FB}} = \frac{(R_{\mathrm{H}} + R_{\mathrm{S}})^2}{R_{\mathrm{H}} G_1 H_{\mathrm{a}} H_{\mathrm{c}}^2} - 1, \tag{6.50}$$

where G_1 is the ratio of the output signal of the irradiated detector and of the power level causing the output signal (i.e., its mean responsivity).

6.5.10 The Area of the Radiometer Aperture

For measuring the irradiance in the plane of the radiometer aperture, it is necessary to know the area of that aperture to an accuracy commensurate with the uncertainties in the other corrections. The calculation of the aperture area is usually based on measurements of the aperture diameter averaged over values measured at various angular intervals across it. Before being used, such apertures should be inspected under a microscope for scratches and other imperfections and their roundness should be determined on a roundness-measuring machine. The aperture should be bevelled and the remaining cylindrical edge should be normal to the front surface and as narrow as possible while still producing a smooth edge of acceptable quality. For an aperture diameter of 10 mm, an uncertainty of 0.01 % in the aperture area is equivalent to an uncertainty of $\pm 0.5\ \mu m$ for the mean diameter. At that level of accuracy, the greatest care must be taken not to apply any force to the aperture when mounting it in its designated position as this could affect the aperture area. Even if the aperture is measured in the mounted position, the assembly of the radiometer head may still produce stresses, which could invalidate the value found for it. Brusa and Fröhlich (1986) reported changes of up to 1 % when using incorrect materials or techniques. Crommelynck (1985) reported unacceptably large systematic differences between diameters determined at five different metrology laboratories and the uncertainty associated with the area of the radiometer aperture was identified as one of the major remaining sources of uncertainty for absolute radiometers (Geist, 1985). The results of a more detailed intercomparison involving a number of national metrology laboratories and organized by Crommelynck are not as yet available.

If the temperature at which the aperture diameter was measured differs from the temperature at which it is used in the absolute radiometer, a correction for thermal expansion may be required. (See Eq. 6.27.) This correction can be a problem in the case of large temperature differences as occur, for example, in cryogenic absolute radiometers (Quinn and Martin, 1985).

6.5.11 The Response Nonlinearity

The detector of an absolute radiometer is an ordinary thermal-radiation detector and it is thus nonlinear in its response to absorbed radiant power to the same degree as other thermal-radiation detectors. This nonlinearity does not affect the results of measurements where electrical power is substituted for radiant power since the comparison is made at the same power level. However, it is sometimes convenient to use the detector output directly, as, for example, when performing correction experiments or even for ordinary measurements. In that case, it is necessary to know the response nonlinearity for correction purposes. As an absolute radiometer has a built-in electrical heating element, its response nonlinearity is easier to determine than that of any other detector.

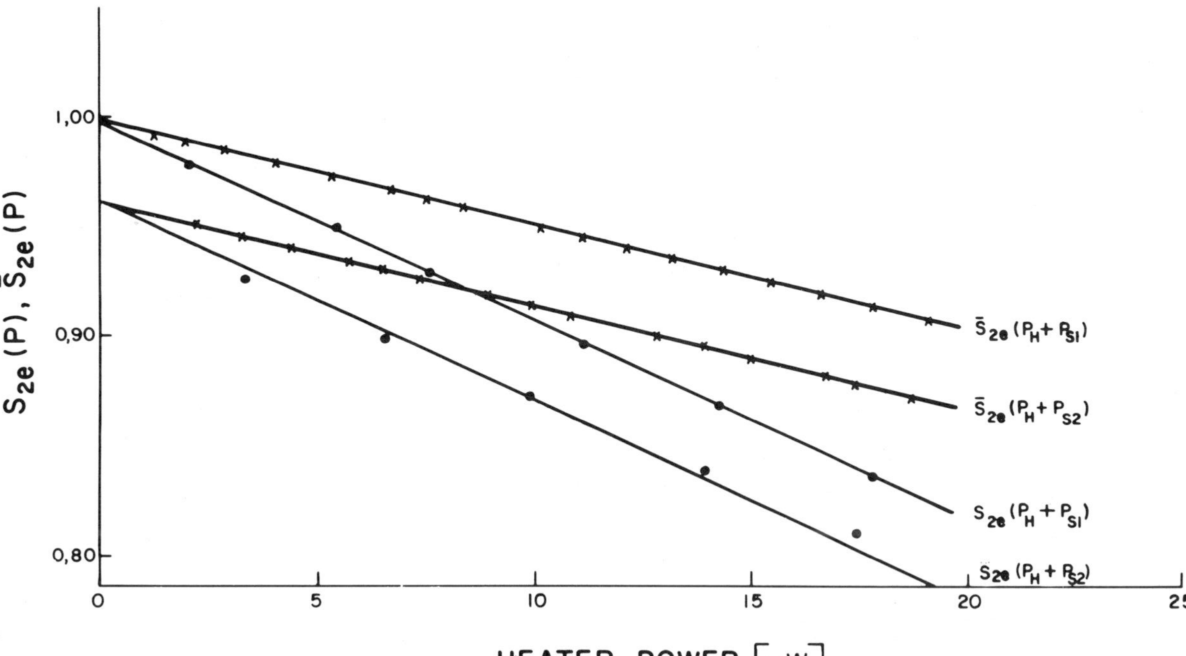

Fig. 6.11 Measured response nonlinearity of an absolute radiometer at the NPRL.

The mean responsivity $\bar{S}_e(E'_e)$ as defined in Eqs. 6.3 and 6.4 can be found for any power level by simply dividing the observed signal by the measured electrical-power level. For radiometer elements in which significant electrical power is dissipated (as, for instance, in bolometric detector elements), the response nonlinearity is a function of the sum of the powers dissipated in the heater and the detector. In addition, it is dependent on the temperature of the radiometer head. Mean responsivities determined for the absolute radiometer at the NPRL as a function of both heater- and detector-power dissipation are shown in Fig. 6.11 (Hengstberger, 1977b).

Also shown in Fig. 6.11 is the power dependence of the responsivity $S_{2e}(P)$, defined analog to the definition of $S_{2r}(P)$ in Eq. (6.23a). It differs from the mean responsivity $\bar{S}_e(E'_e)$ by representing the responsivity at the power level P_e. Therefore, it is not found by dividing the detector signal by the electrical power level, but by making a small power step ΔP_e from the power level P_e (preferably in both directions and averaging the result) and dividing the observed signal step ΔY by ΔP_e. If the mean responsivity is a linear function of power, as is the case in Fig. 6.11, then Hengstberger (1977d) has shown that the slope for the responsivity $S_{2e}(P)$ is exactly double that of the mean responsivity curve and that they intersect the y-axis at the same point. He also pointed out that the mean responsivity can be determined from the data acquired in the first part of the lead-heating experiment, provided it is conducted at a number of different power levels. Therefore, the determination of this function requires no additional measurements.

6.6 FURTHER CONSIDERATIONS

Although modern digital voltmeters usually have very high input impedances and thus do not load the measuring circuit when measuring the heater voltage and the heater current, the specifications should be checked to verify this matter. For those with a low input impedance and for analog voltmeters, if these should be used for some reason, corrections to the measured voltages will be required. The voltmeter used should also have a valid calibration certificate from a recognized calibration authority and the standard resistor used for the measurement of the heater current should be measured with a calibrated ohmmeter.

For accuracies at the 0.01 % level and because of the construction of the detector elements of most absolute radiometers, it seems unlikely that the Peltier and Thomson effects (see Section 8.4) will have any influence on the measurement results. This can be easily verified by reversing the polarity of the heater current and analyzing the results for a systematic difference between the two polarities. Even if there were a difference, it could be corrected by taking the mean of measurements done with both polarities. The latter measure should be adequate right down to the ppm level. Since there could be polarity effects associated with the electronics or with construction faults, a DC reversal experiment should form an integral part of the radiometer-characterization procedure.

As absolute radiometers are often used for the measurement of source parameters, the distance between the source and the radiometer aperture needs to be accurately known. Since the radiometer aperture is generally located inside the radiometer head and is thus difficult to reach, its distance from a reference plane accessible at the outside of the radiometer head should be accurately measured after or during assembly and recorded. Care should be taken not to exert undue force on the aperture in the process and to avoid the possibility of mechanical damage to its precisely machined edges.

When dealing with measurements of radiant power that changes with time (as, for example, in meterorological applications) or when designing response-accelerating circuits, it is necessary to know the frequency response of the absolute radiometer. Because of the long thermal time-constants found in many absolute radiometers, a variation of the heater power in a sinusoidal fashion and simultaneous recording of the detector output require instrumentation operating at frequencies as low as 0.001 Hz. From the data acquired by varying the frequency of the power source and by measuring the amplitude and phase of the detector output, the frequency response can be established. Another technique, one that avoids the low-frequency measurements and instead uses a laser beam chopped in a pseudorandom way followed by Fourier analysis, has also been used for this purpose (Malcorps, 1981).

6.7 CONCLUSION

The material in this chapter represents the most comprehensive discussion and theory of the instrumental corrections in absolute radiometry published anywhere to date (late 1988). It includes all the instrumental corrections applied by absolute radiometrists in a single logical framework and puts this system in perspective relative to other sets of parameters and terminologies used in the field. In most cases, these are shown to be valid subsets of the more comprehensive framework presented here.

In the interest of better communication in absolute radiometry, authors should consider using the symbols, definitions, and terminology presented here in preference to their personal variants.

REFERENCES

Ångstrom, K. (1893). Eine elektrische Kompensationsmethode zur quantitativen Bestimmung strahlender Wärme. *Nova Acta Reg. Soc. Sci. Ups. Ser. III*, 1 (English translation in *Phys. Rev.* **1**, 365 of 1893).

Blevin, W. R., and Brown, W. J. (1965). An infra-red reflectometer with a spheroidal mirror. *J. Sci. Instr.* **42**, 385.

Blevin, W. R., and Brown, W. J. (1966). Black coatings for absolute radiometers. *Metrologia* **2**, 139.

Blevin, W. R., and Geist, J. (1974a). Influence of black coatings on pyroelectric detectors. *Appl. Opt.* **13**, 1171.

Blevin, W. R., and Geist, J. (1974b). Infrared reflectometry with cavity-shaped pyroelectric detector. *Appl. Opt.* **13**, 2212.

Brandenberg, W. M. (1964). Focusing properties of hemispherical and ellipsoidal mirror reflectometers. *J. Opt. Soc. Am.* **54**, 1235.

Brusa, R. W. (1983). *Solar radiometry*. Publication no. 598. Physikalisch-Meteorologisches Observatorium and World Radiation Center, Davos, Switzerland.

Brusa, R. W., and Fröhlich, C. (1975). Realization of the absolute scale of total irradiance. *Scientific Discussion of the IV. International Pyrheliometer Comparisons*. World Radiation Center, Davos, Switzerland.

Brusa, R. W., and Fröhlich, C. (1986). Absolute radiometers (PMO6) and their experimental characterization. *Appl. Opt.* **25**, 4173.

Budde, W. (1983). *Physical Detectors of Optical Radiation*. Academic Press, New York and London.

Coblentz, W. W., and Emerson, W. B. (1916). Studies of instruments for measuring radiant energy in absolute value: an absolute thermopile. *Bull. Bur. Std.* **12**, 503.

Commission Internationale d'Eclairage (1986). *Colorimetric Illuminants*, 1st edition. CIE Publ. No. S001. International Commission on Illumination (CIE), Vienna.

Crommelynck, D. A. (1982). Fundamentals of absolute pyrheliometry and objective characterization. NASA Conference Publication 2239. NASA, Scientific and Technical Information Branch, United States.

Crommelynck, D. A. (1985). The IRMB/SSD absolute radiometric base, objectives and developments. *Proc. of an International Meeting on Advances in Absolute Radiometry, Cambridge, Massachusetts*, P. V. Foukal, ed. Atmospheric and Environmental Research, Cambridge, Mass., pp. 12–15.

Etchechoury, E. M., and Cogno, J. A. (1982). Correccion por calentamiento de conductores en radiometros absolutos. *Carta Metrologica* **5**, 23. Instituto Nacional de Tecnologia Industrial, Buenos Aires, Argentina.

Gao zhizhong, Wang zhenchang, Piao dazhi, Mao shihua, and Yang chiuhong (1983). Realization of the candela by electrically calibrated radiometers. *Metrologia* **19**, 85.

Geist, J. (1971). *Fundamental Principles of Absolute Radiometry and the Philosophy of This NBS Program (1968 to 1971)*. National Bureau of Standards Techn. Note 594-1. U.S. Government Printing Office, Washington, D.C.

Geist, J. (1985). Conference summary. *Proc. of an International Meeting on Advances in Absolute Radiometry, Cambridge, Massachusetts*, P. V. Foukal, ed. Atmospheric and Environmental Research, Cambridge, Mass., pp. 78–79.

Gentile, C., and Rastello, M. L. (1980). *Misure de Sensibilita e dell'Effetto di Riscaldamento dei Conduttori in un Radiometro Assoluto*. Technical Report. Istituto Elettrotecnico, Torino, Italy.

Gillham, E. J. (1953). A method for measuring the spectral reflectivity of a thermopile. *British J. Appl. Phys.* **4**, 151.

Gillham, E. J. (1962). Recent investigations in absolute radiometry. *Proc. Roy. Soc. (London) Ser. A* **269**, 249.

Guild, J. (1937). Investigations in absolute radiometry. *Proc. Roy. Soc. Ser. A* **161**, 1.

Hengstberger, F. (1977a). *The Absolute Radiometer at the National Physical Research Laboratory.* CSIR Research Report 331. Council for Scientific and Industrial Research, Pretoria, South Africa.

Hengstberger, F. (1977b). *A Fresh Look at the Lead-heating Effect in Absolute Radiometry.* CSIR Research Report 333. Council for Scientific and Industrial Research, Pretoria, South Africa.

Hengstberger, F. (1977c). An improved theory of the instrumental corrections for absolute radiometers. *Metrologia* **13**, 69.

Hengstberger, F. (1977d). Mesure des caractéristiques de sensibilité d'un radiomètre absolu: e'tude d'un cas concret. *Proc. CCPR, 9e Session.* Bureau International des Poids et Mesures, Paris, pp. 115–131.

Hengstberger, F. (1979). A correction for spatial responsivity variations in absolute radiometers. *S. Afr. J. Phys.* **2**, 41.

Hengstberger, F., Dressler, R. E., Monard, L. A. G., Kok, C. J., and Turner, R. (1985). Further advances with the fully automated absolute radiometer developed at the NPRL. *Proc. of an International Meeting on Advances in Absolute Radiometry*, Cambridge, Massachusetts, P. V. Foukal, ed. Atmospheric and Environmental Research, Cambridge, Mass., pp. 34–37.

Hengstberger, F., Thain, E., and Turner, R. (1977). *A New Measuring System for Realizing Photometric and Radiometric Scales.* CSIR Research Report 332. Council for Scientific and Industrial Research, Pretoria, South Africa.

Kurlbaum, F. (1894). Notiz über eine Methode zur quantitativen Bestimmung strahlender Wärme. *Wied. Ann.* **51**, 591.

Kurlbaum, F. (1898). Über eine Methode zur Bestimmung der Strahlung in absolutem Maass und die Strahlung des schwarzen Körpers zwischen 0 und 100 Grad. *Wied. Ann.* **65**, 746.

Labsphere (1987). *Integrating Spheres.* Catalogue. Labsphere, Sutton, New Hampshire, United States.

Malcorps, H. (1981). Frequency-response of heat fluxmeters. *J. Phys. E: Sci. Instr.* **14**, 1054.

Quinn, T. J., and Martin, J. E. (1985). A radiometric determination of the Stefan–Boltzmann constant and thermodynamic temperatures between −40°C and 100°C. *Phil. Trans. R. Soc. Lond. A* **316**, 85.

Tkachuk, R., and Kuzina, F. D. (1978). Sulfur as a proposed near infrared reflectance standard. *Appl. Opt.* **17**, 2817.

Willson, R. C. (1973). Active cavity radiometer. *Appl. Opt.* **12**, 810.

Willson, R. C. (1980). Active cavity radiometer type V. *Appl. Opt.* **19**, 3256.

Wood, B. E., Pipes, J. G., Smith, A. M., and Roux, J. A. (1976). Hemi-ellipsoidal mirror infrared reflectometer: development and operation. *Appl. Opt.* **15**, 940.

Zalewski, E. F., Geist, J., and Willson, R. C. (1979). Cavity radiometer reflectance. *Proc. SPIE* **196**, 152.

7

Alternative Optical Power Scales

F. HENGSTBERGER

Council for Scientific and Industrial Research
Pretoria, South Africa

7.1 INTRODUCTION

As already pointed out in Section 1.3.2, absolute radiometry is only one of a number of methods that can be used to realize a radiant power scale. The most prominent alternative techniques (namely, blackbody radiators, silicon photodiodes with a predictable quantum efficiency, and synchrontron radiation) will be reviewed briefly in this chapter. The review itself will merely summarize the physical principles on which the alternative techniques are based and will not treat the subject matters in detail. Separate, relatively comprehensive reference lists covering the recent literature on these subjects are given at the end of this chapter.

When absolute radiometry is included in the list, the four methods discussed in this book represent two source-based and two detector-based possibilities for realizing an absolute radiation scale. Two of the four alternatives (absolute radiometry and blackbody radiators) originated in the last decade of the nineteenth century. As for the other two, synchrotron radiation only became competitive during the 1970s to 1980s, while silicon photodiodes with a predictable quantum efficiency were only developed over the same two decades. As each one of these methods has its strengths and weaknesses, they complement each other to a large extent and should therefore be used according to their relative merits for a given application and, of course, their availability and affordability. Their mutual consistency seems to be well assured by now. (See Section 7.5.) In the light of the complex demands of modern radiometry, each would seem to have a role to play.

ISBN 0-12-340810-5

7.2 BLACKBODY RADIATORS

Blackbody radiation in the strict sense of the word is only present inside an enclosure with a uniform wall temperature (Fig. 7.1).

The theoretical concepts for deriving the properties of blackbody radiation were only developed over the past century. They depended on the identification of light and radiant heat as transverse electromagnetic waves, which exert a pressure equal to one-third of their energy per unit volume. Making use of this property and applying thermodynamic concepts, some general deductions became possible, which led to the formulation of some important radiation laws. One of these is the so-called Stefan–Boltzmann law, which relates the total radiant energy U in an enclosed volume V to the temperature T of its walls as

$$U = \mathrm{K} V T^4, \tag{7.1}$$

where K is a constant.

Another important relationship is Wien's displacement law, which relates the maximum emission wavelength λ_{m} to the wall temperature T in the form

$$\lambda_{\mathrm{m}} T = \text{constant}. \tag{7.2}$$

For both laws, the theoretical derivation was preceded by an empirical derivation on the basis of measured data. The first experimental realization of a good blackbody radiator, consisting of a hollow body surrounded by a constant-temperature bath and with a small hole in the wall to observe the

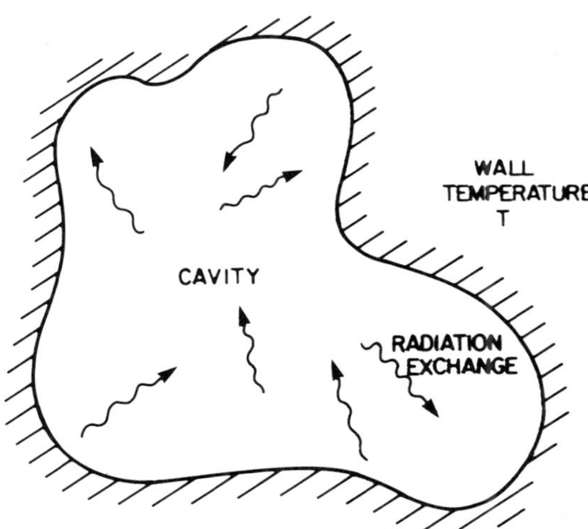

Fig. 7.1 Blackbody radiation in a cavity with a uniform wall temperature T.

radiation, was due to Lummer and Wien around 1890. If the effect of the hole on the radiation is neglected, the radiant exitance M of the hole is given by

$$M = \sigma T^4, \tag{7.3}$$

where σ is known as the Stefan–Boltzmann constant. While the values of the radiation constants were measurable by experiment, neither they nor the exact form of the spectral distribution of blackbody radiation could be derived solely on the basis of electromagnetic theory and thermodynamics, although good approximations for small and large values of the product of wavelength and temperature (λT) were found (Wien and Rayleigh–Jeans formulas) on the basis of some additional assumptions about the radiation mechanism. The fact that no single formula seemed to agree with the experimental data for all values of λT motivated Planck to introduce his quantum hypothesis for the emission and absorption of radiant energy. He succeeded eventually in 1900, at first on a purely empirical basis, to find a formula that satisfied the experimental results. This formula, which is now known as Planck's law, has the form

$$\mathrm{L}_\lambda(\lambda, T) = \frac{\mathrm{c}_1}{\pi}\,\lambda^{-5}\{\exp[\mathrm{c}_2/(\lambda T)] - 1\}^{-1}, \tag{7.4}$$

where $\mathrm{L}_\lambda(\lambda, T)$ is the spectral radiance of the blackbody and c_1 and c_2 are constants.

All previously known radiation laws follow readily from this formula and Planck could also express the constants in it clearly in terms of other known physical constants. The Planck function, which is now thought of as representing the birth of quantum theory, has received extensive experimental verification over a wide range of wavelengths and temperatures and can therefore be regarded as one of the most reliably established relationships in physics.

If blackbody radiation is used as a means to realize a high-accuracy radiation scale, a correction has to be applied for the effect of the opening in the cavity wall on the emitted radiation. This correction is the so-called emissivity of the cavity, which depends mainly on the reflection properties of the cavity wall and on the geometric parameters characterizing the cavity. An excellent review of methods for calculating the emissivity of practical blackbody cavities is given by Quinn (1983); a considerable number of further publications dealing with this subject are listed in the reference section at the end of this chapter. The emissivity of a cavity is used as a multiplier on the right-hand side of Eqs. 7.3 and 7.4 to facilitate the calculation of either the total radiation or of the spectral distribution of the radiation from a real cavity.

For radiation scales based on the Stefan–Boltzmann relationship, a systematic difference between the measured and calculated values of the Stefan–Boltzmann constant was a source of considerable concern for a long time. The uncertainty from this source was only removed relatively recently (Blevin and Brown, 1970; Quinn and Martin, 1985), after diffraction had been identified as a major error source affecting many experimental determinations of the constant. Agreement between the theoretical and measured values now exists at the 0.01% level (Quinn and Martin, 1985).

Further error sources for radiation scales based on cavity radiators are temperature gradients in the cavity and the fact that the thermodynamic temperatures of the fixed points at which the blackbodies are operated, or that are used to calibrate the temperature transducers measuring the blackbody temperature, themselves have uncertainties of a nonnegligible magnitude. Differences between the measured temperatures and the surface temperature of the radiating cavity are also of concern.

Another uncertainty is the fact that Planck's law was derived under the assumptions of the wavelengths of the standing waves being negligibly small compared with the dimensions of the cavity, of loss-free (perfectly reflecting) cavity walls, and of a cavity in the form of a parallelepiped. These factors are still not completely resolved even as of 1987, and modern contributions arising from quantum optics and the theory of partial coherence and stimulated by the impact of various size, shape, and proximity effects on the radiation laws continue to appear in the literature (Baltes, 1976).

Blackbody-based radiation scales have been in use at the National Bureau of Standards (NBS), United States, since about 1913 (Coblentz and Stair, 1933). The NBS scale was originally based on the total radiation emitted by a blackbody cavity via the Stefan–Boltzmann law and was later derived directly from the spectral radiance of the blackbody cavity via Planck's law (Stair, 1970; Walker et al., 1987). A low-temperature blackbody cavity (Bedford, 1960) was used as a standard for total radiation at the NRC (Canada).

7.3 SILICON PHOTODIODES WITH A PREDICTABLE QUANTUM EFFICIENCY

Planar, shallow-junction, oxide-passivated silicon photodiodes of the type oxide-p^+-n-n^+ (Fig. 7.2) were first identified by Geist (1979) as candidates for an absolute radiometric standard.

The typical thickness of a silicon wafer used to make such photodiodes is 300 μm, with the doped p^+ and n^+ regions in the order of 1 μm. Devices of this type, when manufactured with a high degree of perfection, attain an

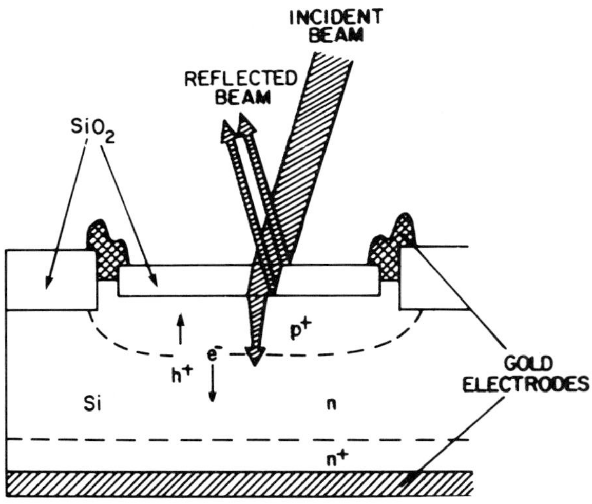

Fig. 7.2 Silicon photodiode oxide -p^+-n-n^+ junction.

internal quantum efficiency in the red region of the visible spectrum within a few tenths of one percent of unity and within a few percentage points of unity over most of the visible region. When the absorption of radiation in the SiO_2 antireflection layer is negligible, which can be achieved with thermally grown SiO_2 layers in the visible region, there are three main mechanisms, which can reduce the internal quantum efficiency to a value of less than unity. These are recombinations at the oxide–silicon interface, in the p^+ region, and in the n region.

The cause for recombinations at the oxide–silicon interface is the positive charge trapped close to the interface in thermally grown silicon dioxide layers. As this charge repels holes from the interface region, an electrical dipole centered on the interface is created. Its field increases the electron concentration there, thus increasing the probability for recombinations mediated by surface states at the interface. This recombination mechanism can be practically eliminated if sufficient charge is supplied to the oxide surface to repel electrons (minority carriers) away from the interface. In practice, this can be achieved by using either tin oxide (Geist et al., 1982b) or water drop (e.g. Zalewski and Geist, 1980) electrodes on the front surface or by applying a corona discharge (Geist et al., 1982b). The measured correction ε_0, which is a function of wavelength, is the ratio of the photocurrent without a bias voltage across the oxide layer (or a corona discharge to the surface) to that with a bias voltage (corona discharge) large enough to saturate the photocurrent. However, it has been found that the probability for recombination increases after each application of a surface charge due to the creation of midgap interface states, which promote recombination (Verdebout and Booker, 1984). The surface reflectance also changes after the application of the electrodes (Verdebout, 1983). Key et al. (1985) described methods for achieving the highest accuracy in oxide bias measurements and also pointed out that corona discharge is not a very satisfactory technique for this purpose.

Recombinations in the p^+ region have been shown to be much less significant than the interface recombinations (Geist, 1980), but the exact magnitude is difficult to determine (Wilkinson et al., 1983). Indications are that it can amount to a few tenths of a percent at short wavelengths.

In the UV region, the penetration depth of the radiation is so short that none of the radiation is absorbed beyond the space charge region (Geist et al., 1982a). The strong electric field associated with this region (in the vicinity of a p–n junction) separates the electron–hole pairs created there by the absorbed photons and sweeps them toward the n-type region (electrons) and the p^+-type region (holes), respectively. This movement is so fast that the probability of recombination is negligibly low, at least at current levels of radiometric accuracy. For photons of longer wavelength, the absorption takes place at larger and larger depths in the n-type material, which is eventually well outside the space charge region, from where the resulting electron–hole pairs now have to diffuse through the n-type material before being collected. This increases their chance of recombination considerably, with a resultant drop in quantum efficiency. Fortunately, it is possible to extend the depth of the space charge region (normally just a few μm) by means of a reverse-bias voltage to the rear electrode. If the reverse bias is made sufficiently large, the space charge region can be made to extend all the way to the rear of the photodiode, a distance typically of some 300 μm. However, electrical breakdown often occurs before this condition is reached. In this way, negligible recombination losses can be achieved for radiations with wavelengths up to 700 nm or more. The required reverse-bias voltage increases both with wavelength and with irradiation level. The wavelength-dependent correction factor ε_R is defined as the ratio of the photocurrent without reverse bias voltage to that with a bias voltage sufficiently large to saturate the photocurrent. It need only be applied if the photodiode is operated with zero reverse bias.

Assuming that all recombination losses other than those characterized by $\varepsilon_0(\lambda)$ and $\varepsilon_R(\lambda)$ are negligible, the internal quantum efficiency $\varepsilon(\lambda)$ of a photodiode can be expressed as (Schaefer et al., 1983)

$$\varepsilon(\lambda) = \frac{\varepsilon_0(\lambda)\varepsilon_R(\lambda)}{1 - [1 - \varepsilon_0(\lambda)][1 - \varepsilon_R(\lambda)]}, \tag{7.5}$$

which is a combination of measurable parameters. The spectral response $R(\lambda)$ of the photodiode is then given by

$$R(\lambda) = \frac{\lambda\varepsilon(\lambda)}{hc}[1 - \rho(\lambda)] \tag{7.6}$$

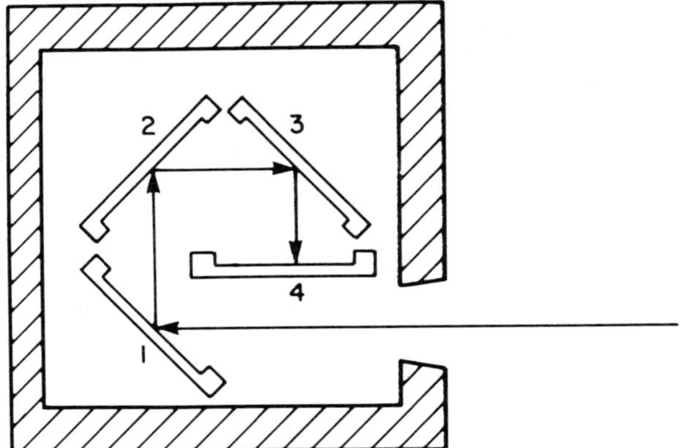

Fig. 7.3 Light-trapping configuration consisting of four photodiodes (numbered 1 to 4).

where $\rho(\lambda)$ is the spectral reflectance of the diode front surface, λ is the wavelength, h is Planck's constant, and c is the speed of light in vacuum.

Since the reflection from the photodiode front surface is highly specular, accurate measurements of the reflectance are not too difficult.

In order to eliminate the need for determining a correction for the recombination at the oxide–silicon interface, induced-junction (inversion-layer) silicon photodiodes of the type oxide-n^+-p were later shown to be suitable devices for producing an accurately predictable quantum efficiency (Zalewski and Duda, 1983; Booker and Geist, 1984). In these devices, the charge trapped in the oxide has the proper polarity to prevent minority carrier recombinations at the oxide–silicon interface. Zalewski and Duda (1983) also devised a light-trapping design for these photodiodes (see Fig. 7.3), with which they achieved an external quantum efficiency (product of internal quantum efficiency and absorptance) of 0.999 at short wavelengths. By employing a reverse bias, they extended the high-quantum-efficiency operating range to the entire visible spectrum and to power levels of up to several milliwatts.

7.4 SYNCHROTRON RADIATION

Although it was known for a long time that accelerated charged particles emit electromagnetic radiation, the phenomenon came under close experimental and theoretical scrutiny with the advent of particle accelerators. In the case of synchrotrons in particular, which are used to accelerate electrons, the loss of radiant energy from the accelerated particles posed a serious limitation to the attainment of high particle energies. The emitted radiation thus eventually became known as "synchrotron radiation." Schwinger (1949) was the first to derive the angular and spectral distribution of the radiation emitted by accelerated electrons. As a special case, he also derived the formula applying to an electron moving in a circular orbit with constant velocity. The radiant power in watts per radian, per electron, and per nanometer emitted in this case is given by

$$P(E, R, \theta, \lambda) = \frac{4e^2 cR}{3\lambda^4} \gamma^{-4}(1 + x^2)^2 \left[K_{2/3}^2(\xi) + \frac{x^2}{1 + x^2} K_{1/3}^2(\xi) \right] \quad (7.7)$$

where R is the radius of the electron orbit, E the energy of the orbiting electron, θ the angle perpendicular to the orbital plane, λ the wavelength of

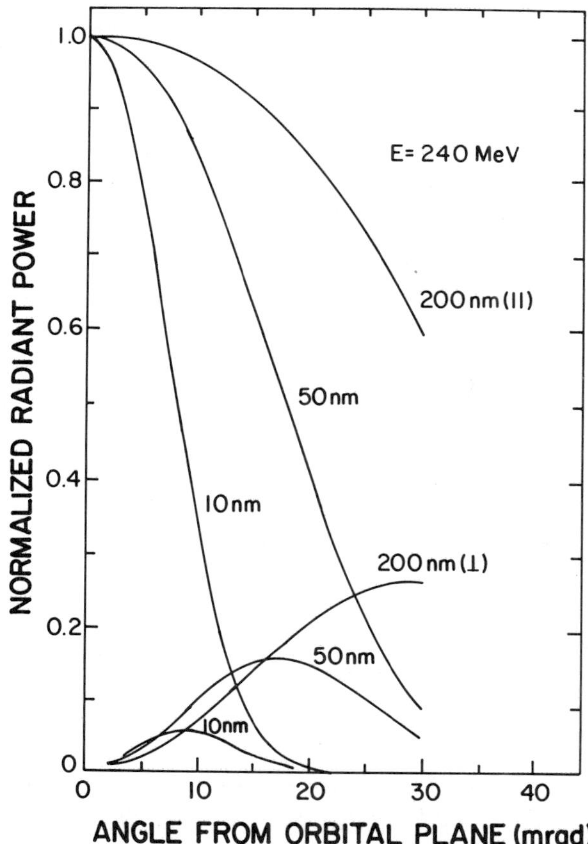

Fig. 7.4 Angular distribution of normal (⊥) and parallel (‖) polarization components from SURF-II for different wavelengths. (After Saloman et al., 1982)

the emitted radiation, e the electron charge, c the velocity of light in vacuum, m_0 the electron rest mass, $\gamma = E/(m_0c^2)$, and $\xi = (2\pi R/3\lambda)\gamma^{-3}(1 + x^2)^{3/2}$ with $x = \gamma\theta$. The functions $K_{2/3}(\xi)$ and $K_{1/3}(\xi)$ are modified Bessel functions of the second kind. The spectral irradiance on a target plane can be calculated as well if the collection geometry is defined.

Synchrotron radiation is emitted in a narrow angular interval around the orbital plane and in the direction of motion of the orbiting electrons. With decreasing wavelength, the radiation is confined to narrower and narrower angles from that plane. It is also highly polarized, with completely linear polarization in the orbital plane (electric vector parallel to the orbital plane). Figure 7.4 shows the distribution of the emitted radiant power with the angle from the orbital plane for both polarizations and for different wavelengths in the case of the NBS electron storage ring SURF-II (Saloman et al., 1982).

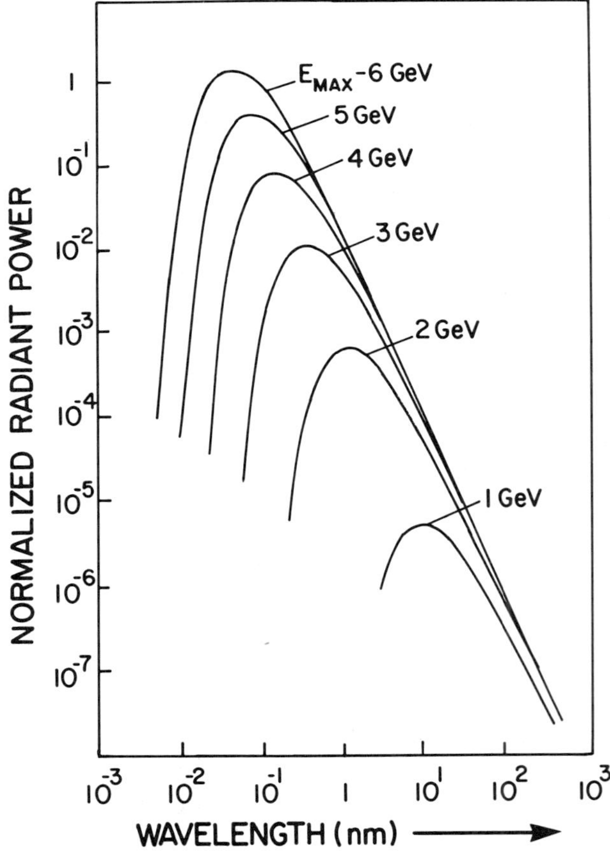

Fig. 7.5 Spectral distribution of synchrotron radiation. (After Lemke and Labs, 1967)

The spectral distribution of the emitted radiant power averaged over the acceleration time for an electron accelerated to different end energies is depicted in Fig. 7.5 (Lemke and Labs, 1967).

The theoretical predictions of the Schwinger theory were confirmed by several experimental studies (Tomboulian and Hartman, 1956; Codling and Madden, 1965; Lemke and Labs, 1967; Key, 1970) and its validity has thus been established with a high degree of confidence.

It soon became clear that the emitted radiation not only limited the attainment of high electron energies in synchrotrons, but also possessed some potentially useful features in other application areas. First, it is a powerful radiation source that surpasses all other artificial radiation-sources in intensity in certain spectral regions as, for example, in the shortwave UV region. This gives it an immediate application potential for spectroscopic, solid-state physics, microelectronic, and other purposes. Figure 7.6 shows a comparison between the spectrums of a 3000-K blackbody radiator and the storage ring BESSY (Kuhne et al., 1983).

The fact that the spectral distribution of the radiation emitted by a synchrotron (or storage ring) radiation source can be predicted theoretically from a set of instrument parameters and fundamental constants also makes it an obvious candidate for use as an absolute radiometric standard. Suggestions to this effect were put forward quite early (Tomboulian and Hartman, 1956), but the experimental realization of these ideas required a lot of further

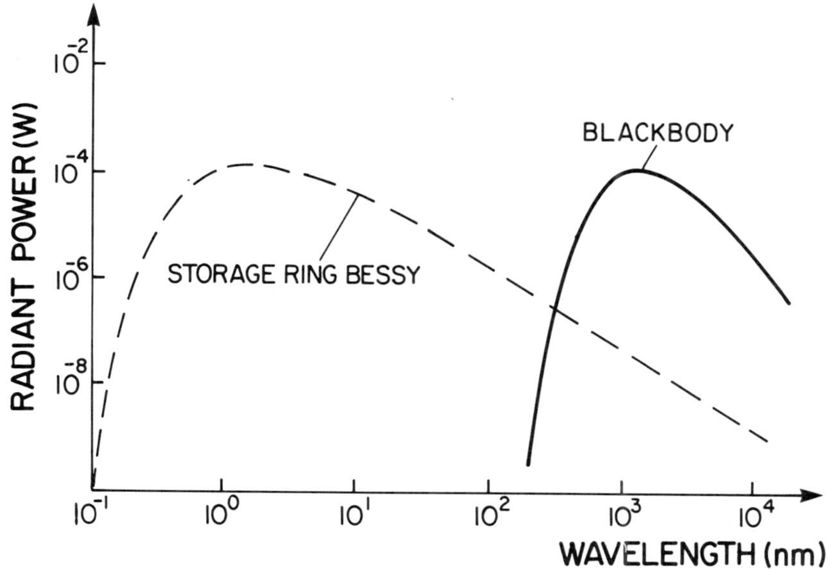

Fig. 7.6 Spectrum of 3000-K blackbody and storage ring BESSY. (After Kühne et al., 1983)

advances in instrumentation and measurement techniques. The efforts were sustained by the incentive to use the synchrotron radiation source for realizing an absolute radiometric scale in the vacuum ultraviolet and soft X-ray regions of the spectrum, where alternative standards are difficult to find. These are required for a variety of applications ranging from plasma physics to astrophysics as well as for the applications mentioned in the previous paragraph. A cross-check with other radiation scales in use in the visible and infrared regions was a further potential benefit. A number of such projects have been successfully concluded (Riehle and Wende, 1986; Schaefer et al., 1986), with uncertainties of the realized synchrotron radiation scales comparable to those of other alternative methods. There would not seem to be any fundamental reasons why this method could not be improved further in the future in line with similar progress with other absolute methods.

7.5 COMPARISONS BETWEEN INDEPENDENT OPTICAL POWER SCALES

Until the 1980s, the only absolute radiation scales in the optical region were based either on absolute radiometry or on blackbody radiators. Thus, comparisons were originally restricted to scales realized by these two methods. Using late-1980s knowledge of the value of the Stefan–Boltzmann constant, one could in principle regard every determination of this constant in the past as a scale comparison. Unfortunately, as pointed out already in Section 7.2, most of the early measurements of this constant were plagued, unbeknown to the experimenters, by diffraction problems, which complicates the interpretation of these data for the purpose of a scale comparison. Guild reported a comparison between his absolute radiometer and the blackbody-based scale of the NBS in 1937 (see Section 1.3.4.4) and found agreement well within the estimated uncertainty of the comparison of 0.2%. A comparison to check the consistency of the radiometric scales in use at various metrology laboratories was conducted by the Comité Consultatif de Photométrie of the International Committee on Weights and Measures (CIPM) in the period between 1962 and 1968. (See Section 1.3.1.5.) Five of the participants realized their scales with absolute radiometers and two by means of blackbody radiators. The comparison provided no conclusive evidence for a systematic difference between the two methods.

With the emergence of further absolute means of realizing radiant power scales, the number of comparisons between the various scales increased considerably.

The National Institute of Metrology (Chengdu, Peoples Republic of China) performed a comparison between a group of three self-calibrated silicon photodiodes and a group of five absolute radiometers (Li Tong-Bao, 1983). The radiation scales represented by the means of these groups agreed to better than 0.2%, with the uncertainty of the comparison being of about the same order of magnitude.

In a comparison between scales based on a silicon photodiode (SPD) with a predictable quantum efficiency and a gold point blackbody (BB), Schaefer and Saunders (1984) observed a scale ratio BB/SPD of 1.0076, with the uncertainty (66% confidence level) for the comparison being in excess of 1.2%. When the comparison was later extended to include a synchrotron source (SS), a scale ratio SS/SPD of 0.9975 was found (Schaefer et al., 1986). The uncertainty of the latter comparison was 0.84% at a 66% confidence level.

As of 1987, the most accurate comparison involving a synchrotron source was performed between the scales realized with the cryogenic absolute radiometer (AR) at the NPL (see Section 1.3.4.6) and with the electron storage ring BESSY of the PTB. (See also Section 7.4.) The wavelengths used in the comparison were 676 and 799 nm, and the mean scale ratio AR/SS measured was 1.0013, with an uncertainty (66% confidence level) of the comparison of 0.35% (Fox et al., 1986).

Boivin and McNeely (1986) reported a comparison between the scale based on three of the absolute radiometers at the NRC (Canada) and on two silicon photodiodes with a predictable quantum efficiency. Using 5 krypton laser lines between 476 and 676 nm at power levels of about 1 mW, they found agreement between the scales at the level of 0.1%.

A further comparison was performed between a silicon photodiode with a predictable quantum efficiency and an absolute radiometer at the National Measurement Laboratory (Australia). The photodiode was of the inversion layer type (see Section 7.3) and measurements were performed at four wavelengths between 458 and 633 nm. The two scales agreed at a level of 0.1% (Gardner and Brown, 1987).

A comparison of radiant power scales at 633 and 488 nm was conducted under the auspices of the Comité Consultatif de Photométrie et Radiométrie of the CIPM between 1985 and 1986 (Zalewski, 1986). Both pn- and np-type photodiodes were used as transfer detectors and the NBS acted as the organizing laboratory. Participating laboratories were the NRC (Canada), NPL (U.K.), PTB (Federal Republic of Germany), NIM (Chengdu, Peoples Republic of China), NML (Australia), Research Institute of Technical Physics of the Hungarian Academy of Sciences, National Office of Measures (Hungary), ETL (Japan), NPRL (South Africa), and INM (France). (For a list of the acronyms used, see Appendix 1.1). At 633 nm, four of the participants used absolute radiometers, 6 (including the NBS) used silicon photodiodes with a predictable quantum efficiency, and one laboratory used both as their absolute scale references. At 400 nm, ETL did not participate, while 5 laboratories used absolute radiometers and 5 laboratories (including NBS) used silicon photodiodes with a predictable quantum efficiency. The internal scatter of both groups (absolute radiometers and silicon photodiodes) was of about the same order of magnitude as the scatter of all the participating laboratories (standard deviation 0.7 %) and neither the type of transfer detector nor the wavelength seemed to have a major influence on the overall results within the uncertainty of the comparison. No statistically significant difference was observed between the radiant power scales based on absolute radiometers and those based on silicon photodiodes with a predictable quantum efficiency.

REFERENCES

Blackbody Radiators

Alfano, G., and Sarnoe, A. (1975). Normal and hemispherical thermal emittances of cylindrical cavities. *J. Heat Transfer* **C97**, 387.

Babushkin, V. V. (1981). Certification of models of an absolutely black body as standard sources for radiometer research. *Izmer. Tekh.* (USSR) **24**, 39.

Babushkin, V. V., Dolgikh, I. I., and Libova, I. V. (1979). A blackbody model as a standard IR facility. *Izmer. Tekh.* (USSR) **22**, 68.

Bacherikov, V. V., Vlasov, L. V., Morozov, N. A., Samoilov, L. N., and Sapritskii, V. I. (1983). High-temperature radiators for standards in energy photometry. *Izmer. Tekh.* (USSR) **26**, 31. Cover-to-cover translation in *Meas. Tech.* (U.S.), Dec. 1983, 988.

Baltes, H. P. (1973). Deviations from the Stefan-Boltzmann law at low temperatures. *Appl. Phys.* **1**, 39.

Baltes, H. P. (1976). Planck's radiation law for finite cavities and related problems. *Infrared Phys.* **16**, 1.

Baltes, H. P., and Kneubuehl, F. K. (1970). Spectral density, thermodynamics and temporal coherence of non-Planckian blackbody radiation for small cavities. *Opt. Comm.* **2**, 14.

Bartell, F. O., and Wolfe, W. L. (1975). New approach for the design of blackbody simulators. *Appl. Opt.* **14**, 249.

Bartell, F. O., and Wolfe, W. L. (1976a). Cavity radiation theory. *Infrared Phys.* **16**, 13.

Bartell, F. O., and Wolfe, W. L. (1976b). Cavity radiators: an ecumenical theory. *Appl. Opt.* **15**, 84.

Bastie, J., Keller, A., and Lacaud, C. (1984). Etude et réalisation d'un corps noir de transfert. *Bulletin BNM* (France) **55**, 5.

Bauer, G., and Bischoff, K. (1970). Hohlraumstrahler mit Temperaturgefälle. *Optik* **31**, 507.

Bauer, G., and Bischoff, K. (1971). Evaluation of the emissivity of a cavity source by reflection measurements. *Appl. Opt.* **10**, 2639.

Bedford, R. E. (1960). A low temperature standard of total radiation. *Can. J. Phys.* **38**, 1256.

Bedford, R. E. (1970). Blackbodies as absolute radiation standards. In *Advances in Geophysics*, H. E. Landsberg and J. van Mieghem, eds., Vol. 14, p. 165. Academic Press, New York and London.

Bedford, R. E., and Ma, C. K. (1974). Emissivities of diffuse cavities: isothermal and nonisothermal cones and cylinders. *J. Opt. Soc. Am.* **64**, 339.

Bedford, R. E., Ma, C. K., Zaixiang Chu, Yuxing Sun, and Shouren Chen (1985). Emissivities of diffuse cavities. IV. Isothermal and nonisothermal cylindro-inner-cones. *Appl. Opt.* **24**, 2971.

Blevin, W. R., and Brown, W. J. (1970). A precise measurement of the Stefan-Boltzmann constant. *Metrologia* **7**, 15.

Chandos, R. J., and Chandos, R. E. (1974). Radiometric properties of isothermal, diffuse wall cavity sources. *Appl. Opt.* **13**, 2142.

Chernin, S. M. (1973). High-temperature source of blackbody radiation for calibrating spectroscopic and radiometric instruments. *Prib. Tekh. Eksp.* (USSR) **16**, 148.

Coblentz, W. W., and Stair, R. (1933). The present status of the standards of thermal radiation maintained by the Bureau of Standards. *J. Res. Nat. Bur. Std.* **11**, 79.

Cussen, A. J. (1982). Overview of blackbody radiation sources. *Proc. SPIE* (*Infrared Sensor Technology*) **344**, 2.

de Vos, J. C. (1954). Evaluation of the quality of a blackbody. *Physica* **20**, 669.

Evans, C. M. (1978). Calculating 'in band' flux density of blackbody sources. *Electro-opt. Syst. Des.* **10**, 39.

Flemming, J. C. (1966). An evaluation of a high temperature blackbody as a working standard of spectral irradiance. *Appl. Opt.* **5**, 195.

Fussell, W. B. (1972). Normal emissivity of an isothermal, diffusely reflecting cylindrical cavity (with top) as a function of inside radius. *J. Res. Nat. Bur. Std.* (U.S.) **76A**, 347.

Geist, J. (1971). Note on the quality of freezing point blackbodies. *Appl. Opt.* **10**, 2188.

Geist, J. (1972). The effect of wall roughness on the spectral density of radiation within symmetric closed cavities in good conductors. *J. Opt. Soc. Am.* **62**, 602.

Globus, M. E. (1979). Effect of the reflectivity of the radiating surface on the quality of a model of an absolute blackbody. *Izmer. Tekh.* (USSR) **22**, 16.

Hochhheimer, B. F. (1977). Radiation pattern for a diffuse wall cavity, nonuniform in temperature and emissivity. *Appl. Opt.* **16**, 2038.

Jones, O. C., and Forno, C. (1971). Reflectance measurements on cavity radiators. *Appl. Opt.* **10**, 2644.

Jones, O. C., and Gordon-Smith, G. W. (1973). Absolute radiometry by means of a blackbody source. *Proc. R. Soc. Lond.* **A335**, 369.

Karoli, A. R. (1970). Experimental blackbody (absolute) radiometry. In *Advances in Geophysics*, H. E. Landsberg and J. van Mieghem, eds., Vol. 14, p. 203. Academic Press, New York and London.

Kelly, F. J. (1966). An equation for the local thermal emissivity at the vertex of a diffuse conical or V-groove cavity. *Appl. Opt.* **5**, 925.

Kelly, F. J., and Moore, D. G. (1965). A test of analytical expressions for the thermal emissivity of shallow cylindrical cavities. *Appl. Opt.* **4**, 31.

Kholopov, G. K. (1973). Radiation of diffuse isothermal cavities. *Inzh.-Fiz. Zh.* (USSR) **25**, 1112.

Kolokolov, A. A., and Skrotskii, G. V. (1974). Equilibrium radiation in a finite-size cavity. *Opt. and Spektrosk.* (USSR) **36**, 217.

Kostkowski, H. J., Erminy, D. E., and Hattenburg, A. T. (1970). High-accuracy spectral radiance calibration of tungsten-strip lamps. In *Advances in Geophysics*, H. E. Landsberg and J. van Mieghem, eds., Vol. 14, p. 111. Academic Press, New York and London.

Kotyuk, A. F., Samoilov, L. N., and Lovinskii, L. S. (1976). A high-temperature black-body model for energy photometry. *Izmer. Tekh.* (USSR) **19**, 63.

Lapworth, K. C., Quinn, T. J., and Allnutt, L. A. (1970). A black-body source of radiation covering a wavelength range from the ultraviolet to the infrared. *J. Phys. E: Sci. Instr.* **3**, 116.

Marette, G. (1975). Angular properties of blackbody simulators. *Appl. Opt.* **14**, 2665.

Marette, G. (1976). Realisation of a blackbody simulator for the absolute calibration of luminous flux detectors. *Opt. Commun.* (the Netherlands) **18**, 576.

Ohwada, Y. (1981). Numerical calculation of effective emissivities of diffuse cones with a series technique. *Appl. Opt.* **20**, 3332.

Palmer, J. M. (1979). Spectral radiant emittance of a nonisothermal cavity radiator. *Appl. Opt.* **18**, 758.

Peavy, B. A. (1966). A note on the numerical evaluation of thermal radiation characteristics of diffuse cylindrical and conical cavities. *J. Res. Nat. Bur. Std.* (U.S.) **70C**, 139.

Powell, W. R. (1974). Evaluation of nonisothermal semi-infinite cylinder with specularly reflecting walls as a blackbody source. *Appl. Opt.* **13**, 593.

Quinn, T. J. (1983). *Temperature.* Monographs in Physical Measurement, A. H. Cook, ed. Acadmic Press, New York and London.

Quinn, T. J., and Martin, J. E. (1985). A radiometric determination of the Stefan–Boltzmann constant and thermodynamic temperatures between −40°C and +100°C. *Phil. Trans. R. Soc. Lond.* **A316**, 85.

Shirley, J. H., and Eberly, J. H. (1979). Local effective emissivity of conical cavities. *Appl. Opt.* **18**, 3810.

Simonov, V. P. (1974). Characteristics of thermal radiation in axisymmetric cavities. *Inzh.-Fiz. Zh.* (USSR) **27**, 208.

Sparrow, E. M., and Jonsson, V. K. (1963). Radiant emission characteristics of diffuse conical cavities. *J. Opt. Soc. Am.* **53**, 816.

Sparrow, E. M., Kruger, P.D., and Heinisch, R. P. (1974). Radiation from cavities with nonisothermal heat conducting walls. *J. Heat Transfer*, Feb. 1974, 15.

Stair, R. (1970). Sources as radiometric standards. In *Advances in Geophysics*, H. E. Landsberg and J. van Mieghem, eds., Vol. 14, p. 83. Academic Press, New York and London.

Stair, R. S., Johnston, R. G., and Halbach, E. W. (1960). Standard of spectral radiance for the region of 0.25 to 2.6 microns. *J. Res. Nat. Bur. Std.* **64A**, 291.

Sydnor, C. L. (1969). Series representation of the solution of the integral equation for emissivity of cavities. *J. Opt. Soc. Am.* **59**, 1288.

Vollmer, J. (1957). Study of the effective thermal emittance of cylindrical cavities. *J. Opt. Soc. Am.* **47**, 926.

Walker, J. H., Saunders, R. D., and Hattenburg, A. T. (1987). *Spectral Radiance Calibrations.* National Bureau of Standards Special Publication 250–1. U.S. Government Printing Office, Washington, D.C.

Williams, C. S. (1969). Specularly vs. diffusely reflecting walls for cavity-type sources of radiant energy. *J. Opt. Soc. Am.* **59**, 249.

Silicon Photodiodes with a Predictable Quantum Efficiency

Booker, R. L., and Geist, J. C. (1982). Photodiode quantum efficiency enhancement at 365 nm: optical and electrical. *Appl. Opt.* **21**, 3987.

Booker, R. L., and Geist, J. (1984). Induced junction (inversion layer) photodiode self-calibration. *Appl. Opt.* **23**, 1940.

Campos, J., Pons, A., and Corrons, A. (1984). Autocalibrado de algunos fotodiodos de Si. *Opt. Pura Y Aplicada* **17**, 141.

Eppeldauer, G. (1983a). Compact selfcalibration setup for high sensitivity absolute light measurements. *Proc. 20th Session* International Commission on Illumination (CIE), p. E19. Amsterdam, the Netherlands.

Eppeldauer, G. (1983b). High sensitivity absolute radiometer. *Proc. 10th Int. Symp. IMEKO Technical Committee on Photon-Detectors*, p. 145, Berlin, Federal Republic of Germany.

Fan, C. L., and Boyd, J. T. (1983). Improvement in the quantum efficiency of silicon photodetectors at near IR wavelengths by edge illumination. *Appl. Opt.* **22**, 3297.

Gardner, J. L., and Brown, W. J. (1987). Silicon radiometry compared to the Australian radiometric scale. *Appl. Opt.* **26**, 2431.

Geist, J. (1979). Quantum efficiency of the p–n junction in silicon as an absolute radiometric standard. *Appl. Opt.* **18**, 760.

Geist, J. (1980). Silicon photodiode front region collection efficiency models, *J. Appl. Phys.* **51**, 3993.

Geist, J. (1983). The quantum yield of silicon in the ultraviolet. *Proc. 10th Int. Symp. International Measurement Confederation* (IMEKO) *Technical Committee on Photon-Detectors*, p. 49. Berlin, Federal Republic of Germany.

Geist, J., and Zalewski, E. F. (1979). The quantum yield of silicon in the visible. *Appl. Phys. Lett.* **35**, 503.

Geist, J., Zalewski, E. F., and Schaefer, A. R. (1980). Spectral response self-calibration and interpolation of silicon photodiodes. *Appl. Opt.* **19**, 3795.

Geist, J., Liang, E., and Schaefer, A. R. (1981). Complete collection of minority carriers from the inversion layer in induced junction diodes. *J. Appl. Phys.* **52**, 4879.

Geist, J., Gladden, W. K., and Zalewski, E. F. (1982a). Physics of photon-flux measurements with silicon photodiodes. *J. Opt. Soc. Am.* **72**, 1068.

Geist, J. C., Farmer, A. J. D., Martin, P. J., Wilkinson, F. J., and Collocott, J. (1982b). Elimination of interface recombination in oxide passivated silicon p^+n photodiodes by storage of negative charge on the oxide surface. *Appl. Opt.* **21**, 1130.

Hughes, C. G. (1982). Silicon photodiode absolute response self-calibration using a filtered tungsten source. *Appl. Opt.* **21**, 2129.

Key, P. J., Fox, N. P., and Rastello, M. L. (1985). Oxide-bias measurements in the silicon photodiode self-calibration technique. *Metrologia* **21**, 81.

Li Tong-Bao (1983). Checking the accuracy of absolute radiometers with technique of silicon photodiode self-calibration. *Proc. 10th Int. Symp. International Measurement Confederation* (IMEKO) *Technical Committee on Photon-Detectors*, p. 137. Berlin, Federal Republic of Germany.

Saito, I. (1986). Measurement of Absolute Responsivity of Silicon Photodiode. *Proc. CCPR*, Document CCPR/86–9. Bureau International des Poids et Mesures, Sevres, France.

Schaefer, A. R., and Saunders, R. D. (1984). Intercomparison between silicon and blackbody based radiometry using a silicon photodiode/filter radiometer. *Appl. Opt.* **23**, 2224.

Schaefer, A. R., Zalewski, E. F., and Geist, J. (1983). Silicon detector nonlinearity and related effects. *Appl. Opt.* **22**, 1232.

Verdebout, J. (1983). Change in the photodiode reflectance when the oxide bias is applied during the self-calibration procedure. *J. Phys. E: Sci. Instr.* **16**, 522.

Verdebout, J., and Booker, R. L. (1984). Degradation of native oxide passivated silicon photodiodes by repeated oxide bias. *J. Appl. Phys.* **55**, 406.

Wilkinson, F. J., Farmer, A. J. D., and Geist, J. (1983). The near ultraviolet quantum yield of silicon. *J. Appl. Phys.* **54**, 1172.

Zalewski, E. F., and Duda, C. R. (1983). Silicon photodiode device with 100 percent external quantum efficiency. *Appl. Opt.* **22**, 2867.

Zalewski, E. F., and Geist, J. (1980). Silicon photodiode absolute spectral response self-calibration. *Appl. Opt.* **19**, 1214.

Zalewski, E. F., and Gladden, W. K. (1984). Absolute spectral irradiance measurements based on the predicted quantum efficiency of a silicon photodiode. *Opt. Pura y Aplicada* **17**, 133.

Zalewski, E., and Tufino, M. (1981). Silicon photodiode self-calibration as a basis for radiometry in the infrared. *Proc. Conf. on Contemporary Infrared Standards and Calibration*, p. 2. San Diego, California.

Synchrotron Radiation

Codling, K., and Madden, R. P. (1965). Characteristics of the "synchrotron light" from the NBS 180-MeV machine. *J. Appl. Phys.* **36**, 380.

Ederer, D. L., Saloman, E. B., Ebner, S. C., and Madden, R. P. (1975). The use of synchrotron radiation as an absolute source of VUV radiation. *J. Res. Nat. Bur. Std.-Phys. and Chem.* **79A**, 761.

Einfield, D., and Stuck, D. (1979). Synchrotron radiation as an absolute standard source. *Proc. National Conference on Synchrotron Radiation Instrumentation*, p. 101. Gaithersburg, Maryland, United States.

Einfield, D., Stuck, D., Behringer, K., and Thoma, P. (1976). Comparison of synchrotron radiation and hydrogen continuum radiation in the VUV by means of a deuterium transfer standard. *Z. Naturforsch.* **31A**, 1131.

Gluskin, E. S., Trakhtenberg, E. M., and Feldman, I. G. (1980). Equipment for the measurement of the absolute efficiency of VUV detectors using synchrotron radiation of the VEPP-2M storage ring. *Space Sci. Instrum.* (the Netherlands) **5**, 129.

Hänsel, R., and Kunz, C. (1967). Experimente mit der Synchrotronstrahlung. *Z. Angew. Phys.* **23**, 276.

Hughey, L. R., and Schaeffer, A. R. (1982). Reduced absolute uncertainty in the irradiance of SURF-II and instrumentation for measuring linearity of X-ray, XUV and UV detectors. *Nucl. Instrum. Methods* **195**, 367.

Kaase, H. (1975). Measurement of the irradiance of UV sources by comparison with synchrotron radiation. *J. Phys. E: Sci. Instr.* **8**, 590.

Kaase, H. (1976). Untersuchungen über die Eignung des PTB-Synchrotrons als Strahlungsnormal und Realisierung einer Skala für Bestrahlungsstärke im UV. *Optik* **46**, 149.

Kaase, H. (1981). A direct radiometric comparison of two radiation standards for the vacuum ultraviolet: synchrotron and Argon cascade arc. *Optik* **59**, 1.

Kaase, H., Stephen, K. H., and Burton, W. M. (1980). Intercomparison of radiometric irradiance scales in the 90–250 nm wavelength range. *Appl. Opt.* **19**, 2529.

Key, P. J. (1970). Synchrotron radiation as a standard of spectral emission. *Metrologia* **6**, 97.

Key, P. J., and Preston, R. C. (1977). Vacuum ultraviolet radiation scales. *Nature* (*GB*) **265**, 717.

Key, P. J., and Ward, T. H. (1978). The establishment of ultraviolet spectral emission scales using synchrotron radiation. *Metrologia* **14**, 17.

Koch, E. E. (1983). *Handbook on Synchrotron Radiation*, Vols. 1A, 1B. North-Holland Physics, Amsterdam, the Netherlands.

Kühne, M., Riehle, F., Tegeler, E., and Wende, B. (1983). The radiometric laboratory of the PTB at BESSY. *Proc. Int. Conf. X-ray and VUV Synchrotron Radiation Instrumentation*, p. 399.

Lemke, D., and Labs, D. (1967). The synchrotron radiation of the 6-GeV DESY machine as a fundamental radiometric standard. *Appl. Opt.* **6**, 1043.

Madden, R. P. (1980). A status report on the SURF-II synchrotron radiation facility at the NBS. *Nucl. Instrum. Methods* **172**, 1.

Masuoka, T., Oshio, T., Iwanaga, R., and Sonoda, H. (1976). Absolute photon-flux and angular distribution of the Tokyo synchrotron radiation in the vacuum ultraviolet. *Jap. J. Appl. Phys.* **15**, 1579.

Meyer, P., and Lagarde, P. (1976). Synchrotron radiation in the infrared. *J. Phys.* (France) **37**, 1387.

Pitz, E. (1969). Absolute calibration of light sources in the VUV by means of the synchrotron radiation of DESY. *Appl. Opt.* **8**, 255.

Riehle, F., and Wende, B. (1986). Establishment of a spectral irradiance scale in the visible and near infrared using the electron storage ring BESSY. *Metrologia* **22**, 75.

Rusbuldt, D., and Thimm, K. (1974). The synchrotron as a radiation standard for the vacuum ultraviolet. *Nucl. Instrum. and Methods* **116**, 125.

Saloman, E. B., Ebner, S. C., and Hughey, L. R. (1982). Vacuum ultraviolet and extreme ultraviolet radiometry using synchrotron radiation at the National Bureau of Standards. *Opt. Eng.* **21**, 951.

Schaefer, A. R., Hughey, L. R., and Fowler, J. B. (1984). Direct determination of the stored electron beam current at the NBS electron storage ring SURF-II. *Metrologia* **19**, 131.

Schaefer, A. R., Saunders, R. D., and Hughey, L. R. (1986). Intercomparison between independent irradiance scales based on silicon photodiode physics, gold point blackbody radiation and synchrotron radiation. *Opt. Eng.* **25**, 892.

Schwinger, J. (1949). On the classical radiation of accelerated electrons. *Phys. Rev.* **75**, 1912.

Stevenson, J. R., Ellis, H., and Bartlett, R. (1973). Synchrotron radiation as an infrared source. *Appl. Opt.* **12**, 2884.

Tomboulian, D. H., and Hartman, P. L. (1956). Spectral and angular distribution of ultraviolet radiation from the 300-Mev Cornell synchrotron. *Phys. Rev.* **102**, 1423.

Williams, G. P. (1982). The National Synchrotron Light Source in the infrared region. *Proc. Second National Conf. on Synchrotron Radiation Instrumentation*, p. 383. Ithaca, New York.

Comparisons Between Independent Optical Power Scales

Boivin, L. P., and McNeely, F. T. (1986). Electrically calibrated absolute radiometer suitable for measurement automation. *Appl. Opt.* **25**, 554.

Fox, N. P., Key, P. J., Riehle, F., and Wende, B. (1986). Intercomparison between two independent primary radiometric standards in the visible and near infrared: a cryogenic radiometer and the electron storage ring BESSY. *Appl. Opt.* **25**, 2409.

Gardner, J. L., and Brown, W. J. (1987). Silicon radiometry compared to the Australian radiometric scale. *Appl. Opt.* **26**, 2341.

Li Tong-Bao (1983). Checking the accuracy of absolute radiometers with technique of silicon photodiode self-calibration. *Proc. 10th Int. Symp. International Measurement Confederation* (IMEKO) *Technical Committee on Photon-Detectors*, p. 137. Berlin, Federal Republic of Germany.

Schaefer, A. R., and Saunders, R. D. (1984). Intercomparison between silicon and blackbody-based radiometry using a silicon photodiode/filter radiometer. *Appl. Opt.* **23**, 2224.

Schaefer, A. R., Saunders, R. D., and Hughey, L. R. (1986). Intercomparison between independent irradiance scales based on silicon photodiode physics, gold-point blackbody radiation, and synchrotron radiation. *Opt. Eng.* **25**, 892.

Zalewski, E. F. (1986). *Addendum to the report on the preliminary round of the comparison of the national standards of absolute spectral responsivity.* Document CCPR/86-10. Bureau International des Poids et Mesures (BIPM), Sevres, France.

8

DC Substitution Methods Used in Other Areas of Metrology

F. HENGSTBERGER
Council for Scientific and Industrial Research
Pretoria, South Africa

8.1 INTRODUCTION

If one considers the whole electromagnetic-frequency spectrum (see Section 1.2.1), it is clear that DC, corresponding to the frequency zero, occupies a special position. With the influence of the time parameter and of quantities varying with it (or with some of its derivatives) removed, it seems natural that DC quantities should be the most accurately measurable of all the electrical quantities. Taking into account the common physical basis of electromagnetic radiation at any frequency and of DC, it is not surprising to find that the measurement of many electromagnetic radiation parameters is done by DC substitution techniques. Examples for such parameters are power, pulse energy, and voltage. Similar techniques are employed throughout the whole electromagnetic frequency domain. They are all based on converting the electromagnetic radiation to heat and then determining the parameter of interest from the corresponding DC parameter that produces the same heating effect.

Apart from their use for determining the power or energy in beams of electromagnetic radiations, DC substitution techniques are also widely used to measure these same parameters in particle beams (as for instance emitted by radioactive substances or in nuclear reactions or, as propagating in particle accelerators). In some cases the energies from the radiative and particle components of a particular reaction are measured with the same instrument.

ISBN 0-12-340810-5

When reviewing DC substitution techniques in these other areas of metrology in the following sections, the coverage of the subject matter will obviously not be as extensive as in the earlier chapters dealing with absolute radiometry in the optical region. It is merely intended to alert the reader to related work being carried out in these areas and to provide a list of key references (arranged by chapter subhead) for more detailed study.

8.2 GAMMA-RAY AND X-RAY REGION

Although the absolute measurement of the radiant flux (also referred to as "energy flux" in this field) or radiant energy of gamma-rays and X-rays is important in its own right, the penetrating nature of these radiations also makes it important to be able to measure the power ("absorbed dose rate") or energy ("absorbed dose") absorbed per unit mass in a particular material. Absorbed-dose calorimeters differ from calorimeters used to measure radiant energy mainly in that the latter are designed to measure the total beam energy while the former only measure the energy absorbed in a limited test mass. The test mass of an absorbed-dose calorimeter absorbs only a fraction of the incident radiation; the rest passes through. For reasons of radiation protection and the application of these radiations in medicine, it is of special importance to know the absorbed dose rate or the absorbed dose in tissuelike materials such as water, carbon, and certain plastics.

Tissue consists to a large extent of water; therefore, absorbed dose in water is very important in radiation dosimetry. As measurements of absorbed dose in a large enough volume of water were not considered feasible for practical reasons, substitute materials such as graphite were generally preferred. Later, polystyrene was also used due to its more waterlike radiation-absorption properties. In both cases, the conversion of the measured absorbed dose to the equivalent absorbed dose in water necessitates corrections that depend on the spectrum of the incident radiation. In the 1980s, water calorimeters have been shown to be quite feasible (e.g. Domen, 1980, 1982), but their ultimate performance and usefulness is still the subject of detailed investigations. Polystyrene–water (e.g. Domen, 1983a, 1983b, 1983c) and graphite–water calorimeters (e.g., Sundara and Naik, 1980) are also being studied.

Considerable progress has been made over the past few decades to reduce uncertainties associated with instrumental imperfections, to achieve higher sensitivities, and to make the instruments amenable to automation.

Calorimetric techniques employed in radiation dosimetry are reviewed by Laughlin and Genna (1966) and Radak and Markovic (1970).

8.3 MICROWAVE AND RADIO-FREQUENCY REGION

Instruments and techniques employed in this area have many similarities with those used in the optical region. This is exemplified by a DC substitution radiometer for radio-frequency (RF) power at the NPRL (Pretoria, South Africa), which uses exactly the same electronic operating system as the absolute radiometer developed at the same institute (Dressler, 1985). Even the housing for the RF radiometer head is the same as that of its optical equivalent. The system is depicted in Fig. 8.1. (See Fig. 1.38 for comparison.)

Voltage (V), current (I), and impedance (Z) are the basic quantities used to characterize electromagnetic waves and their transmission at low frequencies (see Section 8.4), and power is a derived quantity that can be calculated from these parameters as (Klonz, 1987)

$$P = \frac{1}{T}\int_0^T \mathrm{VI}\,dt = \mathrm{Re}(Z)\frac{1}{T}\int_0^T I^2\,dt = \frac{\mathrm{Re}(Z)}{|Z|^2}\frac{1}{T}\int_0^T V^2\,dt, \tag{8.1}$$

where $\mathrm{Re}(z)$ is the real part of the complex impedance Z and $|Z|$ is its numerical magnitude (modulus). At microwave frequencies, circuit constants are distributed quantities and the transmission line lengths are appreciable fractions of the wavelength. (At lower frequencies, the wavelength is generally much larger than the length of the transmission line; e.g., 10 MHz corresponds to a wavelength of 30 m.) Consequently, it becomes difficult at high frequencies to measure the impedance at a particular point and any impedance mismatch between a source and a load results in standing waves along the transmission line. These standing waves make voltage and current measurements highly arbitrary unless an exact reference plane is specified for the measurement. Power, however, remains invariant with position along a lossless transmission line. This makes power the preferred parameter to be measured at high frequencies, with voltage and current being derived from it if necessary. Also, in the case of waveguides, voltage and current are no longer essential and convenient concepts, and power becomes the more fundamental quantity.

For coaxial transmission lines, the line is usually terminated with a 50Ω load resistor, which dissipates either the RF power or substituted DC power. Dual-load instruments are often employed for temperature compensation as well as for the simultaneous dissipation of RF power in one and DC power in the other load. The NPRL instrument shown in Fig. 8.1 is of this dual-load type. The mechanical construction of its RF measuring head is depicted in Fig. 8.2. It can measure RF power at frequencies of up to 3 GHz and at power levels between 0.5 and 50 mW. Instead of the thick-film bifilar gold spirals used as temperature sensors in the optical version, it employs thermistor beads, which are incorporated in the same type of bridge circuit as the

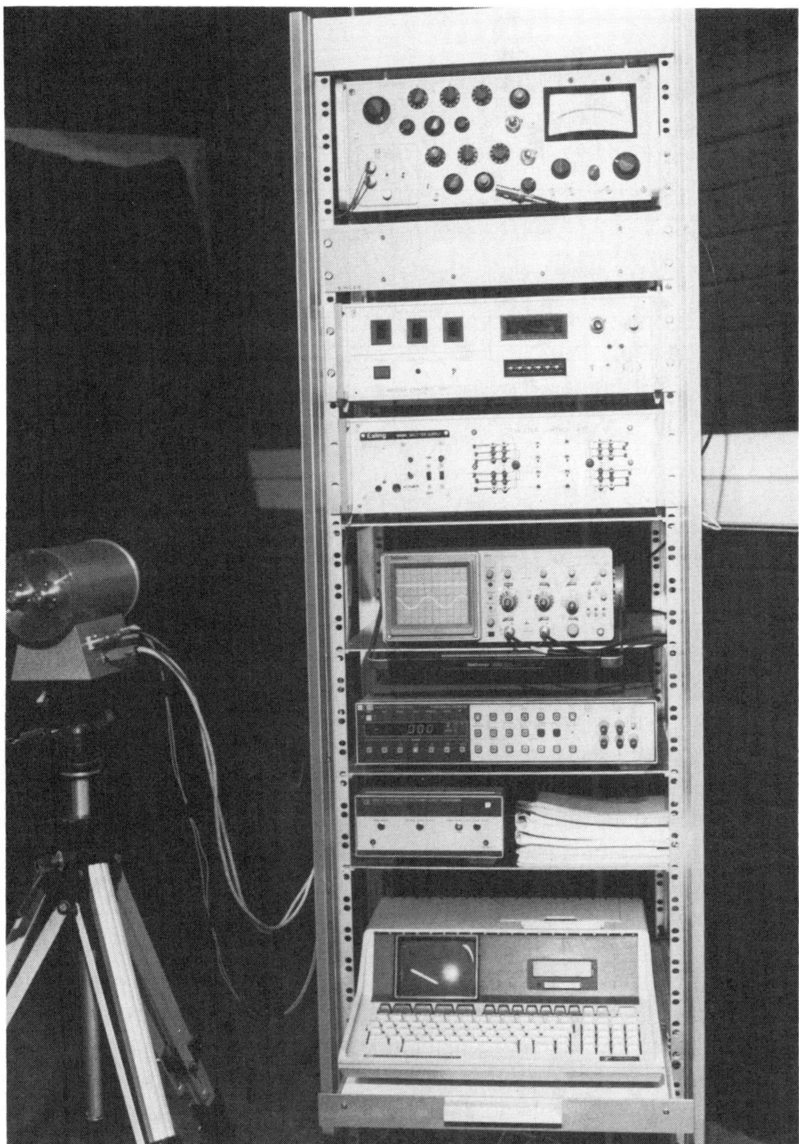

Fig. 8.1 Photograph of the NPRL RF power head and control electronics.

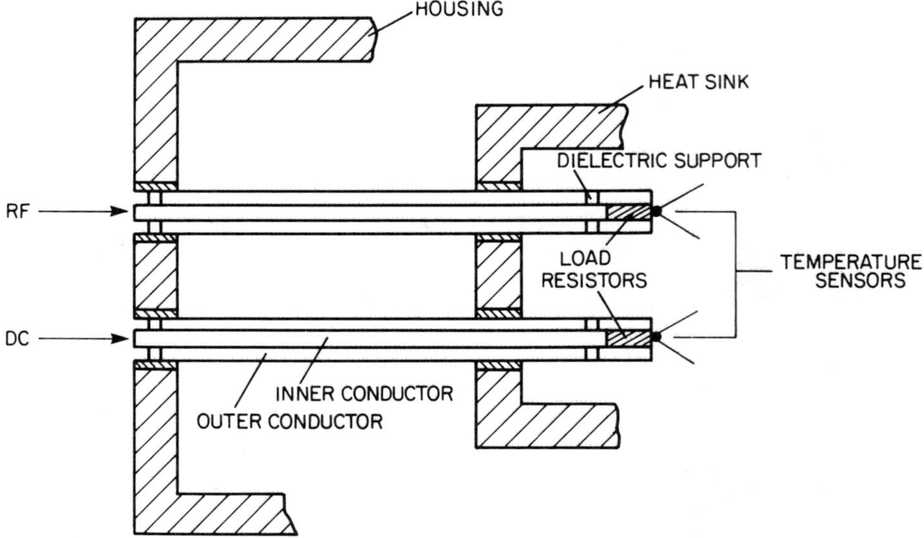

Fig. 8.2 Schematic diagram of the RF power head constructed at the NPRL. [After Dressler (1985)]

absolute radiometer (Fig. 1.32). In contrast to the latter, it is usually operated with one load heated by RF and the other by a feedback-controlled DC power. This operating mode introduces an additional correction term (dual detector correction; see Section 1.3.3). Unfortunately, a simultaneous heating of one load by RF and DC is not feasible, contrary to the case for optical power. An electronic time-constant compensation circuit of the type in Fig. 1.49c reduces the stabilization period following a power change by a factor of 3.5 compared with the uncompensated case—i.e., to about two minutes. (The thermal time-constant is about one minute.)

In order to be able to heat the same detector element simultaneously by DC and RF, Fantom (1979) used auxiliary heaters around each load, with the heaters electrically isolated from it. However, instead of controlling the DC power applied to the detector element used for the RF measurement, he connected both auxiliary heaters in series to a constant voltage source and controlled the potential of the connection between the two heaters with a feedback loop. He showed (Fantom, 1978) that this results in a linear relationship between the measured power and the controlling voltage. This way of linearizing the feedback loop avoids the use of a special square-root element as employed in many optical substitution radiometers. On the negative side, this is achieved at the cost of a slightly different operating temperature of the detector element with and without RF heating. The circuit is depicted schematically in Fig. 8.3.

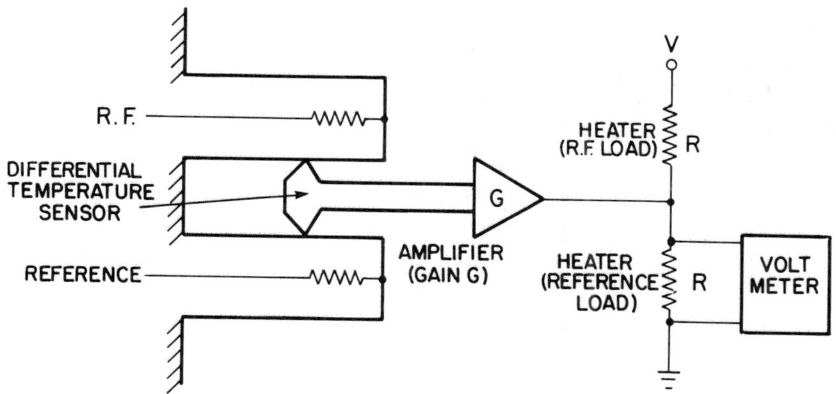

Fig. 8.3 Linearized feedback system. [After Fantom (1979)]

The feedback operation reduced the response time for 99.9% of the final reading from 25 to 7 minutes. An additional response-shaping circuit (Fantom, 1979) reduced the stabilization period to 3 minutes.

The operating principle of DC substitution instruments used to measure RF power in waveguides at even higher frequencies is similar to that for coaxial transmission lines (Lane, 1972; Vowinkel, 1980).

The most important corrections that must be applied to the substituted DC power to derive the correct amount of RF power are for the RF power loss in the transmission line between the input connector and the load resistor (equivalent to the "absorption correction"; see Section 1.3.3), for the magnitude of the frequency-dependent, complex load reflection coefficient (equivalent to the "reflection correction"; see Section 1.3.3), and for the difference in the temperature distributions between DC and RF heating (equivalent to the "nonequivalence correction"; see Section 1.3.3). If one detector is heated by DC and the other one by RF power, a correction for the DC asymmetry (equivalent to the "dual detector correction"; see Section 1.3.3) also has to be applied. A correction for the multiple reflections between load and source could only be made if both the amplitude and phase of the complex source and load reflection coefficients were known. Generally, only the magnitudes of these quantities can be determined, which merely allows the calculation of mismatch uncertainty limits.

8.4 AC–DC TRANSFER

As the instantaneous value of an alternating voltage or current changes all the time in a predictable fashion, the instantaneous value at a particular time is seldom of interest. It is more informative to know, for instance, its peak-to-peak value, the average value of the rectified wave of the so-called root-mean-squared value. The latter is found as the square root of the time average of the square of the AC waveform, i.e.,

$$V_{\mathrm{RMS}} = \left[(1/T) \int_0^T V(t)^2 \, dt \right]^{1/2}. \tag{8.2}$$

The RMS value of an alternating voltage or current corresponds to the value of the corresponding DC quantity, which dissipates the same amount of power in a resistive load. In order to measure the RMS value of a given AC voltage or current, one of the possible (and most accurate) techniques is, therefore, the conversion of the AC quantity to heat and the measurement of the dissipated amount by the substitution of an accurately measurable amount of DC power. The most widely used device for this purpose is a thermal converter consisting of a thin, short heater wire suspended between two sturdy supply leads. A small glass bead with a good thermal conductivity is attached to the middle of the heater wire to provide electrical insulation for the measuring thermocouple or thermistor. The whole structure is sealed in a glass envelope (Fig. 8.4) and evacuated to eliminate air conduction and convection.

Transfer uncertainties of a few parts in 10^6 have been achieved with this technique for frequencies below 100 kHz (Inglis, 1978; Inglis and Franchimon 1985). Corrections are applied for different temperature distributions in the wire under AC and DC heating. These are caused by the Peltier effect (which heats one of the junctions of the heater wire while cooling the other during DC heating) and the Thomson effect, which generates a thermal voltage in a conductor in which there is a temperature gradient. Both effects are cancelled during AC heating and only influence the DC result. The

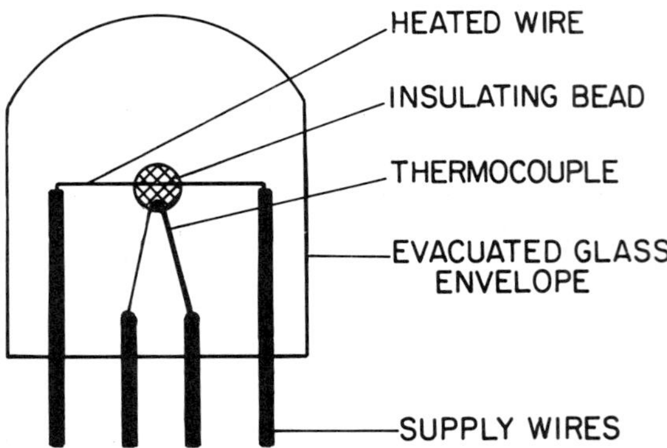

Fig. 8.4 Schematic diagram of a single-element thermal converter.

correct result can be obtained by performing the DC substitution twice, with opposite polarities of the heater voltage, and taking the average of the two values. In order to reduce the DC reversal effect, care is taken to select the materials for the heater and heater supports to produce junctions with the least thermal emf and a heater with a small Thomson coefficient. Good thermal converters of this type have DC reversal differences of the order of one part in 10^4 or less.

Using a twisted and folded (bifilar) heater wire, with the two folded halves in good thermal contact and with multiple thermocouples attached to it as well as special connection techniques for the supply wires, DC reversal differences in the order of 2 ppm or less can be achieved (Klonz, 1987). This results in AC–DC transfer uncertainties for voltage and current in the order of some parts in 10^7 in a frequency range from 10 to 100 kHz (Klonz, 1987), which is comparable to the uncertainties in the realization of the electrical units.

As pointed out in Section 8.3, AC voltage and current measurements become progressively more difficult and eventually even meaningless at frequencies of 1 GHz and higher. In the region between 100 MHz and about 1 GHz, it is possible to use coaxial power meters of the type discussed in Section 8.3 and to calculate the RF voltage at the reference plane of the input connector from the RF power measured in the terminating load resistor. For this purpose, the complex impedance of the load as well as of the coaxial line between connector and load must be known (Janik, 1978).

At low frequencies (<1000 Hz), the thermal-integration capability of the thermal-converter element is reduced as its temperature begins to follow the instantaneous value of the dissipated power. This in turn changes the heater resistance and, together with changes in the effective impedance with decreasing frequency, leads to increased transfer uncertainties (Hermach and Flach, 1976; Klonz, 1987).

8.5 PARTICLE BEAMS

The calorimetric techniques and instruments used for particle beams are in principle identical to those used for X-rays and gamma-rays. (See Section 8.2.)

8.6 CONCLUSION

In this chapter, DC substitution methods used in the non-optical part of the electromagnetic spectrum (and even for particle beams) are reviewed. This completes the circle, which starts with a discussion of units and standards of measurement (Section 1.1), continues with a perspective of the electromagnetic spectrum and of radiometry in general, and then outlines the optical part of the electromagnetic spectrum and optical radiometry (Section 1.2). The main body of the book deals with a subfield of optical radiometry, namely the use of electrically calibrated thermal detectors of optical radiation for the realization of an optical power scale. Alternative methods in use for this purpose are reviewed in Chapter 7.

It is hoped that Chapter 8, through its wider perspective, will promote the interaction and interchange of information between areas of metrology, basically using the same techniques but separated by different application areas, publication channels, and terminology.

REFERENCES

Gamma-Ray and X-Ray Region

Bewley, D. K. (1963). The measurement of locally absorbed dose of megavoltage X-rays by means of a carbon calorimeter. *British J. Radiology* **36**, 865.

Bewley, D. K. (1984). Absorbed dose determination with a water calorimeter. *Phys. Med. Biol.* **29**, 604.

Bregadze, Y. I., Isaev, B. M., Nagl, J., Tultaev, A. V., and Haider, J. (1972). Comparison of the results of absolute absorbed dose measurements performed with two calorimeters of different construction in a Co-60 gamma radiation beam (in Russian). *Izmeritelnaya Teckhnika* **6**, 66.

Burckhart, H., Diehl, R., and Ziegler, B. (1979). A calorimeter for high energy photons. *Nucl. Instr. Methods* **159**, 1.

Busulini, L., Cescon, P., Lora, S., and Palma, G. (1968). Dosimétrie des rayons γ de Co-60 par une simple méthode calorimétrique. *Int. J. Appl. Radiat. Isotop.* **19**, 657.

Calverd, A. M. (1985). A very-low-power AC bridge for high-resolution remote resistance thermometry. *J. Phys. E: Sci. Instr.* **15**, 414.

Degnan, J. H. (1979). Fast, large-signal, free-standing foil bolometer for measuring ultrasoft X-ray burst fluence. *Rev. Sci. Instr.* **50**, 1223.

Domen, S. R. (1969). A heat loss compensated calorimeter and related theorems. *J. Res. Nat. Bur. Stand.* (U.S.) **73C**, 17.

Domen, S. R. (1980). Absorbed dose water calorimeter. *Med. Phys.* **7**, 157.

Domen, S. R. (1981). An improved absorbed dose water calorimeter. *Med. Phys.* **8**, 552.

Domen, S. R. (1982). An absorbed dose water calorimeter: theory, design, and performance. *J. Res. Nat. Bur. Stand.* (U.S.) **87**, 211.

Domen, S. R. (1983a). A polystyrene-water calorimeter. *Int. J. Appl. Radiat. Isot.* **34**, 643.

Domen, S. R. (1983b). A polystyrene-water calorimeter. *J. Res. Nat. Bur. Stand.* (U.S.) **88**, 373.

Domen, S. R. (1983c). A temperature-drift balancer for calorimetry. *Int. J. Appl. Radiat. Isot.* **34**, 927.

Domen, S. R. (1987). Advances in calorimetry for radiation dosimetry. In *The Dosimetry of Ionizing Radiation*, Vol. II, K. R. Kase, B. E. Bjärngard and F. H. Attix, eds. pp. 245–320. Academic Press, New York and London.

Domen, S. R., and Lamperti, P. J. (1974). A heat-loss-compensated calorimeter: theory, design, and performance. *J. Res. Nat. Bur. Stand.* (U.S.) **78A**, 595.

Engelke, B. A., and Hohlfeld, K. (1971). Ein Kalorimeter als Energiedosis Standardmeßeinrichtung und Bestimmung des mittleren Energieaufwandes zur Erzeugung eines Ionenpaares in Luft. *PTB Mitt.* **5**, 336.

Guiho, J.-P., and Simoen, J.-P. (1978). Comparison of BNM-LMRI and NBS absorbed-dose standards for ^{60}Co gamma rays. *Metrologia* **14**, 63.

Hochandel, C. J., and Ghormley, J. A. (1953). A calorimetric calibration of gamma-ray actinometers. *J. Chem. Phys.* **21**, 880.

Janssens, A. (1986). Heat losses occurring in quasi-isothermal absorbed-dose calorimeters. *Metrologia* **22**, 297.

Janssens, A., Cottens, E., Paulsen, A., and Poffijn, A. (1986). Equilibrium of a graphite absorbed-dose calorimeter and the quasi-isothermal mode of operation. *Metrologia* **22**, 265.

Kemp, L. A. W., Marsh, A. R. S., and Baker, M. J. (1971). Pyrolitic graphite microcalorimeter for the measurement of X-ray absorbed dose. *Nature* **230**, 41.

Kubo, H. (1983). Absorbed dose determination with a water calorimeter in comparison with an ionisation chamber. *Phys. Med. Biol.* **28**, 1391.

Kubo, H. (1985). Water calorimetric determination of absorbed dose by 280 kVp orthovoltage X-rays. *Radiotherapy and Oncology* **4**, 275.

Kubo, H., and Brown, D. E. (1984). Calorimeter dose determinations by direct voltage measurements on a Wheatstone-type bridge circuit. *Phys. Med. Biol.* **29**, 885.

Kubo, H., and Mento, D. (1984). An instant dose obtainable *in situ* calorimeter. *Phys. Med. Biol.* **29**, 1433.

Laughlin, J. S., and Genna, S. (1966). Calorimetry. In *Radiation Dosimetry*, 2nd edition, F. H. Attix and W. C. Roesch, eds., p. 389. Academic Press, New York and London.

McDonald, J. C., Laughlin, J. S., and Freeman, R. E. (1976). Portable tissue equivalent calorimeter. *Med. Phys.* **3**, 80.

Myers, I. T., Le Blanc, W. H., and Fleming, D. M. (1961). Precision adiabatic gamma-ray calorimeter using thermistor thermometry. *Rev. Sci. Instr.* **32**, 1013.

Pruit, J. S., Domen, S. R., and Loevinger, R. (1981). The graphite calorimeter as a standard of absorbed dose for cobalt-60 gamma radiation. *J. Res. Nat. Bur. Stand.* (U.S.) **86**, 495.

Radak, B., and Markovic, V. (1970). Calorimetry. In *Manual on Radiation Dosimetry*, N. W. Holm and R. J. Berry, eds., p. 45. Marcel Dekker, New York.

Reid, W. B., and Johns, H. E. (1961). Measurement of absorbed dose with calorimeter and determination of W. *Radiation Research* **14**, 1.

Schulz, R. J., and Weinhous, M. S. (1985). Convection currents in a water calorimeter. *Phys. Med. Biol.* **30**, 1093.

Sundara Rao, I. S., and Naik, S. B. (1980). Graphite calorimeter in water phantom and calibration of ionization chambers in dose to water for ^{60}Co gamma radiation. *Med. Phys.* **7**, 196.

Witzani, J., Duftschmid, K. E., Strachotinsky, C., and Leitner, A. (1984). A graphite absorbed-dose calorimeter in the quasi-isothermal mode of operation. *Metrologia* **20**, 73.

Microwave and Radio-Frequency Region

Clark, R. F., and Jurkus, A. (1968). Ten watt coaxial calorimeter for RF power measurements. *Rev. Sci. Instr.* **39**, 660.

Clark, R. F., Griffin, E. J., Inoue, T., and Weidman, M. P. (1981). An international intercomparison of power standards in WR-28 waveguide. *Metrologia* **17**, 27.

Crawford, M. L. (1968). A new RF-DC substitution calorimeter with automatically controlled reference power. *IEEE Trans. Instr. & Meas.* **IM-17**, 378.

Dressler, E. (1985). *The New RF Power Standard at the NPRL.* CSIR Special Report SFIS 4. Council for Scientific and Industrial Research, Pretoria, South Africa.

Fantom, A. E. (1978). A simple linear feedback method for twin load calorimetric rf power meters. *Metrologia* **14**, 16.

Fantom, A. E. (1979). Improved coaxial calorimetric r.f. power meter for use as a primary standard. *Proc. IEE* **126**, 849.

Hewlett Packard (1969). *Microwave Power Measurement.* Application Note 64. Hewlett Packard, Palo Alto, California.

Hewlett Packard (1977). *Fundamentals of RF and Microwave Power Measurements.* Application Note 64-1. Hewlett Packard, Palo Alto, California.

Jurkus, A. (1966). A coaxial radio-frequency power standard. *IEEE Trans. Instrum. Meas.* **IM-15**, 338.

Klonz, M. (1987). *Entwicklung von Vielfachthermokonvertern zur genauen Rückführung von Wechselgrößen auf äquivalente Gleichgrößen.* Dr-Ing. thesis, Technical University Braunschweig, Federal Republic of Germany.

Lane, J. A. (1972). *Microwave Power Measurement.* IEE Monograph Series 12. Peter Peregrinus, London.

Vowinkel, B. (1980). Broad-band calorimeter for precision measurement of millimeter- and submillimeter-wave power. *IEEE Trans. Instr. & Meas.* **IM-29**, 183.

AC–DC Transfer

Hermach, F. L. (1985). *An Investigation of the Uncertainties of the NBS Thermal Voltage and Current Converters.* Final Report, NBS Contract NB81 SBCA0711. U.S. Department of Commerce, National Bureau of Standards, Washington, D.C.

Hermach, R. L., and Flach, D. R. (1976). An investigation of multijunction thermal converters. *IEEE Trans. Instrum. Meas.* **IM-25**, 524.

Inglis, B. D. (1978). A method for the determination of AC–DC transfer errors in thermoelements. *IEEE Trans. Instrum. Meas.* **IM-27**, 440.

Inglis, B. D. (1980). Evaluation of AC–DC transfer errors for thermal converter–multiplier combinations. *Metrologia* **16**, 177.

Inglis, B. D. (1981). Errors in AC–DC transfer arising from a DC reversal difference. *Metrologia* **17**, 111.

Inglis, B. D. (1982). AC–DC transfer measurements with a precision of < 0.1 ppm. *Metrologia* **18**, 133.

Inglis, B. D., and Franchimon, C. C. (1985). Current-independent AC–DC transfer error components in single-junction thermal converters. *IEEE Trans. Instr. Meas.* **IM-34**, 294.

Janik, D. (1978). "Spannungsmessung im Frequenzbereich bis 1 GHz." In PTB Report E-10, p. 127. PTB, Braunschweig, Federal Republic of Germany.

Janik, D. (1983). Über die Spannungsabhängigkeit der relativen Wechselspannungs-Gleichspannungsdifferenz bei HF-Spannungs-Transfernormalen. PTB Report E-24, PTB, Braunschweig, Federal Republic of Germany.

Klonz, M. (1978). "Wechselspannungs-Gleichspannungs-Transfer bis 100 kHz." In PTB Report E-10, p. 45. PTB, Braunschweig, Federal Republic of Germany.

Klonz, M. (1982). Verfahren zur Herstellung von Vielfachthermokonvertern. In *PTB Annual Report* 1982, p. 135. PTB, Braunschweig, Federal Republic of Germany.

Klonz, M. (1983). The PTB multijunction thermal converter. *Proc. CCE*, 17e Session, Document CCE/83-10. Bureau International des Poids et Mesures, Sevres, France.

Klonz, M. (1984). Transferdifferenz von Vielfachthermokonvertern infolge des Thomsoneffektes. In PTB Annual Report 1984, p. 150. PTB, Braunschweig, Federal Republic of Germany.

Klonz, M. (1987). *Entwicklung von Vielfachthermokonvertern zur genauen Rückführung von Wechselgrößen auf äquivalente Gleichgrößen.* Dr-Ing. thesis, Technical University Braunschweig, Federal Republic of Germany.

Klonz, M., and Wilkens, F. J. (1980). Multijunction thermal converter with adjustable output voltage/current characteristics. *IEEE Trans. Instr. Meas.* **IM-29**, 409.

Schuster, G. (1980). Thermal instrument for measurement of voltage, current, power and energy at power frequencies. *IEEE Trans. Instr. Meas.* **IM-28**, 153.

Widdis, F. C. (1959). *Problems Associated with AC–DC Transfer Devices for the Precise Measurement of Alternating Currents and Voltages.* Thesis, Northampton College of Advanced Technology, London.

Widdis, F. C. (1962). *The Theory of Peltier and Thomson Effects in Thermal AC–DC Transfer Devices.* IEE Monograph 497M. IEE, U.K.

Wilkins, F. J. (1968). Vielfach-Thermokonverter als Wechselstrom/Gleichstrom Transfer-Instrument. *Meßtechnik* **10**, 258.

Wilkins, F. J. (1972). Theoretical analysis of the AC/DC transfer difference of the NPL multijunction thermal converter over the frequency range DC to 100 kHz. *IEEE Trans. Instr. Meas.* **IM-21**, 334.

Wilkins, F. J., Deacon, T. A., and Becker, R. S. (1965). Multijunction thermal converter. *Proc. IEE* **112**, 794.

Williams, E. S. (1971). Thermal voltage converters and comparators for very accurate AC voltage measurements. *J. Res. Nat. Bur. Stand.* (U.S.) **57C**, 145.

Zhang Deh-Shi and Zhang Zhong-Hua (1980). Method for reduction of AC–DC transfer error caused by the Thomson effect for the multijunction thermal converter. *IEEE Trans. Instr. Meas.* **IM-29**, 412.

Particle Beams

Bewley, D. K., McCullough, E. C., Page, B. C., and Sakata, S. (1974). Neutron dosimetry with a calorimeter. *Phys. Med. Biol.* **19**, 831.

Beyer, N. S., Lewis, R. N., and Perry, R. B. (1974). Fast-response fuel rod calorimeter with 36 inch fuel columns. *Nucl. Mater. Mgmt.* **3**, 118.

Bradshaw, A. L. (1965). Calorimetric measurement of absorbed dose with 15 MeV electrons. *Phys. Med. Biol.* **10**, 355.

Caumes, J., and Simoen, J.-P. (1984). A TE-calorimeter as a primary standard for neutron absorbed dose calibrations. *Journal Européen de Radiothérapie* **5**, 235.

Cottens, E. (1980). *Geabsorbeerde dosis kalorimetrie bij hoge energie elektronenbundels en onderzoek van de ijzersulfaat dosimeter.* Ph.D. thesis. Rijksuniversiteit Gent, Belgium.

de Marles, A. E. M. (1981). *Comparison of Measurements of Absorbed Dose to Water Using a Water Calorimeter and Ionization Chambers for Clinical Radiotherapy Photon and Electron Beams.* Ph.D. thesis. University of Texas, Houston.

Gunn, S. R., and Rupert, V. C. (1977). Calorimeters for measurement of ions, X-rays, and scattered radiation in laser-fusion experiments. *Rev. Sci. Instr.* **48**, 1375.

Laughlin, J. S., and Genna, S. (1966). Calorimetry. In *Radiation Dosimetry*, 2nd edition, F. H. Attix and W. C. Roesch, eds., p. 389. Academic Press, New York and London.

Miller, A., and Kovacs, A. (1985). Calorimetry at industrial electron accelerators. *Nucl. Instrum. Methods* **B10/11**, 994.

Thomann, C., and Benn, J. E. (1976). A new type of double-compensated calorimeter for absolute beam intensity measurements. *Nucl. Instrum. Methods* **138**, 193.

Verhey, L. J., Koehler, A. M., McDonald, J. C., Goitein, M., Ma, I., Schneider, R. J., and Wagner, M. (1979). The determination of absorbed dose in a proton beam for purposes of charged-particle radiation therapy. *Radiation Res.* **79**, 34.

Willis, C., Boyd, A. W., and Miller, O. A. (1971). The absolute dosimetry of high intensity, 600 kV pulsed electron accelerator used for radiation chemistry studies of gaseous samples. *Radiation Res.* **46**, 428.

Index